METHODS IN MOLECULAR BIOLOGY™

Series Editor
John M. Walker
School of Life Sciences
University of Hertfordshire
Hatfield, Hertfordshire, AL10 9AB, UK

For further volumes:
http://www.springer.com/series/7651

Chemical Proteomics

Methods and Protocols

Edited by

Gerard Drewes and Marcus Bantscheff

Cellzome AG, Heidelberg, Germany

Humana Press

Editors
Dr. habil. Gerard Drewes
Cellzome AG
Heidelberg, Germany
gerard.drewes@cellzome.com

Dr. Marcus Bantscheff
Cellzome AG
Heidelberg, Germany
marcus.bantscheff@cellzome.com

ISSN 1064-3745 e-ISSN 1940-6029
ISBN 978-1-4939-5800-9 ISBN 978-1-61779-364-6 (eBook)
DOI 10.1007/978-1-61779-364-6
Springer New York Dordrecht Heidelberg London

Printed on acid-free paper

Humana Press is part of Springer Science+Business Media (www.springer.com)

Preface

An Overview of Chemical Proteomics: Methods and Applications

The multidisciplinary science of chemical proteomics studies how small molecules of synthetic or natural origin bind to proteins and modulate their function. Scientists in the field have different backgrounds including molecular and cell biology, biochemistry, pharmacology, organic chemistry, and physics. *Chemical Proteomics: Methods and Applications* is directed at molecular biologists and biochemists with either an interest in small molecules themselves, e.g., in drug discovery projects, or in using small-molecule probes as research tools to study protein function. The book may also be useful for organic chemists with an interest in biology and for specialists in protein mass spectrometry.

In the introductory chapters, we discuss analytical strategies for chemical proteomics projects, with a focus on the current state-of-the-art in protein mass spectrometry, and describe several examples how chemical proteomics can impact the field of drug discovery. The consecutive chapters provide detailed experimental protocols. Most chemical proteomics projects consist of three parts. In the first part, the chemical probe is selected or designed, and then synthesized. In the second part, the probe compound is exposed to the protein sample or cell extract. In the final part, proteins binding to the probe compound are identified and often quantified. In simple applications, this is often achieved by antibody-based detection, but if the aim is the discovery of targets in an unbiased fashion, mass spectrometry is the method of choice. The following chapters cover all of these aspects.

The first set of chapters describes how probes are generated from commercially available reagents without elaborate chemical synthesis procedures, and how the proteins binding to the probes can be analyzed by immunodetection or by mass spectrometry. Rix et al. and Saxena provide protocols for direct noncovalent affinity capture using protein kinase inhibitors as an example, which serves to profile the targets of these compounds and provides probes for kinase expression and activity. Ge and Sem have developed a target class-specific probe for the labeling of dehydrogenase enzymes. Kawamura and Mihai, and Codreanu et al. describe the use of biotin-conjugated probes which form covalent adducts with defined subproteomes, here consisting of adenine-binding proteins and targets of lipid electrophiles. Lenz et al. combine features of noncovalent and covalent capturing strategies in their bifunctional ligands designed to target methyltransferases, an emerging class of drug targets.

The second set of chapters is concerned with techniques for the discovery of small-molecule targets and the probing of target function. Ong et al. describe the use of stable isotope labeling of amino acids in cell culture (SILAC) in identifying proteins that bind small-molecule probes in cell extracts. Hopf et al. perform affinity enrichment of target proteins on a probe matrix in the presence of competing free test compound in solution, thus enabling determination of binding potencies of the free test compound to affinity-captured proteins from cell extracts. The method employs quantitative mass spectrometry with isobaric mass tags to determine the potencies for a large number of targets in a single analysis. Ge and Sem have developed a protocol for the detection and purification of dehydrogenase enzymes, which may represent targets, but also unwanted off-targets, for certain types of drugs. Kovanich et al. describe a combination of cAMP-based affinity chromatography with quantitative mass spectrometry to investigate protein kinase A complexes in extracts of cells and tissues. De Jong et al. use activity-based chemical probes to profile the activity of the proteasome, which has recently emerged as an important cancer target, in cells and tissues. The next three chapters provide innovative protocols for the study of potential drug targets by chemical cross-linking and mass spectrometry. Mueller et al. provide a method to study protein–drug interactions, and Gasilova et al. employ cross-linking and MALDI-mass spectrometry to study ligand modulation of protein–protein interactions. Jeon et al. provide a protocol for in vivo cross-linking via time-controlled transcardiac perfusion, which in principle enables the direct analysis of protein targets in animal models.

The final set of chapters is concerned with small-molecule ligand and drug discovery. Casalena et al. describe the discovery of probe compounds by utilizing compound libraries immobilized on microarrays. Wolf et al. delineate general guidelines for working with small molecules, including aspects like storage, the preparation of solutions, and the determination of solubility. The chapter by de Matos et al. provides guidelines for the use of the ChEBI database, which should be very helpful for researchers tasked with the selection of a particular probe or with building a small molecule collection to purpose. They describe the Chemical Entities of Biological Interest (ChEBI) database which helps to find probe molecules with the desired structural or biological features. Finally, many researchers will consider whether their research tool compound might have the potential to be developed into a drug. Zhang delivers a lucid analysis of the features that distinguish drugs from probe molecules, and lays out a set of rules for "drug likeness."

Affinity- and activity-based chemical probes, combined with quantitative immunodetection and mass spectrometry techniques, are increasingly gaining appreciation as powerful strategies for the molecular analysis of complex biological systems in homeostasis and disease. We hope that the methodologies described in this volume will contribute to a wider application of chemical proteomics methods in biochemical and cell biological laboratories.

Heidelberg, Germany

Gerard Drewes
Marcus Bantscheff

Contents

Contributors

NICO ADAMS • *Department of Genetics, University of Cambridge, Cambridge, UK*
THIN THIN AYE • *Biomolecular Mass Spectrometry and Proteomics Group, Utrecht University and Netherlands Proteomics Centre, Utrecht, The Netherlands*
MARCUS BANTSCHEFF • *Cellzome AG, Heidelberg, Germany*
SONJA BASTUCK • *Cellzome AG, Heidelberg, Germany*
CELIA R. BERKERS • *Division of Cell Biology II, The Netherlands Cancer Institute, Amsterdam, The Netherlands*
MARKUS BOESCHE • *Cellzome AG, Heidelberg, Germany*
DOMINICK E. CASALENA • *Chemical Biology Platform, The Broad Institute of MIT and Harvard, Cambridge, MA, USA*
STEVEN A. CARR • *Proteomics platform, The Broad Institute of MIT and Harvard, Cambridge, MA, USA*
SIMONA G. CODREANU • *Vanderbilt University School of Medicine, Nashville, TN, USA*
ANNEMIEKE DE JONG • *Division of Cell Biology II, The Netherlands Cancer Institute, Amsterdam, The Netherlands*
PAULA DE MATOS • *European Bioinformatics Institute, Hinxton, UK*
MATHIAS DREGER • *caprotec bioanalytics GmbH, Berlin, Germany*
GERARD DREWES • *Cellzome AG, Heidelberg, Germany*
BIRGIT DÜMPELFELD • *Cellzome AG, Heidelberg, Germany*
DIRK EBERHARD • *Cellzome AG, Heidelberg, Germany*
NATALIA GASILOVA • *CovalX AG, Schlieren, Switzerland; Department of Chemistry and Applied Biosciences, ETH, Zürich, Switzerland*
XIA GE • *Chemical Proteomics Facility at Marquette, Department of Chemistry, Marquette University, Milwaukee, WI, USA*
MANUELA GRIDLING • *CeMM Research Center for Molecular Medicine of the Austrian Academy of Sciences, Vienna, Austria*
JANNA HASTINGS • *European Bioinformatics Institute, Hinxton, UK*
ALBERT J.R. HECK • *Biomolecular Mass Spectrometry and Proteomics Group, Utrecht University and Netherlands Proteomics Centre, Utrecht, The Netherlands*
CARSTEN HOPF • *Cellzome AG, Heidelberg, Germany*
AMY HYE WON JEON • *Tanz Centre for Research in Neurodegenerative Diseases and Department of Laboratory Medicine and Pathobiology, University of Toronto, Toronto, ON, Canada*
AKIRA KAWAMURA • *Department of Chemistry, Hunter College of CUNY, New York, NY, USA*
HYE-YOUNG H. KIM • *Vanderbilt University School of Medicine, Nashville, TN, USA*
ANGELA N. KOEHLER • *Chemical Biology Platform, The Broad Institute of MIT and Harvard, Cambridge, MA, USA*
DUANGNAPA KOVANICH • *Biomolecular Mass Spectrometry and Proteomics Group, Utrecht University and Netherlands Proteomics Centre, Utrecht, The Netherlands*

THOMAS LENZ • *caprotec bioanalytics GmbH, Berlin, Germany*
XIAOYU LI • *Chemical Biology platform, The Broad Institute of MIT and Harvard, Cambridge, MA, USA*
DANIEL C. LIEBLER • *Vanderbilt University School of Medicine, Nashville, TN, USA*
DOINA M. MIHAI • *Department of Chemistry, Hunter College of CUNY, New York, NY, USA*
PABLO MORENO • *European Bioinformatics Institute, Hinxton, UK*
MATHIAS Q. MÜLLER • *Department of Pharmaceutical Chemistry and Bioanalytics, Institute of Pharmacy, Martin-Luther-Universität Halle-Wittenberg, Halle (Saale), Germany*
ALEXIS NAZABAL • *CovalX AG, Schlieren, Switzerland*
SHAO-EN ONG • *Proteomics Platform, The Broad Institute of MIT and Harvard, Cambridge, MA, USA*
HUIB OVAA • *Division of Cell Biology II, The Netherlands Cancer Institute, Amsterdam, The Netherlands*
PETER POOT • *Institute of Organic Chemistry, RWTH University, Aachen, Germany*
NED A. PORTER • *Vanderbilt University School of Medicine, Nashville, TN, USA*
FRIEDRICH B.M. REINHARD • *Cellzome AG, Heidelberg, Germany*
UWE RIX • *CeMM Research Center for Molecular Medicine of the Austrian Academy of Sciences, Vienna, Austria*
BORIS RODENKO • *Division of Cell Biology II, The Netherlands Cancer Institute, Amsterdam, The Netherlands*
CHAITANYA SAXENA • *Shantani Proteome Analytics Pvt. Ltd., Pune, MH, India*
MONICA SCHENONE • *Proteomics platform, The Broad Institute of MIT and Harvard, Cambridge, MA, USA*
GEROLD SCHMITT-ULMS • *Tanz Centre for Research in Neurodegenerative Diseases and Department of Laboratory Medicine and Pathobiology, University of Toronto, Toronto, ON, Canada*
ARJEN SCHOLTEN • *Biomolecular Mass Spectrometry and Proteomics Group, Utrecht University andNetherlands Proteomics Centre, Utrecht, The Netherlands*
STUART L. SCHREIBER • *Chemical Biology Platform & Chemical Biology Program, The Broad Institute of MIT and Harvard, Cambridge, MA, USA*
KARIANNE G. SCHUURMAN • *Division of Cell Biology II, The Netherlands Cancer Institute, Amsterdam, The Netherlands*
DANIEL S. SEM • *Department of Pharmaceutical Sciences, Concordia University Wisconsin, Mequon, WI, USA*
SATOKO SHIMAMURA • *Cellzome AG, Heidelberg, Germany*
ANDREA SINZ • *Department of Pharmaceutical Chemistry and Bioanalytics, Institute of Pharmacy, Martin-Luther-Universität Halle-Wittenberg, Halle (Saale), Germany*
CHRISTOPH STEINBECK • *European Bioinformatics Institute, Hinxton, UK*
GIULIO SUPERTI-FURGA • *CeMM Research Center for Molecular Medicine of the Austrian Academy of Sciences, Vienna, Austria*
DINA WASSAF • *Chemical Biology Platform, The Broad Institute of MIT and Harvard, Cambridge, MA, USA*

Elmar Weinhold • *Institute of Organic Chemistry, RWTH University, Aachen, Germany*
Andrea Wolf • *Cellzome AG, Heidelberg, Germany*
Ming-Qiang Zhang • *Merck Sharp & Dohme (MSD), Chaoyang District, Beijing, PR, China*

Chapter 1

Mass Spectrometry-Based Chemoproteomic Approaches

Marcus Bantscheff

Abstract

The term "chemical proteomics" refers to a research area at the interface of chemistry, biochemistry, and cell biology that focuses on studying the mechanism of action of bioactive small molecule compounds, which comprises the mapping of their target proteins and their impact on protein expression and posttranslational modifications in target cells or tissues of interest on a proteome-wide level. For this purpose, a large arsenal of approaches has emerged in recent years, many of which employing quantitative mass spectrometry. This review briefly summarizes major experiment types employed in current chemical proteomics research.

Key words: Affinity chromatography, Chemical proteomics, Drug profiling, Mass spectrometry, Posttranslational modifications, Quantitative proteomics, Stable isotope labeling, Targeted proteomics

1. Introduction

Experimental procedures commonly employed in chemical proteomics approaches can be categorized into two major groups (Fig. 1): (1) global proteomics approaches, aiming at the cell-wide characterization of cellular response to drug treatment, e.g., altered protein expression levels and posttranslational modifications; and (2) activity- or affinity-based protein profiling. The latter summarizing targeted chemoproteomic approaches which employ small molecular probes engineered to selectively capture protein targets/subproteomes via a specific binding mode (e.g., by targeting co-factor binding sites) (1–8).

Gerard Drewes and Marcus Bantscheff (eds.), *Chemical Proteomics: Methods and Protocols*, Methods in Molecular Biology, vol. 803, DOI 10.1007/978-1-61779-364-6_1,

on affinity chromatography, typically using immobilized drugs or tool compounds (1, 8, 26) or covalent active site-labeling probes (3, 6, 27).

Activity-based protein profiling (ABPP) via reactive probes designed to specifically bind to active sites of target enzymes was pioneered by the Cravatt laboratory. In this approach, the reactive probe is typically fused to an affinity tag, such as biotin, via a spacer. In a first step, the small molecule probe is incubated with the biological sample and allowed to covalently attach to proteins it has affinity for. Subsequently, the formed probe–protein conjugates are captured using the affinity tag. Probes have been developed for a variety of enzyme classes including hydrolases (28), proteases (29, 30), kinases (31, 32), phosphatases (33), histone deacetylases (34), and glycosidases (35). For example, fluorescent activity-based probes were reported that enable substrate-free identification of inhibitors of uncharacterized enzymes. By using fluorescence polarization as a read-out, the approach is compatible to high-throughput screening with recombinant enzymes (36). A recent report demonstrated how ABBP can be used to screen compound libraries against an entire target class (37). The authors first synthesized a probe to selectively capture 80% of mammalian serine hydrolases, then screened 70 SHs against 140 structurally diverse carbamates and assessed the selectivity of hits using the very same approach. Patricelli and co-workers (31) used acyl phosphate-containing nucleotides, prepared from a biotin derivative and ATP or ADP to covalently modify ATP-binding proteins directly in cell extracts. Activity-based probes can be adapted for in situ and in vivo labeling by introducing a bio-orthogonal chemical handle, such as an alkyne. Probe-labeled enzymes are then captured by click chemistry conjugation to azide-containing reporter tags (38–40).

Probe design is a crucial step when developing ABPP-based assays and often requires detailed structural information to ensure selective binding of the probe to the catalytic site of target enzymes and to enable the efficient reaction of the active probe with suitable amino acid residues within or close to the catalytic site. In order to streamline probe development, a trifunctional probe design has been suggested consisting of a reversibly interacting selectivity group, a reactive group, and a sorting function (41). In this strategy, small molecules generating selectivity for individual enzyme classes are attached to a common building block containing reactive group and sorting function (i.e., biotin) that is common to all probes. Recent reports demonstrated examples for profiling of cAMP-binding proteins (42), kinases (43), and methyltransferases (44).

For probes interacting with target enzymes with high affinity, however, probe design can be drastically simplified by omitting the reactive group and directly immobilizing the compound of interest on a solid support. The thus generated probe matrix is then incubated with cell extracts and mass spectrometry can be employed in order to identify capture proteins (5, 45). Recent reports demonstrated successful applications of this approach for enrichment of protein kinases and characterization of binding profiles of kinase inhibitors (13, 23, 46–51), proteins binding to ATP/ADP (52), phosphatidylinositols (53, 54), cyclic nucleotides (55, 56), histone deacetylases (57), and peptides (58–60). Further, in several recent mode-of-action studies, immobilized analogs of lead compounds enabled target identification (61–64). Several large-scale studies have been performed for target class specific enrichment of kinase inhibitors using both reactive probes targeting the ATP binding pocket (31, 43) and immobilized unselective kinase inhibitors (46, 65) for affinity enrichment. The kinome coverage achieved with kinase inhibitor probes compared quite favorably to results obtained with reactive probes, indicating a limited impact of the bond-forming reaction with kinases on target class coverage.

4. Impact of Experiment Design and Quantitative Read-Out for Target Identification

Qualitative binding profiles obtained in simple activity/affinity enrichment studies give only limited information about binding potencies of targets and off-targets detected. Consequently, the pharmacological relevance of detected off-target protein has to be validated using the standard repertoire of enzymatic assays (13, 47, 49, 50). While this might be feasible for very selective probes, capturing only few proteins, with recent mass spectrometric equipment often hundreds of proteins can be identified from a single affinity enrichment experiment. Further, the analysis of complex mass spectrometry data is impaired by the fact that the amounts of individual proteins captured do not represent the affinities of these proteins to the immobilized compound. Hence, additional evidence is required to distinguish low-abundant high-affinity interactors from low-affinity abundant ones, e.g., albumin and hemoglobin are known to have low affinity for a range of small molecules and many NADH/NADPH binding proteins bind to immobilized ATP-mimetics (13, 47, 49–51). In addition, proteins might bind to the resin or additional groups introduced to compounds for probe generation, e.g., linkers, reactive groups, biotin, etc. Some of these issues can be addressed by excluding those proteins from further analysis that have been frequently observed in independent experiments using different probe matrices (66). An elegant way to increase over-all specificity of the experiment is to design two probes/probe matrices, one containing the active compound of interest and one containing an inactive analog (45). Experiments with both matrices are then performed in parallel and candidate target proteins can be shortlisted upon differential display of results, e.g., using quantitative mass spectrometric methods.

However, inactive analogs are often not available and designing two independent probes is quite laborious.

Utilizing reactive or immobilized analogs of a small molecule compound in order to identify its targets is a rather indirect approach and modifications introduced for probe generation might alter potency and selectivity of the compound under investigation. This limitation can be addressed by a competition-based experiment design. In this approach, probe/probe matrix binding is performed in the presence or absence of an excess of free, underivatized compound. Experiments are performed in parallel and quantitative mass spectrometry can be applied to identify those proteins for which captured amounts are strongly reduced in the presence of free compound as compared to experiments performed with vehicle control. In a recent application of this approach, Borawski et al. generated an affinity matrix consisting of a derivatized, bioactive PI4KB inhibitor linked to Sepharose beads used to purify cellular and/or viral proteins from a Huh7 HCV replicon cell lysate (61). To determine and quantify specific binding, the experiment was performed in a competition format by adding 10 μM PIK4B inhibitor or DMSO alone into the replicon cell lysate prior to affinity purification. Bound proteins were eluted, digested with trypsin, and labeled with isobaric mass tags for relative and absolute quantification (iTRAQ, (67)) and combined prior to LC-MS/MS analysis. This enabled to quantify binding displacement in the presence of free inhibitor relative to the vehicle. Several hundred proteins were identified and quantified in this experiment but only the binding of class III PI4 kinases, PI4KA and PI4KB, was significantly reduced in the presence of free inhibitor. A similar, SILAC-based strategy was described by Ong et al. for the identification of targets of kinase inhibitors and imunophilin binders (68).

Haystead and co-workers suggested a competition-based approach to estimate relative potencies of target proteins. ATP was linked to Sepharose-beads through the gamma phosphate group, thus generating a probe matrix selective for ATP-binding proteins including protein kinases as well as a variety of other proteins utilizing purine co-factors (52). After enrichment of ATP-binding proteins from cell extracts, the matrix was incubated with increasing concentrations of a set of antimalarial compounds and target proteins were eluted in a dose-dependent manner. In a variation of this approach, Patricelli and co-workers (31) used acyl phosphate-containing nucleotides, prepared from a biotin derivative and ATP or ADP to covalently modify ATP-binding proteins in the co-factor binding site. Biotinylated peptide fragments from labeled proteomes were subsequently captured and identified by mass spectrometry. Target kinases of kinase inhibitors were then determined by comparing MS signals of probe-captured kinases in the absence and presence of excess free kinase inhibitors.

Stable isotope labeling-based quantitative mass spectrometry techniques for precise and accurate relative quantification of proteins (14, 16) have become an indispensable tool in chemical proteomic experiments, since identification of target proteins and their binding affinities (IC_{50}s) largely depends on the ability to quantify differences between vehicle control samples and samples incubated with different amounts of inhibitor. For example, iTRAQ labeling and LC-MS/MS were combined with a mixed kinase inhibitor probe matrix ("kinobeads") for selectivity profiling of three inhibitors of the tyrosine kinase ABL developed for the treatment of chronic myeologeneous leukemia (CML); the phase II compound SKI-606 and the marketed drugs imatinib (Glivec) and dasatinib (46). In this study, cells or cell extracts were treated with inhibitor compounds at varying concentrations followed by incubation with the mixed kinase inhibitor matrix. Kinase inhibitors blocked the ATP binding pockets of target and off-target proteins as a function of concentration and affinity and consequently caused reduced abundance of these target proteins on the kinobeads matrix. Quantitative mass spectrometric analyses revealed binding profiles comprising >500 proteins including ~150 kinases. While dasatinib and SKI-606 revealed very broad target profiles (46 and 42 proteins, respectively, showed >50% competition at 1 μM), imatinib was much more selective. In addition to the primary imatinib targets ABL/BCR-ABL, ARG two novel target candidates, the receptor tyrosine kinase DDR1 (90 nM), and the quinone oxidoreductase NQO2 (43 nM) were identified and validated in independent studies (69, 70). Similarly, SILAC (24) has been employed in conjunction with kinase inhibitor matrices in order to determine the target profile of gefitinib (71). In a very recent report, such a chemoproteomics approach was applied to the analysis of inhibitors binding to native megadalton HDAC complexes. A total of 16 structurally diverse HDAC inhibitors were characterized for their selectivity in targeting multiple HDAC complexes scaffolded by ELM-SANT domain subunits, including a novel mitotic deacetylase complex (MiDAC). Inhibitors clustered according to their target profiles, with stronger binding of aminobenzamides to the HDAC NCoR complex than to the HDAC Sin3 complex, thus suggesting that the selectivity of HDAC inhibitors should be evaluated in the context of HDAC complexes rather than purified catalytic subunits (57).

When cells are treated with an inhibitor, direct targets will be revealed by their reduced binding to the affinity matrix, however, e.g., protein kinases downstream of the respective target kinases will display an altered phosphorylation state due to the reduced signaling by the target kinase. For example, the case of imatinib-treated K562 cells, RSK1, and RSK3 were prominent examples for proteins exhibiting a significant downregulated phosphorylation state (46, 72).

In conclusion, mass spectrometry-based chemoproteomics approaches provide versatile tools to map direct and indirect targets of a compound in a single set of experiments. It is anticipated that these approaches will prove valuable at various stages of drug discovery including validation and selectivity assessment of screening hits and of molecules developed during lead optimization phase as well as in translational studies.

References

1. Bantscheff, M., Scholten, A., and Heck, A. J. (2009) Revealing promiscuous drug-target interactions by chemical proteomics, *Drug Discov Today 14*, 1021–1029.
2. Carlson, S. M., and White, F. M. (2010) Using Small Molecules and Chemical Genetics To Interrogate Signaling Networks, *ACS Chem Biol* 6(1):75–85.
3. Cravatt, B. F., Wright, A. T., and Kozarich, J. W. (2008) Activity-based protein profiling: from enzyme chemistry to proteomic chemistry, *Annu Rev Biochem 77*, 383–414.
4. Kruse, U., Bantscheff, M., Drewes, G., and Hopf, C. (2008) Chemical and pathway proteomics: powerful tools for oncology drug discovery and personalized health care, *Mol Cell Proteomics 7*, 1887–1901.
5. Lolli, G., Thaler, F., Valsasina, B., Roletto, F., Knapp, S., Uggeri, M., Bachi, A., Matafora, V., Storici, P., Stewart, A., Kalisz, H. M., and Isacchi, A. (2003) Inhibitor affinity chromatography: profiling the specific reactivity of the proteome with immobilized molecules, *Proteomics 3*, 1287–1298.
6. Sadaghiani, A. M., Verhelst, S. H., and Bogyo, M. (2007) Tagging and detection strategies for activity-based proteomics, *Curr Opin Chem Biol 11*, 20–28.
7. Yamamoto, K., Yamazaki, A., Takeuchi, M., and Tanaka, A. (2006) A versatile method of identifying specific binding proteins on affinity resins, *Anal Biochem 352*, 15–23.
8. Rix, U., and Superti-Furga, G. (2009) Target profiling of small molecules by chemical proteomics, *Nat Chem Biol 5*, 616–624.
9. Cohen, A. A., Geva-Zatorsky, N., Eden, E., Frenkel-Morgenstern, M., Issaeva, I., Sigal, A., Milo, R., Cohen-Saidon, C., Liron, Y., Kam, Z., Cohen, L., Danon, T., Perzov, N., and Alon, U. (2008) Dynamic proteomics of individual cancer cells in response to a drug, *Science 322*, 1511–1516.
10. Lee, A. Y., Paweletz, C. P., Pollock, R. M., Settlage, R. E., Cruz, J. C., Secrist, J. P., Miller, T. A., Stanton, M. G., Kral, A. M., Ozerova, N. D., Meng, F., Yates, N. A., Richon, V., and Hendrickson, R. C. (2008) Quantitative analysis of histone deacetylase-1 selective histone modifications by differential mass spectrometry, *J Proteome Res 7*, 5177–5186.
11. Liang, X., Hajivandi, M., Veach, D., Wisniewski, D., Clarkson, B., Resh, M. D., and Pope, R. M. (2006) Quantification of change in phosphorylation of BCR-ABL kinase and its substrates in response to Imatinib treatment in human chronic myelogenous leukemia cells, *Proteomics 6*, 4554–4564.
12. Song, D., Chaerkady, R., Tan, A. C., Garcia-Garcia, E., Nalli, A., Suarez-Gauthier, A., Lopez-Rios, F., Zhang, X. F., Solomon, A., Tong, J., Read, M., Fritz, C., Jimeno, A., Pandey, A., and Hidalgo, M. (2008) Antitumor activity and molecular effects of the novel heat shock protein 90 inhibitor, IPI-504, in pancreatic cancer, *Mol Cancer Ther 7*, 3275–3284.
13. Brehmer, D., Godl, K., Zech, B., Wissing, J., and Daub, H. (2004) Proteome-wide identification of cellular targets affected by bisindolylmaleimide-type protein kinase C inhibitors, *Mol Cell Proteomics 3*, 490–500.
14. Bantscheff, M., Schirle, M., Sweetman, G., Rick, J., and Kuster, B. (2007) Quantitative mass spectrometry in proteomics: a critical review, *Anal Bioanal Chem 389*, 1017–1031.
15. Domon, B., and Aebersold, R. (2010) Options and considerations when selecting a quantitative proteomics strategy, *Nat Biotechnol 28*, 710–721.
16. Mallick, P., and Kuster, B. (2010) Proteomics: a pragmatic perspective, *Nat Biotechnol 28*, 695–709.
17. Schirle, M., Heurtier, M. A., and Kuster, B. (2003) Profiling core proteomes of human cell lines by one-dimensional PAGE and liquid chromatography-tandem mass spectrometry, *Mol Cell Proteomics 2*, 1297–1305.
18. Sui, J., Zhang, J., Tan, T. L., Ching, C. B., and Chen, W. N. (2008) Comparative proteomics analysis of vascular smooth muscle cells incubated with S- and R-enantiomers of atenolol

using iTRAQ-coupled two-dimensional LC-MS/MS, *Mol Cell Proteomics* *7*, 1007–1018.

19. Zhang, J., Sui, J., Ching, C. B., and Chen, W. N. (2008) Protein profile in neuroblastoma cells incubated with S- and R-enantiomers of ibuprofen by iTRAQ-coupled 2-D LC-MS/MS analysis: possible action of induced proteins on Alzheimer's disease, *Proteomics* *8*, 1595–1607.
20. Yamanaka, H., Yakabe, Y., Saito, K., Sekijima, M., and Shirai, T. (2007) Quantitative proteomic analysis of rat liver for carcinogenicity prediction in a 28-day repeated dose study, *Proteomics 7*, 781–795.
21. Mackeen, M. M., Kramer, H. B., Chang, K. H., Coleman, M. L., Hopkinson, R. J., Schofield, C. J., and Kessler, B. M. (2010) Small-molecule-based inhibition of histone demethylation in cells assessed by quantitative mass spectrometry, *J Proteome Res 9*, 4082–4092.
22. Pan, C., Olsen, J. V., Daub, H., and Mann, M. (2009) Global effects of kinase inhibitors on signaling networks revealed by quantitative phosphoproteomics, *Mol Cell Proteomics* ***8***, 2796–2808.
23. Li, J., Rix, U., Fang, B., Bai, Y., Edwards, A., Colinge, J., Bennett, K. L., Gao, J., Song, L., Eschrich, S., Superti-Furga, G., Koomen, J., and Haura, E. B. (2010) A chemical and phosphoproteomic characterization of dasatinib action in lung cancer, *Nat Chem Biol 6*, 291–299.
24. Ong, S. E., Blagoev, B., Kratchmarova, I., Kristensen, D. B., Steen, H., Pandey, A., and Mann, M. (2002) Stable isotope labeling by amino acids in cell culture, SILAC, as a simple and accurate approach to expression proteomics, *Mol Cell Proteomics 1*, 376–386.
25. Choudhary, C., Kumar, C., Gnad, F., Nielsen, M. L., Rehman, M., Walther, T. C., Olsen, J. V., and Mann, M. (2009) Lysine acetylation targets protein complexes and co-regulates major cellular functions, *Science* ***325***, 834–840.
26. Hall, S. E. (2006) Chemoproteomics-driven drug discovery: addressing high attrition rates, *Drug Discov Today* ***11***, 495–502.
27. Nomura, D. K., Dix, M. M., and Cravatt, B. F. (2010) Activity-based protein profiling for biochemical pathway discovery in cancer, *Nat Rev Cancer* ***10***, 630–638.
28. Liu, Y., Patricelli, M. P., and Cravatt, B. F. (1999) Activity-based protein profiling: the serine hydrolases, *Proc Natl Acad Sci USA* ***96***, 14694–14699.
29. Kato, D., Boatright, K. M., Berger, A. B., Nazif, T., Blum, G., Ryan, C., Chehade, K. A., Salvesen, G. S., and Bogyo, M. (2005) Activity-based probes that target diverse cysteine protease families, *Nat Chem Biol 1*, 33–38.
30. Saghatelian, A., Jessani, N., Joseph, A., Humphrey, M., and Cravatt, B. F. (2004) Activity-based probes for the proteomic profiling of metalloproteases, *Proc Natl Acad Sci USA* ***101***, 10000–10005.
31. Patricelli, M. P., Szardenings, A. K., Liyanage, M., Nomanbhoy, T. K., Wu, M., Weissig, H., Aban, A., Chun, D., Tanner, S., and Kozarich, J. W. (2007) Functional interrogation of the kinome using nucleotide acyl phosphates, *Biochemistry* ***46***, 350–358.
32. Yee, M. C., Fas, S. C., Stohlmeyer, M. M., Wandless, T. J., and Cimprich, K. A. (2005) A cell-permeable, activity-based probe for protein and lipid kinases, *J Biol Chem* ***280***, 29053–29059.
33. Kumar, S., Zhou, B., Liang, F., Wang, W. Q., Huang, Z., and Zhang, Z. Y. (2004) Activity-based probes for protein tyrosine phosphatases, *Proc Natl Acad Sci USA* ***101***, 7943–7948.
34. Salisbury, C. M., and Cravatt, B. F. (2007) Activity-based probes for proteomic profiling of histone deacetylase complexes, *Proc Natl Acad Sci USA* ***104***, 1171–1176.
35. Vocadlo, D. J., and Bertozzi, C. R. (2004) A strategy for functional proteomic analysis of glycosidase activity from cell lysates, *Angew Chem Int Ed Engl* ***43***, 5338–5342.
36. Bachovchin, D. A., Brown, S. J., Rosen, H., and Cravatt, B. F. (2009) Identification of selective inhibitors of uncharacterized enzymes by high-throughput screening with fluorescent activity-based probes, *Nat Biotechnol* *27*, 387–394.
37. Bachovchin, D. A., Ji, T., Li, W., Simon, G. M., Blankman, J. L., Adibekian, A., Hoover, H., Niessen, S., and Cravatt, B. F. (2010) Superfamily-wide portrait of serine hydrolase inhibition achieved by library-versus-library screening, *Proc Natl Acad Sci USA* ***107***, 20941–20946.
38. Wright, A. T., and Cravatt, B. F. (2007) Chemical proteomic probes for profiling cytochrome p450 activities and drug interactions in vivo, *Chem Biol* ***14***, 1043–1051.
39. Speers, A. E., Adam, G. C., and Cravatt, B. F. (2003) Activity-based protein profiling in vivo using a copper(i)-catalyzed azide-alkyne [3+2] cycloaddition, *J Am Chem Soc* ***125***, 4686–4687.
40. Speers, A. E., and Cravatt, B. F. (2004) Profiling enzyme activities in vivo using click chemistry methods, *Chem Biol* ***11***, 535–546.
41. Koster, H., Little, D. P., Luan, P., Muller, R., Siddiqi, S. M., Marappan, S., and Yip, P. (2007)

Capture compound mass spectrometry: a technology for the investigation of small molecule protein interactions, *Assay Drug Dev Technol 5*, 381–390.

42. Luo, Y., Blex, C., Baessler, O., Glinski, M., Dreger, M., Sefkow, M., and Koster, H. (2009) The cAMP capture compound mass spectrometry as a novel tool for targeting cAMP-binding proteins: from protein kinase A to potassium/sodium hyperpolarization-activated cyclic nucleotide-gated channels, *Mol Cell Proteomics 8*, 2843–2856.
43. Fischer, J. J., Graebner Baessler, O. Y., Dalhoff, C., Michaelis, S., Schrey, A. K., Ungewiss, J., Andrich, K., Jeske, D., Kroll, F., Glinski, M., Sefkow, M., Dreger, M., and Koester, H. (2010) Comprehensive identification of staurosporine-binding kinases in the hepatocyte cell line HepG2 using Capture Compound Mass Spectrometry (CCMS), *J Proteome Res 9*, 806–817.
44. Dalhoff, C., Huben, M., Lenz, T., Poot, P., Nordhoff, E., Koster, H., and Weinhold, E. (2010) Synthesis of S-adenosyl-L-homocysteine capture compounds for selective photoinduced isolation of methyltransferases, *Chembiochem 11*, 256–265.
45. Oda, Y., Owa, T., Sato, T., Boucher, B., Daniels, S., Yamanaka, H., Shinohara, Y., Yokoi, A., Kuromitsu, J., and Nagasu, T. (2003) Quantitative chemical proteomics for identifying candidate drug targets, *Anal Chem 75*, 2159–2165.
46. Bantscheff, M., Eberhard, D., Abraham, Y., Bastuck, S., Boesche, M., Hobson, S., Mathieson, T., Perrin, J., Raida, M., Rau, C., Reader, V., Sweetman, G., Bauer, A., Bouwmeester, T., Hopf, C., Kruse, U., Neubauer, G., Ramsden, N., Rick, J., Kuster, B., and Drewes, G. (2007) Quantitative chemical proteomics reveals mechanisms of action of clinical ABL kinase inhibitors, *Nat Biotechnol 25*, 1035–1044.
47. Brehmer, D., Greff, Z., Godl, K., Blencke, S., Kurtenbach, A., Weber, M., Muller, S., Klebl, B., Cotten, M., Keri, G., Wissing, J., and Daub, H. (2005) Cellular targets of gefitinib, *Cancer Res 65*, 379–382.
48. Daub, H., Olsen, J. V., Bairlein, M., Gnad, F., Oppermann, F. S., Korner, R., Greff, Z., Keri, G., Stemmann, O., and Mann, M. (2008) Kinase-selective enrichment enables quantitative phosphoproteomics of the kinome across the cell cycle, *Mol Cell 31*, 438–448.
49. Godl, K., Gruss, O. J., Eickhoff, J., Wissing, J., Blencke, S., Weber, M., Degen, H., Brehmer, D., Orfi, L., Horvath, Z., Keri, G., Muller, S., Cotten, M., Ullrich, A., and Daub, H. (2005) Proteomic characterization of the angiogenesis inhibitor SU6668 reveals multiple impacts on cellular kinase signaling, *Cancer Res 65*, 6919–6926.
50. Remsing Rix, L. L., Rix, U., Colinge, J., Hantschel, O., Bennett, K. L., Stranzl, T., Muller, A., Baumgartner, C., Valent, P., Augustin, M., Till, J. H., and Superti-Furga, G. (2009) Global target profile of the kinase inhibitor bosutinib in primary chronic myeloid leukemia cells, *Leukemia 23*, 477–485.
51. Rix, U., Hantschel, O., Durnberger, G., Remsing Rix, L. L., Planyavsky, M., Fernbach, N. V., Kaupe, I., Bennett, K. L., Valent, P., Colinge, J., Kocher, T., and Superti-Furga, G. (2007) Chemical proteomic profiles of the BCR-ABL inhibitors imatinib, nilotinib, and dasatinib reveal novel kinase and nonkinase targets, *Blood 110*, 4055–4063.
52. Graves, P. R., Kwiek, J. J., Fadden, P., Ray, R., Hardeman, K., Coley, A. M., Foley, M., and Haystead, T. A. (2002) Discovery of novel targets of quinoline drugs in the human purine binding proteome, *Mol Pharmacol 62*, 1364–1372.
53. Gharbi, S. I., Zvelebil, M. J., Shuttleworth, S. J., Hancox, T., Saghir, N., Timms, J. F., and Waterfield, M. D. (2007) Exploring the specificity of the PI3K family inhibitor LY294002, *Biochem J 404*, 15–21.
54. Krugmann, S., Anderson, K. E., Ridley, S. H., Risso, N., McGregor, A., Coadwell, J., Davidson, K., Eguinoa, A., Ellson, C. D., Lipp, P., Manifava, M., Ktistakis, N., Painter, G., Thuring, J. W., Cooper, M. A., Lim, Z. Y., Holmes, A. B., Dove, S. K., Michell, R. H., Grewal, A., Nazarian, A., Erdjument-Bromage, H., Tempst, P., Stephens, L. R., and Hawkins, P. T. (2002) Identification of ARAP3, a novel PI3K effector regulating both Arf and Rho GTPases, by selective capture on phosphoinositide affinity matrices, *Mol Cell 9*, 95–108.
55. Hanke, S. E., Bertinetti, D., Badel, A., Schweinsberg, S., Genieser, H. G., and Herberg, F. W. (2010) Cyclic nucleotides as affinity tools: phosphorothioate cAMP analogues address specific PKA subproteomes, *Nat Biotechnol 28*(4): 294–301.
56. Scholten, A., Poh, M. K., van Veen, T. A., van Breukelen, B., Vos, M. A., and Heck, A. J. (2006) Analysis of the cGMP/cAMP interactome using a chemical proteomics approach in mammalian heart tissue validates sphingosine kinase type 1-interacting protein as a genuine and highly abundant AKAP, *J Proteome Res 5*, 1435–1447.
57. Bantscheff, M., Hopf, C., Savitski, M. M., Dittmann, A., Grandi, P., Michon, A. M., Schlegl, J., Abraham, Y., Becher, I., Bergamini, G., Boesche, M., Delling, M., Dumpelfeld, B.,

Eberhard, D., Huthmacher, C., Mathieson, T., Poeckel, D., Reader, V., Strunk, K., Sweetman, G., Kruse, U., Neubauer, G., Ramsden, N. G., and Drewes, G. (2011) Chemoproteomics profiling of HDAC inhibitors reveals selective targeting of HDAC complexes, *Nat Biotechnol* *29*, 255–265.

58. Hanke, S., and Mann, M. (2009) The phosphotyrosine interactome of the insulin receptor family and its substrates IRS-1 and IRS-2, *Mol Cell Proteomics* *8*, 519–534.

59. Schulze, W. X., Deng, L., and Mann, M. (2005) Phosphotyrosine interactome of the ErbB-receptor kinase family, *Mol Syst Biol* *1*, 2005 0008.

60. Vermeulen, M., Eberl, H. C., Matarese, F., Marks, H., Denissov, S., Butter, F., Lee, K. K., Olsen, J. V., Hyman, A. A., Stunnenberg, H. G., and Mann, M. (2010) Quantitative interaction proteomics and genome-wide profiling of epigenetic histone marks and their readers, *Cell* *142*, 967–980.

61. Borawski, J., Troke, P., Puyang, X., Gibaja, V., Zhao, S., Mickanin, C., Leighton-Davies, J., Wilson, C. J., Myer, V., Cornellataracido, I., Baryza, J., Tallarico, J., Joberty, G., Bantscheff, M., Schirle, M., Bouwmeester, T., Mathy, J. E., Lin, K., Compton, T., Labow, M., Wiedmann, B., and Gaither, L. A. (2009) Class III phosphatidylinositol 4-kinase alpha and beta are novel host factor regulators of hepatitis C virus replication, *J Virol* *83*, 10058–10074.

62. Fleischer, T. C., Murphy, B. R., Flick, J. S., Terry-Lorenzo, R. T., Gao, Z. H., Davis, T., McKinnon, R., Ostanin, K., Willardsen, J. A., and Boniface, J. J. (2010) Chemical proteomics identifies Nampt as the target of CB30865, an orphan cytotoxic compound, *Chem Biol* *17*, 659–664.

63. Huang, S. M., Mishina, Y. M., Liu, S., Cheung, A., Stegmeier, F., Michaud, G. A., Charlat, O., Wiellette, E., Zhang, Y., Wiessner, S., Hild, M., Shi, X., Wilson, C. J., Mickanin, C., Myer, V., Fazal, A., Tomlinson, R., Serluca, F., Shao, W., Cheng, H., Shultz, M., Rau, C., Schirle, M., Schlegl, J., Ghidelli, S., Fawell, S., Lu, C., Curtis, D., Kirschner, M. W., Lengauer, C., Finan, P. M., Tallarico, J. A., Bouwmeester, T., Porter, J. A., Bauer, A., and Cong, F. (2009) Tankyrase inhibition stabilizes axin and antagonizes Wnt signalling, *Nature* *461*, 614–620.

64. Raijmakers, R., Dadvar, P., Pelletier, S., Gouw, J., Rumpel, K., and Heck, A. J. (2010) Target profiling of a small library of phosphodiesterase 5 (PDE5) inhibitors using chemical proteomics, *ChemMedChem* *5*, 1927–1936.

65. Oppermann, F. S., Gnad, F., Olsen, J. V., Hornberger, R., Greff, Z., Keri, G., Mann, M., and Daub, H. (2009) Large-scale proteomics analysis of the human kinome, *Mol Cell Proteomics* *8*, 1751–1764.

66. Trinkle-Mulcahy, L., Boulon, S., Lam, Y. W., Urcia, R., Boisvert, F. M., Vandermoere, F., Morrice, N. A., Swift, S., Rothbauer, U., Leonhardt, H., and Lamond, A. (2008) Identifying specific protein interaction partners using quantitative mass spectrometry and bead proteomes, *J Cell Biol* *183*, 223–239.

67. Ross, P. L., Huang, Y. N., Marchese, J. N., Williamson, B., Parker, K., Hattan, S., Khainovski, N., Pillai, S., Dey, S., Daniels, S., Purkayastha, S., Juhasz, P., Martin, S., Bartlet-Jones, M., He, F., Jacobson, A., and Pappin, D. J. (2004) Multiplexed protein quantitation in Saccharomyces cerevisiae using amine-reactive isobaric tagging reagents, *Mol Cell Proteomics* *3*, 1154–1169.

68. Ong, S. E., Schenone, M., Margolin, A. A., Li, X., Do, K., Doud, M. K., Mani, D. R., Kuai, L., Wang, X., Wood, J. L., Tolliday, N. J., Koehler, A. N., Marcaurelle, L. A., Golub, T. R., Gould, R. J., Schreiber, S. L., and Carr, S. A. (2009) Identifying the proteins to which small-molecule probes and drugs bind in cells, *Proc Natl Acad Sci USA* *106*, 4617–4622.

69. Day, E., Waters, B., Spiegel, K., Alnadaf, T., Manley, P. W., Buchdunger, E., Walker, C., and Jarai, G. (2008) Inhibition of collagen-induced discoidin domain receptor 1 and 2 activation by imatinib, nilotinib and dasatinib, *Eur J Pharmacol* *599*, 44–53.

70. Winger, J. A., Hantschel, O., Superti-Furga, G., and Kuriyan, J. (2009) The structure of the leukemia drug imatinib bound to human quinone reductase 2 (NQO2), *BMC Struct Biol* *9*, 7.

71. Sharma, K., Weber, C., Bairlein, M., Greff, Z., Keri, G., Cox, J., Olsen, J. V., and Daub, H. (2009) Proteomics strategy for quantitative protein interaction profiling in cell extracts, *Nat Methods* *6*, 741–744.

72. Breitkopf, S. B., Oppermann, F. S., Keri, G., Grammel, M., and Daub, H. (2010) Proteomics analysis of cellular imatinib targets and their candidate downstream effectors, *J Proteome Res* *9*, 6033–6043.

Chapter 2

Chemical Proteomics in Drug Discovery

Gerard Drewes

Abstract

Real-world drug discovery and development remains a notoriously unproductive and increasingly uneconomical process even in the Omics era. The dominating paradigm in the industry continues to be target-based drug design, with an increased perception of the role of signaling pathways in homeostasis and in disease. Since proteins represent the major type of drug targets, proteomics-based approaches, which study proteins under relatively physiological conditions, have great potential if they can be reduced to practice such that they successfully complement the arsenal of drug discovery techniques. This chapter discusses examples of drug discovery processes where chemical proteomics-based assays using native endogenous proteins should have substantial impact.

Key words: Chemical Proteomics, Drug target, Target deconvolution, Target validation, Drug discovery, Selectivity profiling

1. Introduction

Despite the dawn of the Omics era, drug discovery and development remains a notoriously unproductive and increasingly uneconomical process (1, 2). The dominating paradigm in the industry is still target-based drug design, with an increased perception of the role of signaling pathways in homeostasis and in disease (3). Because proteins represent the major type of drug targets, proteomics-based approaches, which allow to study a wide variety of proteins under relatively physiological conditions, have great potential if they can be reduced to practice such that they successfully complement the arsenal of drug discovery techniques (4). Industry standard assays of drug action typically assess the biochemical activity of the purified target protein in isolation. Frequently, recombinant enzymes or protein fragments are used instead of the full-length endogenous proteins. The activity of a

Gerard Drewes and Marcus Bantscheff (eds.), *Chemical Proteomics: Methods and Protocols*, Methods in Molecular Biology, vol. 803, DOI 10.1007/978-1-61779-364-6_2,

compound determined in this type of assays is often not predictive for its pharmacodynamic efficacy. One reason for this discrepancy is that an isolated recombinant protein, or protein fragment, does not necessarily reflect the native conformation and activity of the target in its physiological context, because of the lack of regulatory domains, expression of alternative splice variants, interacting regulatory proteins, or incorrect protein folding or posttranslational modifications. As a consequence, data generated in such assays may not be predictive for the effects of a compound or drug in cell-based or in vivo models. Ideally, assays should be developed to generate data on native proteins in cell extracts or cell fractions, under conditions carefully optimized to preserve protein integrity, folding, posttranslational modification state, and interactions with other proteins. Both activity-based and affinity-based chemical proteomics techniques, as described in this volume, should complement, or in some instances replace the traditional recombinant protein-based assays.

2. Chemical Proteomics Can Aid More Informed Selection and Validation of Targets

In target-based drug discovery, a project begins with the nomination of a target. The target is typically defined as a protein which should be

1. Tractable: Its biochemical activity can be modulated by the desired therapeutic agent (e.g., a small molecule) in a dose-dependent fashion.
2. Validated: It mediates a pathophysiological process such that its modulation reverses a disease-relevant parameter, which can be measured in disease-related cell-based or animal models, and is expected to be predictive of human disease.

Targets are often referred to as "druggable" and "clinically validated" when the modulation of the target was demonstrated to lead to the desired clinical outcome. Historically almost all druggable targets belong to a small number of target classes, biased toward cell surface proteins (e.g., G protein-coupled receptors, ion channels, or transporters) and a small number of intracellular protein classes (e.g., nuclear receptors, metabolic enzymes, kinases, or phosphodiesterases). A recent study estimated that the entirety of approved small molecule drugs acts through approximately 200 human proteins as targets (5), obviously a small number when compared to the 20,000–25,000 protein-coding genes in the human genome (6). It has been estimated that ten times as many suitable drug targets may exist, waiting to be discovered (7). In fact there are numerous proteins in pathways with a strong disease implication, e.g., based on pathobiochemical and human genetic evidence, which are not tractable by current small molecule-based approaches. Chemical proteomics approaches should serve to expand the number of accessible drug targets by aiding the identification of tractable targets without the heavy bias toward the traditional target classes. This type of "target deconvolution" approaches was pioneered by the Schreiber laboratory in the classical studies which identified the molecular targets of immunosuppressants (8, 9). More recent exemplary approaches employed a combination of screening of diverse compound libraries in cell-based assays, which are not biased toward a particular family of targets, with chemoproteomics-based target identification. Huang et al. discovered the tankyrase proteins as tractable targets in the Wnt signaling pathway, which plays a central role in colon cancer but was characterized by a dearth of tractable drug targets (10). Using a related strategy, Fleischer et al. found that the potent and selective cytotoxic agent CB30865 exerts its effects by inhibition of nicotinamide phosphoribosyltransferase, an enzyme in the NAD biosynthetic pathway which helps cancer cells to sustain their increased energy metabolism (11). In another recent study, cell-based screening was performed for the upregulation of apolipoprotein AI production, and the proteomic profiling of hit compounds led to the unexpected discovery of bromodomain proteins as tractable targets for the modulation of the expression of apolipoprotein AI and certain proinflammatory genes (12). These bromodomain inhibitors exhibit a novel mechanism of action by blocking a protein–protein interaction formed between acetylated histones and BET-family bromodomains, which were not previously regarded as tractable targets. These and other successful studies support the notion that there is a general need for small molecules as research tools to study protein function, particularly for proteins which are not classical drug targets. Both the Structural Genomics Consortium (http://www.thesgc.org) and the Center for Protein Research (http://www.cpr.ku.dk) have recently initiated extensive programs for the development of chemical probes which will be made available to the scientific community.

3. Chemical Proteomics-Based Screening of Compound Libraries

Many drug discovery assays rely on the ability to express and purify the target protein in active form in the substantial amounts – typically milligrams of pure protein – necessary for the screening of compound libraries. The drug industry has encountered many so-called "difficult" target classes where this is not easily achieved, for instance, because the target protein is very large or requires additional factors like interacting proteins for proper activity. Therefore, methods based on immobilized probe compounds to capture the target directly from a cell or tissue extract without further purification can represent a viable alternative strategy. This approach was used by Fadden et al. who captured purine-binding proteins from porcine tissue with ATP-derivatized Sepharose and performed affinity elutions with 5,000 different compounds, resulting in the identification of 463 small molecule compounds eluting a total of 77 distinct proteins. Among these, novel and structurally diverse inhibitors of the cancer target Hsp90 were identified, which were further optimized to enter clinical development (13). A different strategy was used by Bantscheff et al. who screened a compound library for histone deacetylase (HDAC) inhibitors in a human cell line extract, using an immobilized hydroxamate-based probe. Here, compounds were added directly to the cell extract rather than using them for elution, such that each compound was assayed for the inhibition of the binding of HDACs to the immobilized probe (14). An important feature of both approaches is that the entire complement of proteins binding selectively to the immobilized probe is screened simultaneously. This represents a major advantage over traditional screening approaches, in particular, for target classes with a substantial number of structurally related targets, like protein kinases or deacetylases, because possible "off-targets" (undesired additional proteins, which typically share a related active site with the target) are revealed early in the project. In conventional approaches, one is left to resort to educated guesses regarding possible "off-targets," and distinct assays have to be configured for each individual protein.

4. Chemical Proteomics for Drug Target Profiling

Despite the fact that drugs are usually optimized against a single target, many compounds exhibit polypharmacology, i.e., they act on multiple targets. These "off-targets" can increase the therapeutic potential of a drug, but they might also cause toxic side effects, which represent a major reason why drugs fail in clinical development (15). An important recent example was the chemoproteomics-based identification of cereblon (CRBN) as a target of the drug thalidomide which mediates the drug's teratogenic effects (16). However, for oncology drugs, polypharmacology is the rule rather than the exception, as they often target proteins from large target classes with a high degree of structural conservation around the active site, like protein and lipid kinases, HDACs, or heat shock proteins. Compared to a truly selective drug, such a spectrum of targets is more likely to produce toxic side effects, but in oncology the increase in therapeutic potential may outweigh this disadvantage (17). Conventional strategies typically rely on assay panels comprising 10–100 purified enzymes to address compound potency, selectivity, and potential off-target liabilities (18). The recent progress in affinity-based proteomic techniques has enabled the direct determination of protein-binding profiles of small molecule drugs under close to physiological conditions. These techniques utilize immobilized compounds as noncovalent affinity baits (14, 19–22) or covalent active-site labeling probes (23, 24). The affinity probes are designed to selectively enrich a larger set of up to several hundreds of proteins defined by structurally related active sites, which can be viewed as chemically tractable subproteomes (25). Noncovalent probes are used either immobilized to an affinity matrix like sepharose or conjugated to biotin, and have been used successfully for purine-binding proteins (26), protein kinases including transmembrane receptor kinases (21, 22), lipid kinases (27, 28), phosphodiesterases (29), and HDACs (14). Covalent active-site labeling probes are typically biotin conjugates and have been applied to kinases (30), GTPases (31), methylases (32), dehydrogenases (33), serine-, cysteine-, metallo-, and proteasomal proteases (23, 34, 35), and HDACs (36). These methodologies typically generate protein affinity profiles for the immobilized compounds, which may reveal novel target candidates, but precautions must be taken to avoid false positives due to the background problems caused by nonspecific interactions with abundant proteins. Moreover, for the application to drug discovery, e.g., in screening or affinity/selectivity profiling assays, the generation of robust quantitative data for hit and lead compounds is an absolute necessity. These problems can be managed if the affinity capture protocols are formatted as quantitative competition-binding assays. This can be achieved by adding the compound of interest in its free form in the tissue extract, before or together with the affinity matrix or the active site label, such that the free compound binds to its targets in the lysate, thereby effectively competing with the capturing probe. By assaying the free compound in the cell extract over a range of concentrations, dose–response binding curves are generated for as many proteins as can be captured by the probe compound and robustly quantified. In case of the "Kinobeads" matrix for protein kinases and the hydroxamate matrix for HDACs developed by Bantscheff et al., more than 1,000 proteins were found to bind to the matrix and were routinely quantified in drug-profiling experiments using a competition binding assay format coupled to protein quantification by isobaric tagging and high-resolution LC-MS/MS peptide sequencing (14, 22). For a more detailed discussion of qualitative and quantitative small molecule target profiling, the reader is referred to recent comprehensive reviews (20, 23, 37). Finally, in addition to the in vitro applications described above, many chemical proteomics strategies can potentially be adapted to the identification and activity profiling of targets in living cells and in animal models (38).

In conclusion, the recent advances in chemical proteomics and in analytical instrumentation have promoted new drug discovery strategies based on assays with increased content and better appreciation of the molecular context of the targets. These methodologies are providing complementary approaches to drug screening, drug target identification, and selectivity profiling, and have the potential to substantially contribute to in vivo studies and clinical studies of drug–target interactions.

References

1. Brown, D. (2007) Unfinished business: target-based drug discovery. *Drug Discov. Today* **12**, 1007–1012.
2. Kola, I., and Landis, J. (2004) Can the pharmaceutical industry reduce attrition rates? *Nat. Rev. Drug Discov.* **3**, 711–715.
3. Fishman, M. C., and Porter, J. A. (2005) Pharmaceuticals: a new grammar for drug discovery. *Nature* **437**, 491–493.
4. Hall, S. E. (2006) Chemoproteomics-driven drug discovery: addressing high attrition rates. *Drug Discov. Today* **11**, 495–502.
5. Overington, J. P., Al-Lazikani, B., and Hopkins, A. L. (2006) How many drug targets are there? *Nat. Rev. Drug Discov.* **5**, 993–996.
6. Clamp, M., Fry, B., Kamal, M., Xie, X., Cuff, J., Lin, M. F., Kellis, M., Lindblad-Toh, K., and Lander, E. S. (2007) Distinguishing protein-coding and noncoding genes in the human genome. *Proc. Natl. Acad. Sci. USA* **104**, 19428–19433.
7. Hopkins, A. L., and Groom, C. R. (2002) The druggable genome. *Nat. Rev. Drug Discov.* **1**, 727–730.
8. Brown, E. J., Albers, M. W., Shin, T. B., Ichikawa, K., Keith, C. T., Lane, W. S., and Schreiber, S. L. (1994) A mammalian protein targeted by G1-arresting rapamycin-receptor complex. *Nature* **369**, 756–758.
9. Harding, M. W., Galat, A., Uehling, D. E., and Schreiber, S. L. (1989) A receptor for the immunosuppressant FK506 is a cis-trans peptidyl-prolyl isomerase. *Nature* **341**, 758–760.
10. Huang, S. M., Mishina, Y. M., Liu, S., Cheung, A., Stegmeier, F., Michaud, G. A., Charlat, O., Wiellette, E., Zhang, Y., Wiessner, S., Hild, M., Shi, X., Wilson, C. J., Mickanin, C., Myer, V., Fazal, A., Tomlinson, R., Serluca, F., Shao, W., Cheng, H., Shultz, M., Rau, C., Schirle, M., Schlegl, J., Ghidelli, S., Fawell, S., Lu, C., Curtis, D., Kirschner, M. W., Lengauer, C., Finan, P. M., Tallarico, J. A., Bouwmeester, T., Porter, J. A., Bauer, A., and Cong, F. (2009) Tankyrase inhibition stabilizes axin and antagonizes Wnt signalling. *Nature* **461**, 614–620.
11. Fleischer, T. C., Murphy, B. R., Flick, J. S., Terry-Lorenzo, R. T., Gao, Z. H., Davis, T., McKinnon, R., Ostanin, K., Willardsen, J. A., and Boniface, J. J. (2010) Chemical proteomics identifies Nampt as the target of CB30865, an orphan cytotoxic compound. *Chem. Biol.* **17**, 659–664.
12. Nicodeme, E., Jeffrey, K. L., Schaefer, U., Beinke, S., Dewell, S., Chung, C. W., Chandwani, R., Marazzi, I., Wilson, P., Coste, H., White, J., Kirilovsky, J., Rice, C. M., Lora, J. M., Prinjha, R. K., Lee, K., and Tarakhovsky, A. (2010) Suppression of inflammation by a synthetic histone mimic. *Nature* **468**, 1119–1123.
13. Fadden, P., Huang, K. H., Veal, J. M., Steed, P. M., Barabasz, A. F., Foley, B., Hu, M., Partridge, J. M., Rice, J., Scott, A., Dubois, L. G., Freed, T. A., Silinski, M. A., Barta, T. E., Hughes, P. F., Ommen, A., Ma, W., Smith, E. D., Spangenberg, A. W., Eaves, J., Hanson, G. J., Hinkley, L., Jenks, M., Lewis, M., Otto, J., Pronk, G. J., Verleysen, K., Haystead, T. A., and Hall, S. E. (2010) Application of chemoproteomics to drug discovery: identification of a clinical candidate targeting hsp90. *Chem. Biol.* **17**, 686–694.
14. Bantscheff, M., Hopf, C., Savitzki, M. M., Dittmann, A., Grandi, P., Michon, A. M. M., Schlegl, J., Abraham, Y., Becher, I., Bergamini, G., Boesche, M., Delling, M., Dümpelfeld, B., Eberhard, D., Huthmacher, C., Mathieson, T., Poeckel, D., Strunk, K., Sweetman, G., Kruse, U., Neubauer, G., Ramsden, N., and Drewes, G. (2011) Chemoproteomics profiling of HDAC inhibitors reveals selective targeting of HDAC complexes. *Nat. Biotechnol.* **29**, 255–265.
15. Morphy, R., Kay, C., and Rankovic, Z. (2004) From magic bullets to designed multiple ligands. *Drug Discov. Today* **9**, 641–651.
16. Kruse, U., Bantscheff, M., Drewes, G., and Hopf, C. (2008) Chemical and pathway proteomics: powerful tools for oncology drug discovery and personalized health care. Mol. *Cell Proteomics.* 7, 1887–1901.
17. Ito, T., Ando, H., Suzuki, T., Ogura, T., Hotta, K., Imamura, Y., Yamaguchi, Y., and Handa H. (2010) Identification of a primary

target of thalidomide teratogenicity. *Science.* 327, 1345–1350.

18. Fliri, A. F., Loging, W. T., Thadeio, P. F., and Volkmann, R. A. (2005) Analysis of drug-induced effect patterns to link structure and side effects of medicines. *Nat. Chem. Biol.* **1**, 389–397.
19. Ong, S. E., Schenone, M., Margolin, A. A., Li, X., Do, K., Doud, M. K., Mani, D. R., Kuai, L., Wang, X., Wood, J. L., Tolliday, N. J., Koehler, A. N., Marcaurelle, L. A., Golub, T. R., Gould, R. J., Schreiber, S. L., and Carr, S. A. (2009) Identifying the proteins to which small-molecule probes and drugs bind in cells. *Proc. Natl. Acad. Sci. USA* **106**, 4617–4622.
20. Rix, U., and Superti-Furga, G. (2009) Target profiling of small molecules by chemical proteomics. *Nat. Chem. Biol.* **5**, 616–624.
21. Sharma, K., Weber, C., Bairlein, M., Greff, Z., Keri, G., Cox, J., Olsen, J. V., and Daub, H. (2009) Proteomics strategy for quantitative protein interaction profiling in cell extracts. *Nat. Methods* **6**, 741–744.
22. Bantscheff, M., Eberhard, D., Abraham, Y., Bastuck, S., Boesche, M., Hobson, S., Mathieson, T., Perrin, J., Raida, M., Rau, C., Reader, V., Sweetman, G., Bauer, A., Bouwmeester, T., Hopf, C., Kruse, U., Neubauer, G., Ramsden, N., Rick, J., Kuster, B., and Drewes, G. (2007) Quantitative chemical proteomics reveals mechanisms of action of clinical ABL kinase inhibitors. *Nat. Biotechnol.* **25**, 1035–1044.
23. Cravatt, B. F., Wright, A. T., and Kozarich, J. W. (2008) Activity-based protein profiling: from enzyme chemistry to proteomic chemistry. *Annu. Rev. Biochem.* 77, 383–414.
24. Paulick, M. G., and Bogyo, M. (2008) Application of activity-based probes to the study of enzymes involved in cancer progression. *Curr. Opin. Genet. Dev.* **18**, 97–106.
25. Haystead, T. A. (2006) The purinome, a complex mix of drug and toxicity targets. *Curr. Top. Med. Chem.* **6**, 1117–1127.
26. Graves, P. R., Kwiek, J. J., Fadden, P., Ray, R., Hardeman, K., Coley, A. M., Foley, M., and Haystead, T. A. (2002) Discovery of novel targets of quinoline drugs in the human purine binding proteome. *Mol. Pharmacol.* **62**, 1364–1372.
27. Borawski, J., Troke, P., Puyang, X., Gibaja, V., Zhao, S., Mickanin, C., Leighton-Davies, J., Wilson, C. J., Myer, V., Cornellataracido, I., Baryza, J., Tallarico, J., Joberty, G., Bantscheff, M., Schirle, M., Bouwmeester, T., Mathy, J. E., Lin, K., Compton, T., Labow, M., Wiedmann, B., and Gaither, L. A. (2009) Class III phosphatidylinositol 4-kinase alpha and beta are novel host factor regulators of hepatitis C virus replication. *J. Virol.* **83**, 10058–10074.
28. Gharbi, S. I., Zvelebil, M. J., Shuttleworth, S. J., Hancox, T., Saghir, N., Timms, J. F., and Waterfield, M. D. (2007) Exploring the specificity of the PI3K family inhibitor LY294002. *Biochem. J.* **404**, 15–21.
29. Raijmakers, R., Dadvar, P., Pelletier, S., Gouw, J., Rumpel, K., and Heck, A. J. (2010) Target profiling of a small library of phosphodiesterase 5 (PDE5) inhibitors using chemical proteomics. *ChemMedChem.* **5**, 1927–1936.
30. Patricelli, M. P., Szardenings, A. K., Liyanage, M., Nomanbhoy, T. K., Wu, M., Weissig, H., Aban, A., Chun, D., Tanner, S., and Kozarich, J. W. (2007) Functional interrogation of the kinome using nucleotide acyl phosphates. *Biochemistry* **46**, 350–358.
31. Luo, Y., Fischer, J. J., Baessler, O. Y., Schrey, A. K., Ungewiss, J., Glinski, M., Sefkow, M., Dreger, M., and Koester, H. (2010) GDP-capture compound – a novel tool for the profiling of GTPases in pro- and eukaryotes by capture compound mass spectrometry (CCMS). *J. Proteomics.* 73, 815–819.
32. Dalhoff, C., Huben, M., Lenz, T., Poot, P., Nordhoff, E., Koster, H., and Weinhold, E. (2010) Synthesis of S-adenosyl-L-homocysteine capture compounds for selective photoinduced isolation of methyltransferases. *Chembiochem.* **11**, 256–265.
33. Ge, X., Wakim, B., and Sem, D. S. (2008) Chemical proteomics-based drug design: target and antitarget fishing with a catechol-rhodanine privileged scaffold for NAD(P)(H) binding proteins. *J. Med. Chem.* **51**, 4571–4580.
34. Kato, D., Boatright, K. M., Berger, A. B., Nazif, T., Blum, G., Ryan, C., Chehade, K. A., Salvesen, G. S., and Bogyo, M. (2005) Activity-based probes that target diverse cysteine protease families. *Nat. Chem. Biol.* **1**, 33–38.
35. Berkers, C. R., Verdoes, M., Lichtman, E., Fiebiger, E., Kessler, B. M., Anderson, K. C., Ploegh, H. L., Ovaa, H., and Galardy, P. J. (2005) Activity probe for in vivo profiling of the specificity of proteasome inhibitor bortezomib. *Nat. Methods* **2**, 357–362.
36. Salisbury, C. M., and Cravatt, B. F. (2007) Activity-based probes for proteomic profiling of histone deacetylase complexes. *Proc. Natl. Acad. Sci. USA* **104**, 1171–1176.
37. Nomura, D. K., Dix, M. M., and Cravatt, B. F. (2010) Activity-based protein profiling for biochemical pathway discovery in cancer. *Nat. Rev. Cancer* **10**, 630–638.
38. Edgington, L. E., Berger, A. B., Blum, G., Albrow, V. E., Paulick, M. G., Lineberry, N., and Bogyo, M. (2009) Noninvasive optical imaging of apoptosis by caspase-targeted activity-based probes. *Nat. Med.* **15**, 967–973.

Part II

Small Molecules and Probe Design

Chapter 3

Compound Immobilization and Drug-Affinity Chromatography

Uwe Rix, Manuela Gridling, and Giulio Superti-Furga

Abstract

Bioactive small molecules act through modulating a yet unpredictable number of targets. It is therefore of critical importance to define the cellular target proteins of a compound as an entry point to understanding its mechanism of action. Often, this can be achieved in a direct fashion by chemical proteomics. As with any affinity chromatography, immobilization of the bait to a solid support is one of the earliest and most crucial steps in the process. Interfering with structural features that are important for identification of a target protein will be detrimental to binding affinity. Also, many molecules are sensitive to heat or to certain chemicals, such as acid or base, and might be destroyed during the process of immobilization, which therefore needs to be not only efficient, but also mild. The subsequent affinity chromatography step needs to preserve molecular and conformational integrity of both bait compound and proteins in order to result in the desired specific enrichment while ensuring a high level of compatibility with downstream analysis by mass spectrometry. Thus, the right choice of detergent, buffer, and protease inhibitors is also essential. This chapter describes a widely applicable procedure for the immobilization of small molecule drugs and for drug-affinity chromatography with subsequent protein identification by mass spectrometry.

Key words: Drug, Target, Affinity chromatography, Immobilization, Chemical proteomics, Mechanism of action

1. Introduction

Promiscuity of small molecule drugs and tool compounds has important implications for understanding their cellular mechanisms, side effects, and clinical potential. Chemical proteomics is a postgenomic, mass spectrometry (MS) and bioinformatics-enabled version of previous versions of drug-affinity chromatography that can identify a drug's cellular target profile, thus aiding in the subsequent

Gerard Drewes and Marcus Bantscheff (eds.), *Chemical Proteomics: Methods and Protocols*, Methods in Molecular Biology, vol. 803, DOI 10.1007/978-1-61779-364-6_3, © Springer Science+Business Media, LLC 2012

characterization of its molecular mechanism of action (1, 2). This approach requires the immobilization of the compound of interest on solid support as an early step. The strategy for immobilization has to be designed very carefully as chemical modification of any bioactive compound harbors the risk of impairing or abrogating its biological activity through disruption of the protein interaction interface. Therefore, one should utilize all available structure–activity relationship information, which can, for instance, be obtained by activity studies with structural analogs or from X-ray co-crystal structures. If such information is unavailable, it is recommended to immobilize the bait compound in separate experiments via two or more different linker attachment points. In some cases, when a cellular or biochemical readout is at hand, these analogs can be tested for retention of activity. When employing cellular assays, though, it has to be kept in mind that modifying a compound might also alter its cell permeability properties. In addition, some molecules are sensitive to various commonly employed conditions, such as acid, base, or heat. Therefore, any broadly applicable modification or immobilization reaction has to be efficient and mild. This is the case for several different reactions employing, e.g., epoxide or amide chemistry (3). The formation of an amide bond from an activated *N*-hydroxy-succinimide (NHS) ester and the amino group (primary or secondary) of a bioactive compound (analog) is readily achieved at room temperature and the reaction usually comes to completion within several hours (4). Compounds containing hydroxy- or carboxy-groups can be conveniently converted into amines by Steglich esterification with Boc- or Fmoc-protected amino acids or mono-protected ethylenediamine, respectively, and subsequent deprotection (5). The choice of protecting group should be based on chemical stability of the bait compound under deprotection conditions (6). Other important considerations for drug-affinity chromatography are the choice of buffer and pH. Just as critical is the efficient inhibition of cellular proteases, which would destroy proteins and prevent affinity enrichment, and possibly enzymes that carry out unscheduled posttranslational modifications in vitro (e.g., protein dephosphorylation), as well as the choice and concentration of detergent, which aids in the solubilization of proteins. In the following, we present the detailed workflow for immobilization and affinity chromatography, which we have previously applied to the characterization of several clinical kinase inhibitors by chemical proteomics (7). We exemplify the approach with a study on the BCR-ABL tyrosine kinase inhibitor dasatinib (BMS-354825, trade name Sprycel) (8), a drug in clinical use for chronic myeloid leukemia (CML) and several other malignancies. In this example, drug-affinity chromatography is performed using the CML cell line KU812 (9).

2. Materials

2.1. Coupling to Solid Support

1. NHS-activated Sepharose 4 Fast Flow (GE Healthcare) (see Note 1).
2. Dimethyl sulfoxide, absolute over molecular sieve (DMSO, Fluka).
3. Isopropanol, absolute over molecular sieve (iProp, Fluka).
4. Triethylamine (TEA, Sigma).
5. Ethanolamine (Aldrich).
6. 2 mL microcentrifuge tubes (Eppendorf).

2.2. LC-MS Analysis of the Coupling Reaction

1. Methanol (MeOH), LC-MS grade (Fisher Scientific GmbH).
2. Isopropanol (iProp), LC-MS grade (Fisher Scientific GmbH).
3. Water, LiChrosolv grade (Merck).
4. Formic acid.
5. Chromatography column: SunFire C_{18}, 5 μm, 3.0×50 mm (Waters).
6. HPLC 2795 (Waters).
7. Photo diode array detector 2996 (Waters).
8. ZQ 2000 single quadrupole mass spectrometer (Waters/Micromass).

2.3. Cell Culture, Harvest, Lysis, and Protein Quantification

1. Dulbecco's modified Eagle's medium (DMEM) (Gibco) supplemented with 10% fetal bovine serum (Gibco) and penicillin/streptomycin 100× (PAA Laboratories GmbH).
2. Cell culture dishes, Ø 150 mm (VWR International).
3. 4× Concentrated Laemmli sample buffer containing 10% β-mercaptoethanol.
4. Lysis buffer 1×: 50 mM Tris–HCl, 100 mM NaCl, 0.2% NP-40, 5% glycerol, 1.5 mM $MgCl_2$, 25 mM NaF, 1 mM Na_3VO_4, 1 mM phenylmethylsulfonyl fluoride, 1 mM dithiothreitol (DTT), 10 μg/mL TLCK, 1 μg/mL leupeptin, 1 μg/mL aprotinin, and 10 μg/mL soybean trypsin inhibitor (Sigma), pH 7.5 (see Note 2).
5. γ-Globin (1 μg/μL) (Bio-Rad).
6. Protein Assay Dye Reagent Concentrate (5×) (Bio-Rad).
7. Hypodermic Needle, 20 G×1½ (Terumo®), single-use syringe 10 mL (B. Braun Melsungen AG).
8. Cuvettes half-micro, 2.5 mL, PMMA (VWR International).
9. Polycarbonate ultracentrifuge tubes (Beckman Coulter).
10. Ultracentrifuge and fixed-angle rotor (Beckman Coulter).

2.4. Affinity Pulldown

1. Lysis buffer 1× (see above).
2. 4× Concentrated Laemmli sample buffer (containing 10% β-mercaptoethanol) (see above).
3. Polycarbonate ultracentrifuge tubes (Beckman Coulter).
4. Spin columns Mobicol (MoBiTec GmbH).
5. Spin column lower filters – 90 μm pore size (MoBiTec GmbH).
6. 1.5 and 2 mL microcentrifuge tubes (Eppendorf).

2.5. One-Dimensional SDS-PAGE

1. Iodoacetamide (13 μg/μL) (Sigma).
2. Mini-PROTEAN Tetra electrophoresis system (Bio-Rad Laboratories).
3. 4× Running gel buffer (SDS buffer): 1.5 mM Tris–HCl and 0.54% SDS, pH 8.8.
4. 4× Gel buffer 2: 2 M Tris, 1.6 mM SDS, pH 6.8.
5. Ammoniumperoxide sulfate (APS), acrylamide, tetramethyl-ethylenediamine (TEMED, Sigma).
6. 5× Running buffer: 602 g Tris, 2,880 g glycine, 200 g SDS, 20 L water, pH 8.3. Do not use HCl!
7. Broad range 7–175 kDa prestained protein marker (New England Biolabs).

2.6. Silver Staining

1. Fixing buffer: 40% ethanol, 10% acetic acid.
2. Washing buffer: 35% ethanol or distilled water.
3. Sensitizing buffer: 0.02% $Na_2S_2O_3$.
4. 0.2% $AgNO_3$ solution (refrigerated).
5. Developing buffer: 3% Na_2CO_3, 0.05% formaldehyde.
6. Quenching solution: 5% acetic acid.

2.7. Immunoblot

1. 10× Western blot buffer: for 2 L dissolve 60.7 g Tris and 288.4 g glycine in distilled water.
2. 1× Western blot buffer: 10% (v/v) 10× Western blot buffer, 10% (v/v) methanol.
3. Wash buffer: PBS with 1% Tween-20 (PBS-T).
4. Blocking buffer (in this example, see Note 3): 3% bovine serum albumin (BSA) in PBS-T.
5. Primary antibody dilution (in this example): 1:2,000 monoclonal mouse anti-ABL (21–63) (Santa Cruz Biotechnology) in 3% BSA/PBS-T (see Note 4).
6. Secondary antibody dilution (in this example): 1:2,000 horse radish peroxidase (HRP)-conjugated anti-mouse (Rockland Immunochemicals) in PBS-T.

7. Amersham ECL Plus™ Western Blotting Detection Reagents (GE Healthcare).
8. Nitrocellulose transfer membrane (Protran BA 85, 300 mm × 3 m, 0.45 μm).
9. Saran wrap.
10. Amersham Hyperfilm ECL (GE Healthcare).
11. Hoefer Semi-Phor semi-dry transfer unit (Hoefer, Inc.).

3. Methods

3.1. Coupling to Solid Support

This section is illustrated in Fig. 1.

1. The beads are provided in a slurry of approximately 50% isopropanol. For one pull down, pipette 100 μL of this slurry into a 1.5 mL microcentrifuge tube (see Note 5). Compare the settled bed volume with 50 μL DMSO. Add or remove slurry until 50 μL bed volume is reached (see Note 6).

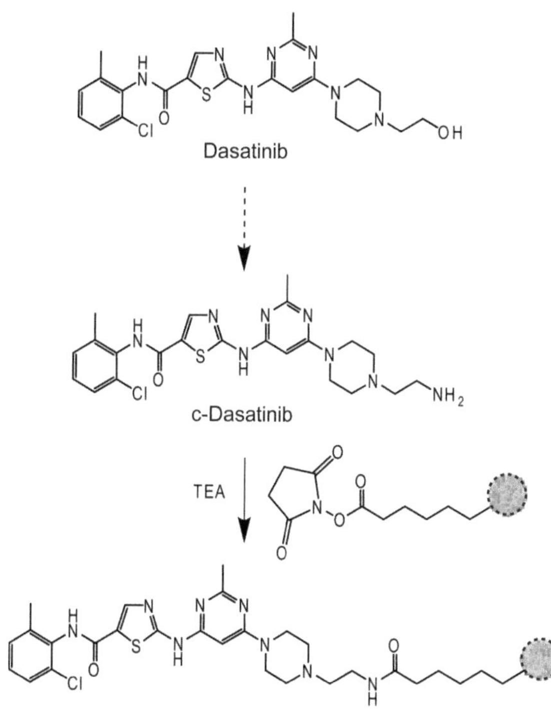

Fig. 1. Immobilization strategy for the BCR-ABL kinase inhibitor dasatinib. The "coupleable" analog c-dasatinib was designed based on the X-ray co-crystal structure of ABL with dasatinib, which indicated that the hydroxy group extends into the solvent space (10). The dasatinib affinity matrix is generated via coupling of c-dasatinib to NHS-activated sepharose.

2. Centrifuge the beads for 3 min at room temperature with 75 × *g* and remove the supernatant.
3. Add 50 μL of DMSO, resuspend gently by inverting several times, centrifuge (as before), and discard the supernatant. Repeat this washing step three times with 0.5 mL DMSO. Then resuspend beads in 50 μL DMSO.
4. Dissolve bait compound to a final concentration of 10 mM in DMSO and add 5 μL of this stock solution and 0.75 μL TEA to the 50% bead slurry, mix carefully and incubate on roto-shaker for 16–24 h (but at least for 8 h) at room temperature with 10 rpm (see Note 7).
5. Centrifuge the beads (as before) and remove 10 μL (≈5 nmol) of the supernatant for the LC-MS control. This is the T_{16} time point for the coupling control.
6. If the drug was successfully coupled to the beads, block the unreacted NHS-ester groups by adding 25 μL ethanolamine and incubate on roto-shaker for further 16–24 h (but at least for 8 h) at room temperature with 10 rpm.
7. Centrifuge beads, remove the supernatant, wash twice with 0.5 mL of DMSO, and discard the supernatant again.
8. Either proceed directly with the drug pulldown and wash the beads with Lysis buffer or add 50 μL isopropanol, centrifuge, discard the supernatant, and repeat this washing step twice with 0.5 mL isopropanol each. The beads can be stored in isopropanol at 4°C (away from light) (see Note 8).

3.2. LC-MS Analysis of the Coupling Reaction

This section is illustrated in Fig. 2.

1. Remove 0.5 μL of a 10 mM bait compound stock solution and add 9.5 μL DMSO. This is the T_0 time point. Add 10 μL MeOH to each sample (T_0 and T_{16}).
2. Inject 5 μL of the samples beginning with the T_{16} time point to avoid carryover.
3. The applied solvent system and LC conditions were:
   (a) Solvent A = 0.1% formic acid in H_2O
   (b) Solvent B = 80%MeOH/20% iProp
   (c) Flow rate = 0.5 mL/min
   (d) 0–6 min 90% A/10% B to 100% B (linear gradient, curve 6)
   (e) 6–8 min 100% B
   (f) 8–8.5 min 100% B to 90% A and 10% B (linear gradient, curve 6)
   (g) 8.5–11 min 90% A and 10% B

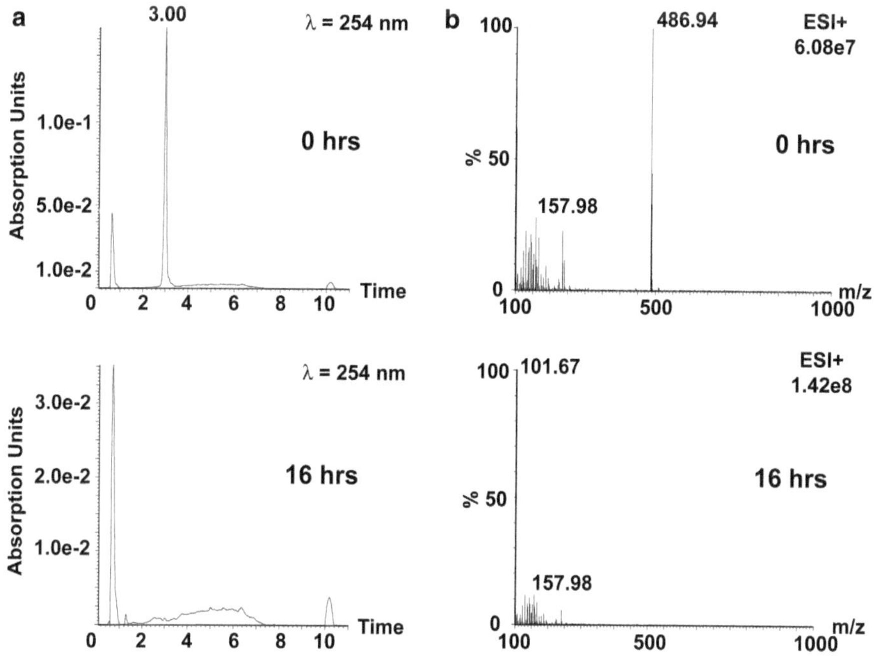

Fig. 2. LC-MS analysis of immobilization reaction of c-dasatinib. (**a**) UV chromatograms of T_0 and T_{16} samples measured at 254 nm. The peak at a retention time of 3.00 min disappears after 16 h of incubation. (**b**) Positive mode electrospray MS spectra at 3.00 min retention time of the T_0 and T_{16} samples confirming that the observed peak represents c-dasatinib, which completely disappears after incubation, i.e., couples to the activated matrix.

4. The column temp was 40°C.
5. The reaction was analyzed on a photodiode array detector (PDA) scanning a range of wavelengths between 210 and 500 nm.
6. The MS was set to scan a mass/charge ratio from 100 to 1,000, switching continuously between the positive and negative electrospray ionization (ESI) mode.
7. The MS parameters were:
   (a) Capillary voltage: 3.2 (ESI+/−)
   (b) Cone voltage: 32 (ESI+), 30 (ESI−)
   (c) Extractor voltage: 3 (ESI+), 4 (ESI−)
   (d) RF lens voltage: 0.3 (ESI+), 1.2 (ESI−)
   (e) Source temperature: 150 (ESI+/−)
   (f) Desolvation temperature: 400 (ESI+/−)
   (g) Desolvation gas (N_2): 450 (ESI+/−)
   (h) Cone gas (N_2): 50 (ESI+/−)
   (i) HM and LM resolution: 15 (ESI+/−)
   (j) Ion energy: 0.6 (ESI+) and 0.1 (ESI−)

3.3. Cell Culture and Harvest

1. Culture KU812 cells in 20 mL DMEM on a 15 cm dish. After cells approach confluence, centrifuge the cell suspension for 3 min with 300 × *g*. Aspirate the medium and resuspend cells in 2 mL fresh medium.
2. Fill 20 mL DMEM in four new 15 cm dishes and add 0.5 mL of the cell suspension. After ca. 2 days cells approach confluence and then need to be split again.
3. To harvest the cells, centrifuge the cell suspension, aspirate the medium, and wash with PBS. Shock freeze the samples in liquid nitrogen and store them at −80°C until further use or use them directly for lysis and the protein quantification.

3.4. Preparation of Total Cell Lysate and Protein Quantification

1. Thaw pellet on ice and resuspend in appropriate amount of Lysis buffer (depending on pellet size) and pull it through a 0.9 mm syringe ten times (see Note 9).
2. Transfer the homogenate to a 15 mL Falcon tube and incubate on ice for 30 min.
3. Transfer lysate to an ultracentrifuge tube, balance tubes, and centrifuge for 10 min at 4°C with 20,000 × *g*.
4. Transfer the supernatant to a fresh polycarbonate ultracentrifuge tube, balance tubes, and centrifuge for 1 h at 4°C with 100,000 × *g*.
5. Transfer supernatant to a fresh 15 mL Falcon tube and keep on ice.
6. Determine the protein concentration by the Bradford method using a freshly determined standard curve. Apply 0, 5, 10, 15, 20, 25, and 30 μL γ-globin and 0.5, 1, and 2 μL lysate to the cuvettes. Add 1 mL of the dye reagent (1×) to each cuvette, mix by careful vortexing, and measure the absorbance at 595 nm.
7. Prepare 25 mg aliquots in microcentrifuge tubes. Use the lysates directly for the drug pulldown or shock freeze the samples in liquid nitrogen and store them at −80°C until use.

3.5. Affinity Pulldown

1. Prepare the 1× Lysis buffer freshly and keep it on ice.
2. Dilute the thawed cell lysate (preferably, prepare cell lysate freshly) with 1× Lysis buffer to a total volume of 3 mL (for 25 mg total protein) transfer to Beckman ultracentrifuge tube, balance tubes, and centrifuge for 20 min at 4°C with 100,000 × *g*.
3. Collect "total cell lysate" (TCL) sample (100 μg per Western blot), add the appropriate amount of 4× Laemmli sample buffer, and denature proteins at 100°C for 3–5 min.
4. Pipette 100 μL 50% drug-bead slurry (i.e., 50 μL settled bead volume = 0.1 μmol drug) into a 2 mL Microcentrifuge tube

(always use cut pipette tips for pipetting beads) and centrifuge beads for 3 min at 4°C with 75 ×*g*, remove supernatant.

5. Add 1 mL 1× Lysis buffer and wash beads by gently resuspending and inverting several times, centrifuge beads, remove supernatant, repeat wash step three times, and remove supernatant.
6. Combine and resuspend the beads with the centrifuged cell lysate.
7. Incubate on roto-shaker for 2 h at 0–4°C with 10 rpm.
8. Centrifuge beads, remove and discard supernatant, but leave ca. 700 μL buffer in microcentrifuge tube (perform following steps in cold room at 0–4°C).
9. Resuspend beads gently and transfer to (plugged!) Mobicol columns, let beads settle by gravity.
10. Drain remaining buffer by gravity flow, fill with fresh 1× Lysis buffer and connect to 30 mL syringe with luer lock tip.
11. Add 7.5 mL 1× Lysis buffer and let the buffer drain by gravity flow (possibly apply gentle pressure with syringe plunger).
12. Place the column in a 2 mL microcentrifuge tube and centrifuge for 1 min at 4°C with 100 ×*g*, plug the column, and place it in a 1.5 mL microcentrifuge tube.
13. Add 30 μL 4× Laemmli sample buffer and denature proteins at 100°C for 3–5 min.
14. Open first the top lid of the column, then unplug the bottom (careful!) and place back into the microcentrifuge tube.
15. Centrifuge for 1 min at room temperature with 400 ×*g*, collect eluate (this is the first elution sample) and replug the column.
16. Place column in another 1.5 mL microcentrifuge tube, add 30 μL 4× Laemmli sample buffer and denature proteins at 100°C for 3–5 min.
17. Unplug the column, centrifuge (as before) and collect eluate (this is the second elution sample) (see Note 10).
18. Store at −20°C until SDS-PAGE analysis.

3.6. One-Dimensional SDS-PAGE

1. For the 7% separating gel, mix 2.5 mL SDS buffer, 2.35 mL 30% acrylamide, 5.17 mL H_2O, 100 μL 10% APS, and 10 μL TEMED and fill ca. 4 mL (leave space for 2.5 mL of stacking gel) of the resulting solution between the two glass plates and overlay with 1 mL isopropanol, wait until the gel has polymerized (ca. 20 min) and pour off the isopropanol.
2. For the stacking gel, mix at first 25 mL Gel buffer 2 with 16.6 mL 30% acrylamide and 58.4 mL H_2O. Take 10 mL of this solution and add 100 μL 10% APS and 15 μL TEMED. Pour ca. 2.5 mL on the polymerized separating gel and insert a comb.

3. After another 20 min remove the comb and place the gel in the electrophoresis chamber, which is filled with running buffer.
4. For immunoblotting, load 10 μL (5 μl each of first and second elution) per well, in the first well inject 5 μL of the protein marker. For silver staining, mix 10 μL of the first and second elutions and alkylate the sample with iodoacetamide (final concentration 13 μg/μL) for 20 min in the dark. Then load the samples on the gel. If possible, fill the first and the last well of the gel with 8 μL of Laemmli sample buffer to ensure an even running profile.
5. Complete the assembly of the gel unit and connect the power supply. The gel can be run at 120 V or higher, if time is short. When the lysates are run through the gel it can be transferred to the nitrocellulose membrane or used for silver staining.

3.7. Silver Staining

This section is illustrated in Fig 3a.

1. Perform all the following steps in a dust-free cabinet to minimize keratin contamination. First, fix the gel with 40% ethanol and 10% acetic acid (always add sufficient solution to completely cover the gel) for 1 h at room temperature on a rocker or at 4°C overnight.
2. Wash the gel two times with 30% ethanol for 20 min and for another 20 min with water (*pro analysi*). In the meantime prepare the sensitizer, the silver nitrate (refrigerator) and the developer.

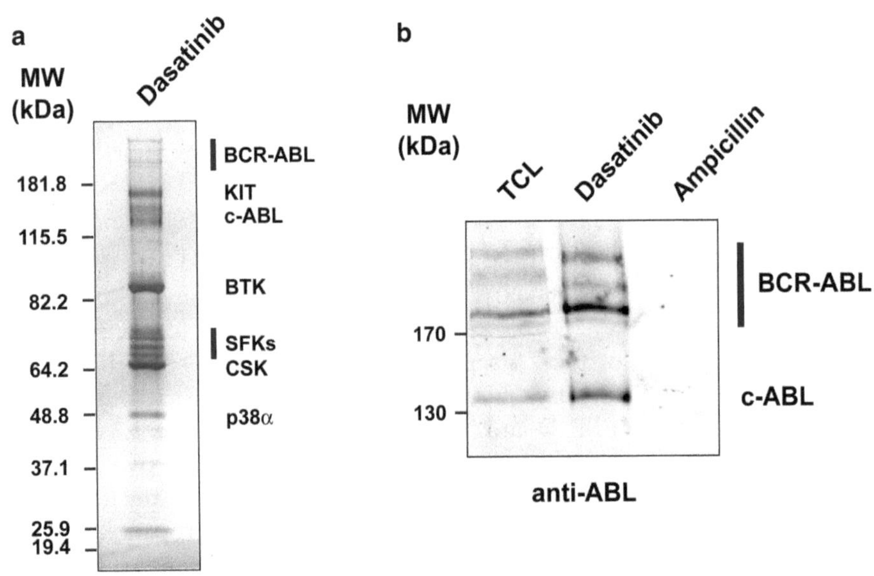

Fig. 3. SDS-PAGE-based analysis of the dasatinib pulldown eluate from KU812 CML cells (9). (**a**) Silver staining shows several bands that subsequent LC-MS/MS analysis reveals to correlate with validated dasatinib target kinases. (**b**) Immunoblot with anti-ABL 21–63 antibody shows the enrichment of BCR-ABL and c-ABL by the dasatinib affinity matrix while an ampicillin control matrix does not display affinity for the dasatinib cognate targets. *SFK* SRC family kinase.

3. Sensitize the gel with 0.02% $Na_2S_2O_3$ for 1.5 min and wash it three times for 20 s with water (*pro analysi*).
4. Incubate the gel with refrigerated 0.2% $AgNO_3$, for 20 min.
5. Discard the silver nitrate in heavy metal waste and wash the gel again (three times for 20 s). Transfer the gel into a clean Petri dish after the second wash step to keep the background as low as possible.
6. Develop the gel in 3% Na_2CO_3 and 0.05% formaldehyde. Rock the Petri dish until the solution changes to a yellow color. Remove the liquid and allow the development without liquid. If the staining is not sufficient, repeat the developing step.
7. When all the bands are visible on the gel, quench the developer by removing the liquid from the Petri dish and adding 5% acetic acid. Leave the gel in the acetic acid for a minimum of 5 min.
8. Remove the acid, place the gel in the lid of the Petri dish, cover it with parafilm, and scan the developed gel for the documentation using a conventional picture-scanner.
9. Cover the gel with water if the gel is stored in the refrigerator for further analysis.

3.8. Immunoblot

This section is illustrated in Fig. 3b.

1. Take three 8 × 9 cm filter papers, soak them in 1× Western blot buffer (containing 10% methanol), and place them in the transfer unit. Then take a 7 × 8 cm membrane, soak it in the buffer, and place it on the filter paper. Cut away the comb of the gel, soak the gel in the buffer, carefully lay it on top of the membrane, and cover it with another three soaked filter papers. Finally ensure that there are no air bubbles in the resulting sandwich and close the transfer cassette. The transfer can be accomplished at 56 mA (1 mA per cm^2) for 1.5 h.
2. Incubate the membrane for 1 h in approximately 50 mL Blocking buffer.
3. Discard the Blocking buffer and incubate the membrane with the primary antibody dilution over night in the cold room. Use gentle agitation.
4. Wash the membrane three times for 5 min with PBS-T.
5. Add 10 mL of the secondary antibody dilution (always use a fresh solution) and incubate for 1 h.
6. Discard the solution and wash the membrane again three times for 5 min.
7. Mix together the ECL Plus solution (1:40) and cover the membrane with 2 mL of the resulting solution. Incubate for 4 min and let the redundant ECL solution drip down. Envelope the membrane with the Saran-Film (ensure that there are no

bubbles) and place it in an X-ray film cassette. Perform the remaining steps in a dark room under safe light conditions.

8. Insert a developing film in the cassette and expose the film for a few seconds to minutes (see Note 11).

4. Notes

1. Always use cut pipette tips for pipetting beads with a clean scalpel or scissors.
2. Use high-purity distilled water of consistent quality to avoid contaminations, such as DNase and RNase free distilled water (Gibco).
3. The primary antibody and therefore the blocking conditions depends on any prior knowledge of validated target proteins of the bait compound. For our example, we have chosen dasatinib as the bait compound, the cognate target of which is ABL. Therefore, we will delineate in the following conditions for the anti-ABL (21–63) antibody (Santa Cruz Biotechnology).
4. The primary antibody can be often reused for subsequent applications. If used in milk/PBS-T, it should be stored at 4°C and be used within a few days. If used in BSA, it can usually be stored at −20°C for several months.
5. Always use aerosol filter pipette tips to avoid cross contamination of samples, as mass spectrometers are highly sensitive instruments, which will detect even trace amounts of contaminants.
6. Contamination of samples with keratin is one of the most common and serious complications for proteomics mass spectrometry analysis, due to the abundance of this family of proteins and dynamic range limits of mass spectrometers. There are several sources of keratin contamination, such as dust, hair, skin, and clothing. It is therefore essential to take the precautions to reduce keratin levels as much as possible.
   (a) Always wear gloves and laboratory coat as much for personal protection as for avoiding keratin contamination.
   (b) Avoid wearing woollen clothing, but rather synthetics or cotton.
   (c) Use a designated dust-free clean bench.
   (d) Use a dedicated set of pipetting and electrophoresis equipment.
   (e) Prepare buffers and reagents freshly from designated stock solutions.
7. The bait compound needs to contain either a primary or secondary amino group, as does c-dasatinib, which was designed

based on the available X-ray co-crystal structure of dasatinib with ABL (10). Alternatively, primary or secondary alcohols can be readily esterified under Steglich conditions, i.e., in the presence of dimethylaminopyridine (DMAP) and dicyclohexylcarbodiimide (DCC) (5). For example, the hydroxyl group of dasatinib can be esterified with N-Boc-glycine. Subsequent removal of the Boc protecting group with trifluoroacetic acid yields a primary amine suitable for immobilization.

8. Duration of storage depends strongly on the stability of the bait compound. Although the dasatinib matrix was sufficiently stable, our experience with other compounds suggests that a storage of maximum 2 weeks should not be exceeded. If possible, drug beads should be used immediately.
9. Some primary cells express high levels of protease activity which may require the addition of additional protease inhibitors to the Lysis buffer, such as the Complete Protease Inhibitor Cocktail Tablets (Roche).
10. Samples from the first and second elution are stored separately, but can be combined directly before loading on SDS-PAGE gel, if desired.
11. Exposure time typically varies.

Acknowledgements

This work was supported by the Austrian Federal Ministry for Science and Research (BMWF) under the GEN-AU program (GZ BMWF-70.081/0018-II/1a/2008) and the Austrian Academy of Sciences (ÖAW).

References

1. Oda, Y., Owa, T., Sato, T., Boucher, B., Daniels, S., Yamanaka, H., Shinohara, Y., Yokoi, A., Kuromitsu, J., and Nagasu, T. (2003) Quantitative chemical proteomics for identifying candidate drug targets, *Anal Chem* 75, 2159–2165.
2. Rix, U., and Superti-Furga, G. (2009) Target profiling of small molecules by chemical proteomics, *Nat Chem Biol* 5, 616–624.
3. Han, S.-Y., and Kim, Y.-A. (2004) Recent development of peptide coupling reagents in organic synthesis, *Tetrahedron* 60, 2447–2467.
4. Anderson, G. W., Zimmerman, J. E., and Callahan, F. M. (1963) N-Hydroxysuccinimide esters in peptide synthesis, *J. Am. Chem. Soc.* 85, 3039.
5. Neises, B., and Steglich, W. (1978) Simple method for the esterification of carboxylic acids, *Angew. Chem. Int. Ed. Engl.* 17, 522–524.
6. Kocienski, P. J. (2005) *Protecting Groups*, 3rd ed., Thieme, Stuttgart.
7. Rix, U., Hantschel, O., Durnberger, G., Remsing Rix, L. L., Planyavsky, M., Fernbach, N. V., Kaupe, I., Bennett, K. L., Valent, P., Colinge, J., Kocher, T., and Superti-Furga, G. (2007) Chemical proteomic profiles of the BCR-ABL inhibitors imatinib, nilotinib, and dasatinib reveal novel kinase and nonkinase targets, *Blood* 110, 4055–4063.
8. Lombardo, L. J., Lee, F. Y., Chen, P., Norris, D., Barrish, J. C., Behnia, K., Castaneda, S., Cornelius, L. A., Das, J., Doweyko, A. M.,

Fairchild, C., Hunt, J. T., Inigo, I., Johnston, K., Kamath, A., Kan, D., Klei, H., Marathe, P., Pang, S., Peterson, R., Pitt, S., Schieven, G. L., Schmidt, R. J., Tokarski, J., Wen, M. L., Wityak, J., and Borzilleri, R. M. (2004) Discovery of N-(2-chloro-6-methyl-phenyl)-2-(6-(4-(2-hydroxyethyl)-piperazin-1-yl)-2-methylpyrimidin-4-ylamino)thiazole-5-carboxamide (BMS-354825), a dual Src/Abl kinase inhibitor with potent antitumor activity in preclinical assays, *J Med Chem* 47, 6658–6661.

9. Kneidinger, M., Schmidt, U., Rix, U., Gleixner, K. V., Vales, A., Baumgartner, C., Lupinek, C., Weghofer, M., Bennett, K. L., Herrmann, H., Schebesta, A., Thomas, W. R., Vrtala, S., Valenta, R., Lee, F. Y., Ellmeier, W., Superti-Furga, G., and Valent, P. (2008) The effects of dasatinib on IgE receptor-dependent activation and histamine release in human basophils, *Blood* 111, 3097–3107.
10. Tokarski, J. S., Newitt, J. A., Chang, C. Y., Cheng, J. D., Wittekind, M., Kiefer, S. E., Kish, K., Lee, F. Y., Borzillerri, R., Lombardo, L. J., Xie, D., Zhang, Y., and Klei, H. E. (2006) The structure of Dasatinib (BMS-354825) bound to activated ABL kinase domain elucidates its inhibitory activity against imatinib-resistant ABL mutants, *Cancer Res* 66, 5790–5797.

Chapter 4

Affinity-Based Chemoproteomics with Small Molecule-Peptide Conjugates

Chaitanya Saxena

Abstract

In affinity-based chemoproteomics strategies, the direct immobilization of small bioactive probe molecules to a solid support may pose problems with respect to the preservation of the functional activity toward the target proteins. Typically, immobilized molecules on solid supports exhibit lower affinity for target proteins compared to the free parent molecule. This may lead to a failure to specifically capture the target proteins or to unacceptable losses during the washing steps. To circumvent these shortcomings, we have devised small molecule-peptide conjugates (SMPCs), which enable wide-ranging experimental strategies for the capturing of protein targets of small molecules from cells or tissues. With the possibilities of synthesizing peptides of tailored biochemical and biophysical properties, SMPCs enable the identification of protein targets of small molecules from cell-lysates and intact cells. Moreover, labeling of these conjugates with fluorophores can provide information on the cellular localization and distribution of the target.

Key words: Peptide-coupled small molecule, Immuno-chemoproteomics, Small molecule-peptide conjugates, Drug target deconvolution, Target protein capture from live cells, Target identification

1. Introduction

Affinity-based chemical proteomics approaches have been successfully used and may provide some level of target deconvolution, but their application typically suffers from limitations of the probes that are immobilized directly to the solid support (1). Problems associated with steric hindrance (2) and limited mobility of the probe molecule may alter its binding to protein targets. The large resin surface area of Sepharose and other matrix types also increases the chance of nonspecific binding and may lead to the identification of false positives. Furthermore, the coupling of small molecules to solid supports

Gerard Drewes and Marcus Bantscheff (eds.), *Chemical Proteomics: Methods and Protocols*, Methods in Molecular Biology, vol. 803, DOI 10.1007/978-1-61779-364-6_4, © Springer Science+Business Media, LLC 2012

complicates the characterization of their affinity toward the protein targets after immobilization. Frequently, immobilized molecules on solid support exhibit lower affinities for their target proteins compared to their free state. This could lead to unacceptable losses of target proteins during the washing steps. Circumventing some of these limitations, small molecule-peptide conjugates (SMPCs) provide an orthogonal strategy to capture the targets of small molecules from biological lysates (3). Moreover, SMPCs can enable capturing of the target from intact biological systems and probe its cellular localization (4). Notably, SMPCs offer several advantages compared to biotinylated probes. The length of the linker moiety (i.e., the peptide) in SMPCs can be kept to a minimum to preserve the biological activity of the probe, in contrast to the extended linkers required for biotinylated probes to dock into the deep binding pocket on avidin. Moreover, by carefully choosing the amino acid composition of the peptide, the hydrophilic properties of the SMPC can be controlled to mimic the behavior of the probe molecule. Conceptually, the design of SMPCs follows the basic design principles of typical fishing probes, which are widely utilized in chemoproteomics experiments. Such probes consist of the bait compound, a linker or spacer, and a "molecular handle" to isolate the small molecule probe–protein complexes from the biological sample. Depending on the isolation strategy, the "molecular handles" are short sequences of carefully selected amino acids of the desired biochemical properties, the choice of which depends on whether the cell lysates or intact cells in culture are used as the protein source. Typically, peptides of hydrophilic nature are preferred to allow diffusion-limited interaction of the SMPC with the target proteins in the cell lysate sample. Cell-permeable SMPCs are utilized to capture the target proteins from intact cells in culture. To obtain additional target localization information, a fluorophore may be coupled to the SMPCs, allowing imaging of the probe inside cells, to generate target location and distribution information using a high-resolution microscope. Once the SMPC has bound to the cellular targets, SMPC–protein complexes are isolated, either using a solid-phase immobilized antibody against the peptide or the fluorophore coupled to the SMPCs. The captured SMPCs–target protein complexes can be eluted via different approaches, and are typically analyzed using SDS-PAGE, immunoblotting, and mass spectrometry-based methods. Mass spectrometry-based protocols are discussed elsewhere in this volume and hence this chapter focuses on the preparation and application of SMPCs for protein target enrichment protocols. In the following protocol, we use the protein kinase C-α (PKCα) and glycogen synthase kinase 3-β (GSK3β) inhibitor Bisindolylmaleimide-III as the bait compound to exemplify the experimental strategy (5).

2. Materials

2.1. Probe Preparation

1. Dimethyl sulfoxide (DMSO).
2. Stock solution of the bait molecule Bisindolylmaleimide-III (Alexis Biochemicals, San Diego, CA), 13 mM in DMSO.
3. Stock solution of the hetero-bifunctional cross-linker succinimidyl-4-(*N*-maleimidomethyl]cyclohexane-1-carboxy-[6-amidocaproate] (LC-SMCC, Pierce, Rockford, IL), 13 mM in DMSO.
4. Modified FLAG® peptide with an extra cysteine residue at the c-terminus, NH_2-DYKDDDDKC-COOH (custom synthesized by GenScript Inc., Piscataway, NJ) (see Note 1). Prepare a 13-mM stock solution of the modified peptide in ultra pure water.
5. Modified 5-FAM-TAT peptide with an extra cysteine residue at the c-terminus, 5-FAM-YGRKKRRQRRRC-COOH (custom synthesized by Anaspec Inc., Fremont, CA) (see Note 1). Prepare a 13-mM stock solution of the modified peptide in ultra pure water.
6. 0.6 M Phosphate buffer, pH 7.2.
7. 10× PBS, pH 7.4 (Sigma, St. Louis, MO). Dilute with water to 2× PBS.
8. Sephadex G-25 (Sigma, St. Louis, MO).
9. 3–5 mL Plastic columns, which can be inserted in long 15-mL plastic tubes (Pierce, Rockford, IL).
10. Orbital shaker.
11. 37°C Incubator.
12. HPLC system, Agilent 1100 series equipped with 250×4.6 mm Kromasil C-18 HPLC column.
13. Applied Biosystems MALDI-TOF 4700 Proteomic analyzer, Spectra max multiwell absorption spectrophotometer, and Spectramax multiwell emission fluorimeter.

2.2. Cell-Culture, Cell-Lysis, and Microscopy

1. Dulbecco's modified Eagle's medium (DMEM) supplemented with 10% fetal bovine serum.
2. Poly-D-lysine coated six-well plates (Millipore, Billerica, MA).
3. Solutions of trypsin (0.25%) and EDTA (1 mM) from Gibco/BRL.
4. Hank's buffered salt solution (HBSS).
5. Cell-lysis buffer: 50 mM Hepes, 150 mM NaCl, 0.25% Triton X-100, 10% glycerol, 1 mM EDTA, 10 mM sodium pyrophosphate,1 mM DTT, and protease inhibitor cocktail (EMD Biosciences, San Diego, CA), pH 7.5.

6. High-salt buffer: 50 mM Hepes, 1 M NaCl, 0.1% NP-40, and 1 mM EDTA, pH 7.5.
7. Teflon cell scrapers (Fisher Scientific).
8. Centrifuge for the removal of cellular debris from cell lysates.

2.3. Affinity Chromatography

1. Equilibration buffer: 50 mM Hepes, 150 mM NaCl, and 1 mM EDTA, pH 7.5.
2. Washing buffer: 50 mM Hepes, 1 M NaCl, 0.1% NP-40, and 1 mM EDTA, pH 7.5.
3. Elution buffer: 50 mM Hepes, 150 mM NaCl, 1 mM EDTA, and 1 mM Bis-III, pH 7.5.
4. Anti-FLAG® M2 affinity resin (Sigma, St. Louis, MO).
5. BioMag® Sheep Anti-Fluorescein IgG (Qiagen, Valencia, CA).
6. Magnetic Separator (Magna-Sep™, Invitrogen, Carlsbad, CA).

2.4. Sample Preparation for SDS-PAGE and Western Blotting

1. Lab grade chloroform and methanol.
2. LDS sample buffer (Invitrogen, Carlsbad, CA).
3. Reducing Agent (Invitrogen, Carlsbad, CA).
4. 4–12% Bis-Tris precast SDS-PAGE gels (NUPAGE®, Invitrogen, Carlsbad, CA).
5. Mini cell apparatus (XCell *Surelock*®, Invitrogen, Carlsbad, CA).
6. MES SDS running buffer (Invitrogen, Carlsbad, CA).
7. Prestained molecular weight marker (SeeBlue, Invitrogen, Carlsbad, CA).
8. Power supply (PowerPac, BioRad, Hercules, CA).
9. Microwave oven.
10. SimplyBlue Safestain (Invitrogen, Carlsbad, CA).
11. Semidry Trans-blot transfer apparatus (BioRad, Hercules, CA).
12. Transfer buffer concentrate (NuPAGE, Invitrogen, Carlsbad, CA). For two gels, use 80 mL ultra pure water, 10 mL methanol, and 10 mL transfer buffer concentrate.
13. Extra thick filter pads (BioRad, Hercules, CA).
14. Nitrocellulose membranes (NuPAGE, Invitrogen, Carlsbad, CA).
15. Protein detector Western blot Kit (KPL, Gaithersburg, MD).
16. Polyclonal anti-PKC-α antibody (Cell Signaling Technology, Beverly, MA).
17. Mouse monoclonal anti-GSK3 α/β antibody (Abcam, Cambridge, MA).

3. Methods

To prepare a SMPC, a molecular attachment site on the bait molecule is chosen to couple the linker and the peptide such that interference with the functional properties of the small molecule is minimized. The presence of a suitable reactive functional group at the "coupling end" of the small molecule, such as a primary or secondary amine or carboxylic acid, facilitates the process of linker attachment and peptide coupling. In cases where suitable reactive functional groups are not present, derivatives of the small molecule with such functional groups must be synthesized. In the following protocol, the preparation of a peptide conjugate of Bisindolylmaleimide-III, a kinase inhibitor compound which already possesses a suitable primary amine function, is described. The functional activity assessment of the target capture probe is the unique feature of the SMPC in chemical proteomics. Hence, once the SMPCs are prepared, they should be thoroughly tested to ensure that the functional activity of the SMPC probe compared to the original small molecule has not been compromised (1, 3). Because this chapter focuses on the SMPC preparation and application protocol, the functional activity assessment assays, which will need to be tailored to any specific small molecule under investigation, are not discussed. For the peptide conjugate of Bisindolylmaleimide-III, examples can be found in refs. 3 and 4.

3.1. Preparation of SMPCs

3.1.1. Preparation of the Modified FLAG®-Coupled Bisindolylmaleimide-III (SMPC for Capturing of Protein Targets from Cell Lysates)

1. Prepare a 500 μL reaction mix of Bis-III and LC-SMCC by diluting the stock solutions to a final concentration of 1 mM in 0.6 M phosphate buffer (see Note 2). The reaction scheme is shown in Fig. 1.
2. Agitate the solution on an orbital shaker for 45 min in the dark at room temperature.
3. After 45 min, add the modified FLAG® peptide to the mixture to a final concentration of 1 mM (see Note 3).
4. Continue to agitate the solution for another 45 min on the orbital shaker.
5. In the meantime, prepare 4 mL of a Sephadex G-25 slurry in water, using a Sephadex to water ratio of 75:25. Carefully pour the slurry into a 3–5 mL plastic column. Wash the resin in the column with two column volumes of water.
6. Carefully apply the mixture from step 4 to the Sephadex G-25 column.
7. Insert the column into a long 15-mL plastic tube and centrifuge for 30 s at 800 ×*g*. Any unreacted small molecule is trapped in the Sephadex. Use the flow-through fraction collected in the plastic tube, which contains unreacted FLAG®, FLAG® reacted

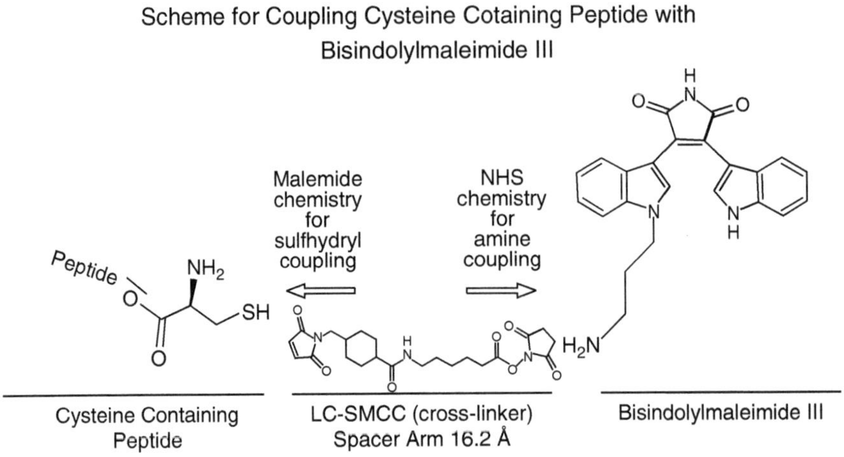

Fig. 1. Reaction scheme. The hetero-bifunctional cross-linker LC-SMCC was utilized to couple the small molecule through the free primary amine to a peptide via a cysteine residue.

to LC-SMCC, and FLAG® coupled to Bis-III through LC-SMCC, for further purification (see Note 4).

8. Load the resultant solution from step 7 on a 250 × 4.6 mm Kromasil C18 HPLC column with a flow rate of 1 mL/min using an Agilent 1100 HPLC pump.
9. Elute the columns using a linear gradient of water:acetonitrile from 95:5 to 55:65 developed over a period of 45 min with a flow rate of 1 mL/min using a HPLC pump.
10. Collect fractions every 3 min and analyze the content using a MALDI-TOF mass spectrometer. A typical elution profile is shown in Fig. 2.
11. Fractions showing peaks at m/z 1,832.8 are pooled and lyophilized (see Note 5).
12. Dissolve the lyophilized SMPC in DMSO and dilute the solution into water or HBSS. Measure the concentration of "FLAG®-coupled-Bis-III" by acquiring absorption spectra of Bis-III, which is the only optically active moiety in the visible spectral range (see Note 6).

3.1.2. Preparation of the Modified 5-FAM-TAT Coupled Bisindolylmaleimide (SMPCs for Capture of Protein Targets from Live Cells)

1. Prepare a 200 μL reaction mix of Bis-III and LC-SMCC by diluting the stock solutions to a final concentration of 1 mM in DMSO (see Note 7).
2. Agitate the solution with an orbital shaker for 45 min in dark at 37°C.
3. After 45 min, add the modified 5-FAM-TAT peptide to the mixture at a final concentration of 500 μM and make up the volume of the reaction to 500 μL by adding 2× PBS (see Note 8).

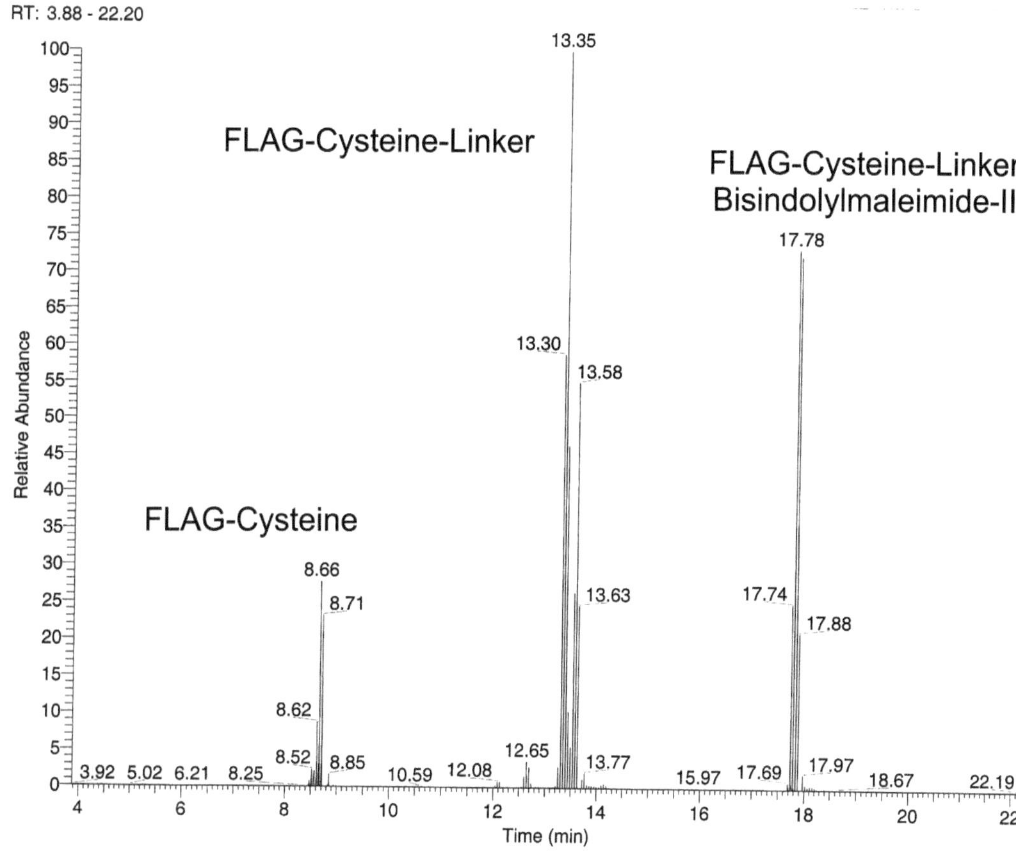

Fig. 2. A typical HPLC elution profile. During the preparation of SMPCs of Bisindolylmaleimide-III, the reaction mixture after the gel filtration step was subjected to HPLC purification. The elution profile shows that the Bisindolylmaleimide-III–FLAG® conjugate was clearly separated from the other two most abundant reaction by-products.

4. Continually agitate the solution for another 45 min.
5. Prepare a slurry of 4 mL Sephadex G-25 in water using a Sephadex to water ratio at 75:25. Carefully pour the slurry into a 3–5 mL plastic column. Wash the resin in the column with two column volumes.
6. Carefully apply the mixture from step 4 to the Sephadex G-25 column.
7. Insert the column into a long 15-mL plastic tube and centrifuge at 800 × *g* for 30 s. Unreacted small molecule compound is trapped in the Sephadex. Use the flow-through fraction (containing unreacted 5-FAM-TAT, 5-FAM-TAT reacted to LC-SMCC, and 5-FAM-TAT coupled to Bis-III through LC-SMCC) collected in the 15 mL tube for further purification.
8. Load the solution obtained in step 7 on a 250 × 4.6 mm C18 column with a flow rate of 500 μL/min using a HPLC pump.

9. Elute the column using a linear gradient of water:acetonitrile from 95:5 to 55:65 developed over a period of 33 min with a flow rate of 500 μL/min.
10. Collect fractions every 30 s and analyze the content using a MALDI-TOF instrument.
11. Fractions showing peaks at *m/z* 2,739.3 are pooled and lyophilized (see Note 9).
12. Dissolve the lyophilized SMPC in DMSO and dilute the solution in water or HBSS. Measure the concentration of "5-FAM-TAT-coupled-Bis-III" by acquiring absorption of 5-FAM at 560 nm. A typical absorption profile of the different species is given in Fig. 3 (see Note 10).

3.2. Application of SMPC for Capture of Protein Target from Cell Lysates

3.2.1. Cell-Lysate Preparation

1. Harvest the HeLa cells and immediately freeze the cells in a −80°C freezer (see Note 11).
2. Slowly thaw the cells 1 day before the lysis in an ice–water mixture.
3. Homogenize the cells using a Dounce homogenizer in cell-lysis buffer.
4. Centrifuge the lysate at 500 × *g* for 15 min at 4°C to remove the cellular debris.
5. Discard the pellets and collect the supernatant.

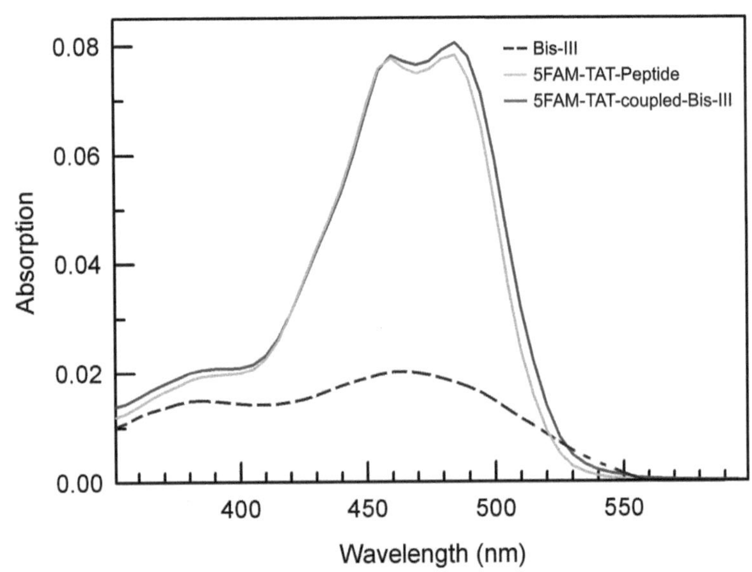

Fig. 3. Calculation of the concentration of 5-FAM-TAT-peptide-coupled Bisindolylmaleimide-III *by* absorption spectroscopy. Typical absorption spectra of Bisindolylmaleimide-III, 5-FAM-TAT peptide, and the 5-FAM-TAT-peptide-coupled-Bisindolylmaleimide-III are shown and were used to calculate the concentration of 5-FAM-TAT-peptide-coupled-Bisindolylmaleimide-III.

6. Centrifuge the supernatant again at 40,000 × *g* at 4°C. Discard the pellets.
7. Measure the protein concentration in the supernatant using the Bradford method (6).
8. Dilute the sample to a final protein concentration of 2 mg/mL in high-salt buffer immediately before the in vitro association experiments.

3.2.2. Affinity Chromatography

1. Incubate 100 μM of the FLAG®-coupled-Bis-III or control-FLAG® peptide (NH_2-DYKDDDDKC-COOH) with 1 mL of high-salt buffer cell-lysate for 2 h at 4°C in a 1.5-mL plastic tube.
2. Pipet 500 μL of Anti-FLAG® M2 affinity resin into a 1.5-mL plastic tube and let the resin settle. Remove the supernatant and add 500 μL of washing buffer to the tube. Gently invert the tube five times to resuspend the resin and allow it to settle again. Remove the supernatant and add 500 μL fresh high-salt buffer. Repeat the washing, let the resin settle and remove the supernatant.
3. Add the mixture from step 1 to the equilibrated resin from step 2 in 1.5-mL plastic tubes. Incubate the mixture overnight at 4°C with constant orbital mixing.
4. The next day, briefly centrifuge the tubes and collect the supernatant.
5. Wash the resin by adding 500 μL of high-salt buffer to the tube containing anti-FLAG® affinity resins. Gently mix the resin and the buffer for 5 min using an orbital shaker (set to minimal speed, just enough to provide a gentle mixing of resin and buffer).
6. Place the tube upright for 2 min, carefully remove and discard the supernatant, and repeat step 5 for two more times.
7. Elute the resin by adding 500 μL of elution buffer containing either (a) 1 mM Bis-III in high-salt buffer, (b) 10 mM FLAG® peptide in high-salt buffer, or (c) 0.1 M glycine (pH 3.5) (see Note 12). Gently mix the resin and the buffer for 5 min using an orbital shaker set to minimal speed.
8. Place the tube upright for 2 min, carefully remove, and collect the supernatant. Repeat step 7 for two more times.
9. Pool the three eluates in a 1.5-mL plastic tube.

3.2.3. Protein Preparation for Visualization and Western Blot Analysis

1. Concentrate the eluted proteins in a speedvac system from 1.5 mL to approximately 100 μL.
2. Precipitate the proteins by chloroform–methanol precipitation.
3. Dissolve precipitated proteins in 100 μL of LDS sample buffer and reducing agent mix and heat the samples at 70°C for 10 min.

4. Place two precasted gel-cassettes in the mini gel system and fill the apparatus with MES SDS running buffer.
5. Load 20 μL each of the sample on a 4–12% SDS/PAGE gel. Load prestained molecular weight marker in one of the wells. The marker will help to assess a smooth run of the gel. Close the lid and turn on the power supply.
6. Separate the proteins by applying 200 V across the gel for 35 min.
7. Carefully remove one of the gels from the cassette, using a cassette opener, and soak the gel in 10 mL of ultra pure water in a petri dish. Wash the gels by slowly rocking the petri dish for 5 min on a rocker platform or manually. Discard the water, add 10 mL of SimplyBlue safestain to the dish and microwave it for 1 min. Remove the dish from the microwave and rock it slowly for 10 min. Discard the staining solution and rinse the gel three times with 20 mL of water or by adding 20 mL of water and rocking it for 10 min. Scan the stained gels, e.g., by placing the gel in a transparent sheet and using a regular document scanner.
8. Carefully remove the second gel from the gel-cassette and soak it in transfer buffer in a petri dish for 20 min (see Note 13).
9. Soak the nitrocellulose membrane in transfer buffer for 5 min.
10. Soak the filter pads in transfer buffer for 10 s.
11. Assemble a sandwich of the gel pad and the membrane on the semidry trans-blotter in the following order: Extra thick filter pad, membrane, gel, extra thin filter pad (see Note 14).
12. Close the trans-blotter and transfer the protein at a constant 15 V for 15 min (see Note 15).
13. After 15 min, carefully open the trans-blotter, remove the filter pads, soak the membrane in 10 mL blocking buffer in a petri dish, and gently rock the dish for 1 h (see Note 16).
14. Drain the blocking buffer and add 10 mL of wash solution to the petri dish. Rock the dish for 5 min.
15. Drain the wash solution, add another 10 mL of wash solution, and rock for another 5 min. Repeat the process one more time and wash the gel at least three times.
16. Drain the final wash solution and add 10 mL of the primary antibody solution (10 μL GSK3-β antibody diluted in 10 mL of buffer) to a petri dish. Rock the dish for 1 h.
17. Drain the primary antibody solution and wash the membrane as in step 14.
18. Drain the wash solution and add 10 mL of secondary antibody solution (10 μL anti-mouse HRP-conjugated antibody in 10 mL of buffer) to a petri dish. Rock the dish for 1 h.
19. Drain the secondary antibody solution and wash the membrane as in step 14.

20. Add 1 mL each of LumiGlo substrate solution A and B to a clean petri dish.
21. Carefully remove the nitrocellulose membrane from the dish using a suitable forceps and transfer it to the LumiGlo solution. Swirl the solution over the membrane for 1 min.
22. Remove the membrane and air dry it for 30 s.
23. Place the membrane between the leaves of an acetate sheet protector cut to the size of an X-ray film cassette (see Note 17).
24. Place the acetate sheet protector with the membrane in an X-ray film cassette equipped with film, for a suitable exposure time (typically from one to a few minutes). Figure 4 shows a typical Western blot analysis of the GSK3-β target capture experiment using the described protocol (see Note 18).

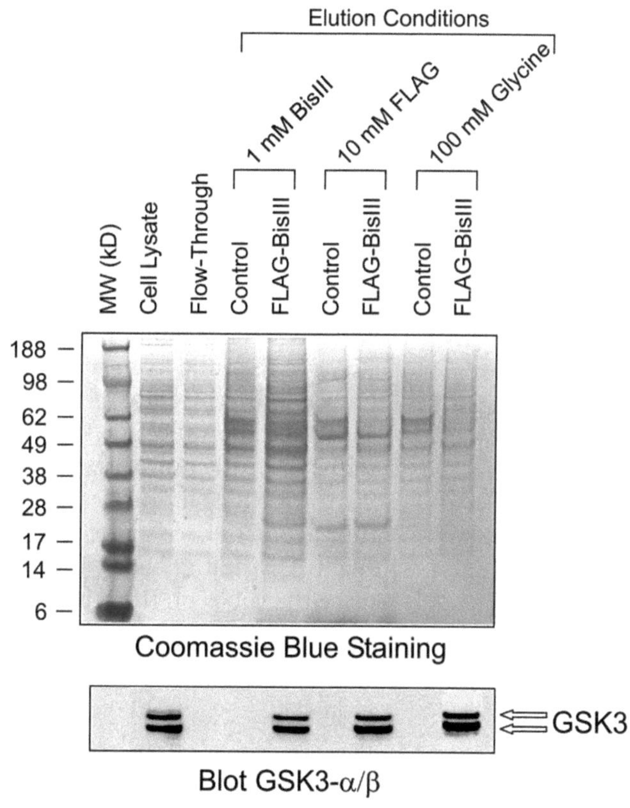

Fig. 4. Elution profile of proteins from anti-FLAG® antibody resins. Different elution conditions were applied for the elution, and the eluted proteins were concentrated before SDS-PAGE analysis. The flow-through represents the fraction which did not bind to FLAG®-coupled-Bis-III. Proteins were separated by SDS-PAGE and visualized using Coomassie blue stain. Aliquots of the samples were analyzed by immunoblotting using a monoclonal antibody against GSK3-α/β protein. The presence of target protein in the elution irrespective of the elution-method used clearly demonstrates that SMPCs were capable of capturing the target protein from the cell lysates.

3.3. Application of SMPCs for Protein Target Capture from Live Cells

3.3.1. Cell-Culture, Probe Incubation, Lysate Preparation, and Affinity Biochemistry

Culture HEK cells on a poly-D-lysine coated six-well plate in DMEM with 10% FBS (and antibiotics, if necessary).

1. Wash the semiconfluent cultures twice with 1 mL of HBSS.
2. Prepare 600 μL of 40 μM 5-FAM-TAT-coupled-Bis-III and 5-FAM-TAT-linker (the control probe) solution each in HBSS.
3. Add 200 μL of 5-FAM-TAT-coupled-Bis-III solutions to the wells of number 1, 2, and 3 and 200 μL of 5-FAM-TAT-linker control probe solution to the wells of number 4, 5, and 6 (see Note 19).
4. Place the plate for 20 min in a 37°C cell-culture incubator.
5. Remove the plate from the incubator and gently remove excess solution by pipetting (see Note 20).
6. Slowly add 500 μL of HBSS to each well and rinse the cells gently by swirling the solution. Remove the solution. Repeat the washing step twice.
7. Add 1 mL of cell-lysis buffer to each well. Scrape the cells from each well and rupture the cell membranes by gently pipetting the cells up and down.
8. Collect the lysed cells from all wells and centrifuge at 800×*g* for 10 min at 4°C to remove debris. Collect the supernatant in 1.5-mL plastic tubes and discard the pellets.
9. Dispense 15 mL of BioMag® Sheep Anti-Fluorescein magnetic bead suspension into a 15 mL tube and place it on a magnetic separator for 5 min. Remove and discard the supernatant. Remove the tube from the separator and add 15 mL of equilibration buffer. Gently mix the magnetic beads with the buffer and place the tube again on the separator for 5 min. Repeat step 2 twice to ensure proper equilibration of the resin. Remove 9 mL of the solution, leaving 6 mL in the tube, remove the tube from the separator, and resuspend the magnetic beads again. Divide the beads into six aliquots in labeled tubes and place the tubes on the separator. Remove the supernatant.
10. Add the cell-lysate obtained in step 9 to the six tubes with anti-fluorescein magnetic beads and incubate for 45 min (see Note 21).
11. Place the tubes on the separator for 2 min, remove, and discard the supernatant.
12. Remove the tubes from the separator and add 1,000 μL of washing buffer.
13. Wash the beads by manually moving the solution up and down for 2 min. Repeat washing steps 12 and 13 twice.
14. Place the tubes on the separator for 2 min, remove, and discard the final wash solution.

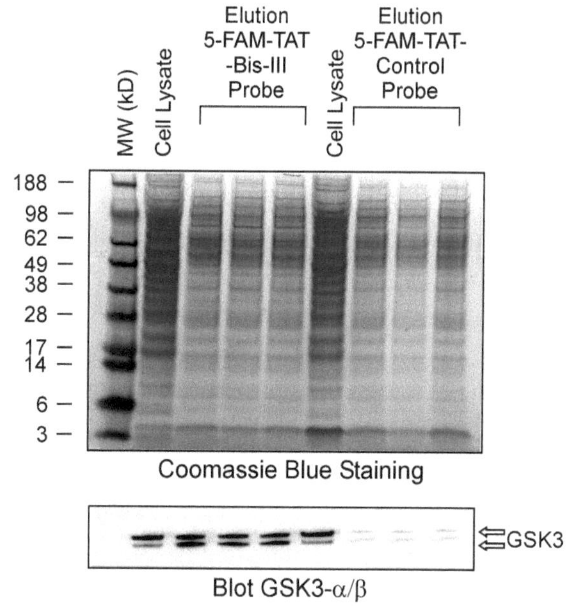

Fig. 5. Elution profile of proteins from anti-FITC® antibody resins. In three independent live cell experiments, eluted proteins were concentrated before SDS-PAGE analysis and visualized using Coomassie blue stain. Corresponding samples were analyzed by immunoblotting using a monoclonal antibody against GSK3-α/β protein. The enrichment of the target protein compared to the control experiment clearly demonstrated that SMPCs efficiently captured the target protein from the live cells.

15. Remove the tubes from the separator and add 500 μL of elution buffer. Elute the proteins by manually moving the elution buffer up and down for 2 min.
16. Place the tubes on the separator for 2 min, remove, and collect the eluate. Repeat step 15 and elute the beads for an additional two times.
17. Pool the three 500 μL eluates in a 1.5-mL plastic tube.
18. For Western blot analysis, prepare the protein samples as described in Subheading 3.2. Figure 5 shows a typical Western blot analysis of the GSK3-β target capture from live cells.

4. Notes

1. We designed the peptides with a C-terminal cysteine residue such that commercially available cross-linkers such as LC-SMCC can be used. 5-FAM is a fluorescein derivative with increased stability compared to fluorescein when coupled to a peptide.
2. Small molecules with a free primary or secondary amine should be reacted first with the cross-linker. If the peptide is added

together with the linker, the free amino group of the peptide will react with the linker and will decrease its availability for the reaction with the small molecule. The reaction can also be performed in DMSO. We did not optimize these reactions with respect to overall SMPC yield (see also Notes 7 and 8).

3. The final concentration of the peptide below 1 mM.
4. Any commercially available quick spin gel filtration columns can be used for this purpose. At this stage, we analyzed the reaction mixture using MALDI-TOF mass spectrometry. We observed multiple peptide species in the sample necessitating HPLC purification of the FLAG®-linker-Bis-III adduct (Fig. 2).
5. The SMPC purification method was initially developed using a TSQ-Quantum series LC-MS system. Based on the elution profile we adopted the parameters which later were used routinely on Agilent HPLC set-up. Mass calculations are as follows: molecular weight (MW) of the peptide = 1,116.46+, MW of the linker (LC-SMCC) = 447.48+, MW of Bisindolylmaleimide-III = 384.4; total = 1,948.34 (NHS reaction with the primary amine yields a byproduct of molecular weight = 115.54 (approx.)) and hence the final molecular weight of the modified FLAG®-coupled Bisindolylmaleimide-III is 1,948.34 − 115.54 = 1,832.8.
6. Instead of determining a molecular extinction coefficient value for Bis-III, we developed a relative scheme to quantify the SMPC concentration by measuring the absorption of unreacted Bis-III (the starting material, which was weighted and quantified). If the small molecule does not absorb in the visible region, nitrogen analysis-based methods can be utilized to measure SMPC concentration.
7. Since we observed that the primary amine of Bis-III can be reacted with LC-SMCC in DMSO, these conditions were used from this stage on.
8. The use of PBS at pH 7.4 resulted in higher yield compared to phosphate buffer at pH 7.2.
9. The SMPC purification method was initially developed using a TSQ-Quantum series LC-MS system and based on the elution profile we adopted the parameters which later were used routinely on Agilent HPLC set-up. Mass calculations are as follows: MW of the peptide = 2,022.96+, MW of the linker (LC-SMCC) = 447.48+, MW of the Bisindolylmaleimide-III = 384.4, total = 2,854.84 (NHS reaction with the primary amine yields a byproduct of molecular weight = 115.54 (approx.)) and hence the final molecular weight of the modified FLAG®-coupled Bisindolylmaleimide-III is 2,854.84 − 115.54 = 2,739.3.
10. Instead of establishing a molecular extinction coefficient value for 5-FAM coupled to the peptide, we developed a relative

scheme to quantify the SMPC concentration by measuring the absorption of uncoupled 5-FAM-TAT-peptide (starting material, which was weighted and quantified).

11. In capturing protein targets from cell-lysate, we typically used a starting protein concentration of 2 mg/mL. Cells can be cultured in large or small quantities to make up to 2 mg/mL concentration. We cultured a large amount of HeLa cells in a bioreactor.
12. One of the advantages of using SMPCs is that either SMPC–protein complexes as a whole or only the proteins can be eluted from the antibody affinity matrix. Because nonspecific binding is an unavoidable feature of this type of experiments, multiple elution methods provide an opportunity to identify the specific SMPC-binding proteins with greater confidence, by comparing the multiple elution profiles and prioritizing the proteins which are commonly identified.
13. The amount of transfer buffer is based on the size of petri dish used to soak the gel. Soaking of the gel is important and any common dish or plastic caps of pipette tip boxes can be used for this purpose.
14. Add a few drops of buffer between the layers to ensure that no bubbles are trapped in between the layers.
15. The transfer efficiency of proteins depends on the size of the protein and may need to be optimized in some cases. A sufficient transfer, which allowed robust Western blot detection, was achieved under the given conditions for GSK3-β.
16. The amount of transfer buffer is based on the size of the petri dish used to soak the membrane.
17. It is advised to use sticky tape to fix the membrane and, after aligning the membrane with the acetate sheet, to mark the edges and sides of the membrane with a permanent ink marker.
18. Before placing the X-ray film over the membrane, we made a cut in the corner of the membrane and used a permanent marker to mark a particular side.
19. The experiment was run in triplicate and hence six wells were used.
20. Do not aspirate the solution to avoid any unwanted removal of loosely attached cells.
21. We found that 45 min incubation was sufficient to capture >90% of the 5-FAM-TAT-peptide. We did not observe a significant increase in target protein capture after an overnight incubation with the antibody.

References

1. Saxena, C., Higgs, R.E., Zhen, E., and Hale, J.E. (2009) Small-molecule affinity chromatography coupled mass spectrometry for drug target deconvolution. *Expert Opinion on Drug Discovery* **4**, 701–714.
2. Hahn, R., Berger, E., Pflegerl, K., and Jungbauer, A. (2003) Directed immobilization of peptide ligands to accessible pore sites by conjugation with a placeholder molecule. *Anal. Chem.* **75**, 543–8.
3. Saxena, C., Zhen, E., Higgs, R.E., and Hale, J.E. (2008) An Immuno-Chemo-Proteomics Method for Drug Target Deconvolution. *J Proteome Res.* 7, 3490–3497.
4. Saxena C., Bonacci, T.M., Huss, K.L., Bloem, L.J., Higgs, R.E., and Hale, J.E. (2009) Capture of Drug Targets from Live Cells Using a Multipurpose Immuno-Chemo-Proteomics Tool. *J Proteome Res.* **8**, 3951–3957.
5. Toullec, D., Pianetti, P., Coste, H., Bellevergue, P., Grand-Perret, T., Ajakane, M., Baudet, V., Boissin, P., Boursier, E., and Loriolle, F. (1991) The bisindolylmaleimide GF 109203X is a potent and selective inhibitor of protein kinase C. *J. Biol. Chem.* **266**, 15771–15781.
6. Bradford, M.M. (1976) A rapid and sensitive method for the quantitation of microgram quantities of protein utilizing the principle of protein dye binding. *Anal. Biochem.* **72**, 248–254.

Chapter 5

A Chemical Proteomic Probe for Detecting Dehydrogenases: *Catechol Rhodanine*

Xia Ge and Daniel S. Sem

Abstract

Inherent complexity of the proteome often demands that it be studied as manageable subsets, termed subproteomes. A subproteome can be defined in a number of ways, although a pragmatic approach is to define it based on common features in an active site that lead to binding of a common small molecule ligand (e.g., a cofactor or a cross-reactive drug lead). The subproteome, so defined, can be purified using that common ligand tethered to a resin, with affinity chromatography. Affinity purification of a subproteome is described in the next chapter. That subproteome can then be analyzed using a common ligand probe, such as a fluorescent common ligand that can be used to stain members of the subproteome in a native gel. Here, we describe such a fluorescent probe, based on a catechol rhodanine acetic acid (CRAA) ligand that binds to dehydrogenases. The CRAA ligand is fluorescent and binds to dehydrogenases at pH > 7, and hence can be used effectively to stain dehydrogenases in native gels to identify what subset of proteins in a mixture are dehydrogenases. Furthermore, if one is designing inhibitors to target one or more of these dehydrogenases, the CRAA staining can be performed in a competitive assay format, with or without inhibitor, to assess the selectivity of the inhibitor for the targeted dehydrogenase. Finally, the CRAA probe is a privileged scaffold for dehydrogenases, and hence can easily be modified to increase affinity for a given dehydrogenase.

Key words: Dehydrogenase, Oxidoreductase, Catechol rhodanine, Chemical proteomics, Subproteome, Staining

1. Introduction

The analysis of families of proteins related by their active site pockets, for instance, by labeling proteins in gels, represents an important chemoproteomics approach (1, 2). This requires a chemical probe that has a detectable group (e.g., a fluorescent label) tethered to an active site ligand (3, 4). The active site ligand might covalently react with a conserved active site residue (5), in which case the

Gerard Drewes and Marcus Bantscheff (eds.), *Chemical Proteomics: Methods and Protocols*, Methods in Molecular Biology, vol. 803, DOI 10.1007/978-1-61779-364-6_5, © Springer Science+Business Media, LLC 2012

probe is called an activity-based probe (6). Such probes have been developed for serine hydrolases and cysteine proteases (7, 8). In contrast, affinity-based probes bind noncovalently, such as ATP tethered to a fluorophor to detect kinases (9, 10). One problematic design constraint for affinity-based probes is that the ligand must be tethered to a detectable group, and the addition of the linker/fluor pair can disrupt binding interactions (11). This is especially problematic since enzyme affinity for cofactors is typically not strong ($K_d > 10\ \mu M$). An improved affinity-based probe would have a high affinity family specific ligand that is itself fluorescent, so that a linker/fluor does not need to be added (12). There are few examples of such probes that can be used in displacement assays, in a native gel (13). The catechol rhodanine acetic acid (CRAA) probe presented herein binds in the NAD(P) cofactor site of dehydrogenases, and its ability to target dehydrogenases broadly has recently been demonstrated (14, 15).

The catechol rhodanine scaffold was also used as a template in a focused combinatorial library, yielding potent (50–200 nM) biligand inhibitors for multiple dehydrogenases (14). Therefore, it serves the dual role of being a fluorescent probe, and a scaffold for a focused library targeting dehydrogenases. It can be used as a stain for dehydrogenases (15), either in its current form or as modified in a focused library. The fluorescent CRAA probe was initially used to inhibit dihydrodipicolinate reductase (DHPR), an anti-infective drug target. It also shows in-gel binding to two lactate dehydrogenase (LDH) isozymes and to 1-deoxy-d-xylulose-5-phosphate reductoisomerase (DXPR), making it a generally useful staining reagent for dehydrogenases. Because binding is noncovalent, it can also be used in a displacement assay performed in a native gel, by monitoring a decrease in fluorescent band intensity as the NAD(P) pocket is occupied by a competing ligand. Finally, the CRAA probe can be used as a stain for mixtures of proteins, to profile the mixture in terms of proteins that are likely to be dehydrogenases. The CRAA probe has been shown, using affinity chromatography of human liver and *Mycobacterium tuberculosis* proteome samples followed by tandem mass spectrometry, to exert cross-reactivity with various dehydrogenases, and yet have some selectivity for the dehydrogenase family (16).

2. Materials

2.1. Proteins and Staining Reagents

1. *Escherichia coli* DHPR expressed in *E. coli* (BL21) and purified following previously described methods (17).
2. l-Lactic acid dehydrogenase from bovine heart (LDH, Sigma-Aldrich).

3. *Escherichia coli* DOXPR expressed and purified from *E. coli* (a gift from Triad Therapeutics).
4. CRAA was prepared and purified as described in the next chapter.
5. CRAA staining buffer: dissolve CRAA (2 mM) in 25 mM Tris–HCl, pH 8.5.
6. CRAA competition buffer: CRAA staining buffer plus 0.5 mM NADH.

2.2. Native Gel Electrophoresis

1. Bio-Rad protein assay reagent.
2. NuPage™ and Novex™ products for native gel and SDS-PAGE (Invitrogen).
3. Native stain: 15.5 mL of 1 M Tris–HCl, 2.5 mL of a 1% solution of Bromophenol blue, 7.0 mL of water, and 25 mL of glycerol, pH 6.8.
4. Tris–Glycine running buffer: 3.0 g of Tris base and 14.4 g of glycine in 1 L of distilled water. Adjust the pH to 8.3.
5. NativeMark™ molecular weight standards (Invitrogen).
6. HP 8452A diode array spectrophotometer for UV-Vis measurements and 1 mL quartz cuvette. Measurements should be taken at 25°C.

2.3. In-Gel Staining of the Dehydrogenase Subproteome Using CRAA

1. Phosphate-buffered saline (PBS): 8 g NaCl, 0.2 g KCl, 1.44 g Na_2HPO_4, 0.24 g KH_2PO_4 in 1 L of distilled water, pH 7.4.
2. Blocking buffer: 2.5% nonfat drug milk in PBS (18).
3. CRAA staining solution: 2.0 mM CRAA in 25 mM Tris–HCl, pH 8.5.
4. Fixing solution: 50% methanol, 10% acetic acid, and 40% deionized water.
5. Coomassie blue staining solution: 0.1% Coomassie Brilliant Blue R-250 in 50% methanol and 10% glacial acetic acid.
6. Destaining solution: 10% glacial acetic acid, 40% methanol, and 50% deionized water.
7. Kodak Image Station 2000MMT System for in-gel fluorescence scanning (IGFS).
8. Canon CanoScan (D1250 U2F) scanner for visible imaging of gel bands.

3. Methods

CRAA is a privileged scaffold, in that it binds to many dehydrogenases. It is visibly colored and is also fluorescent under slightly basic conditions (pH above 7). It binds to dehydrogenases with its

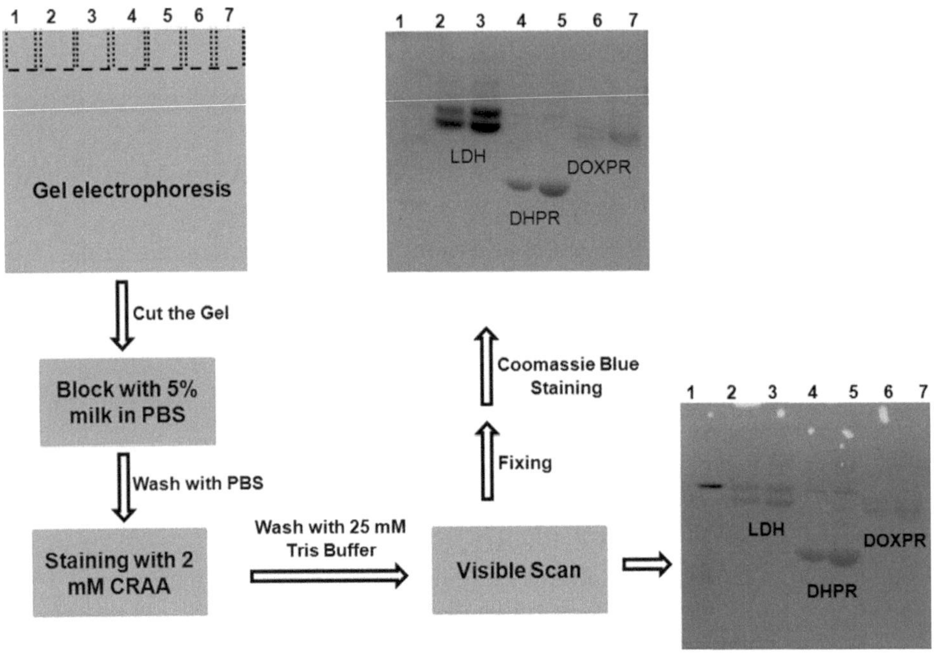

Fig. 1. Schematic description of the process whereby a native gel of dehydrogenase proteins is run, then stained using the CRAA probe.

para-phenol and carboxylic acid in their deprotonated form (Fig. 1). Since the binding is noncovalent, it is reversible so that CRAA can be displaced by a higher affinity ligand like NADH. This makes CRAA a useful reagent for detecting dehydrogenases in native gels using both direct binding and displacement assays (Fig. 2).

3.1. Preparation of the CRAA Staining Reagent

CRAA staining buffer should be prepared immediately before use. For best staining results, it is better to use freshly prepared CRAA buffer, since CRAA can be oxidized during storage (see Note 1).

3.2. Native Gel Electrophoresis

1. Prepare 10 μL aliquots of different concentrations of proteins, using 2.5 μL of native stain. Protein samples are always stored in an ice bath before use (see Note 2).
2. Set two 10% Tris–Glycine native gels up according to the manufacturer's instructions using Tris–Glycine running buffer. Load the molecular weight standard and the above protein samples on each of the two gels. Run the gel at 125 V for ~2.0 h, or until the Bromphenol blue front approaches the bottom of the gel (see Note 3).
3. Remove the gels from the cassette.

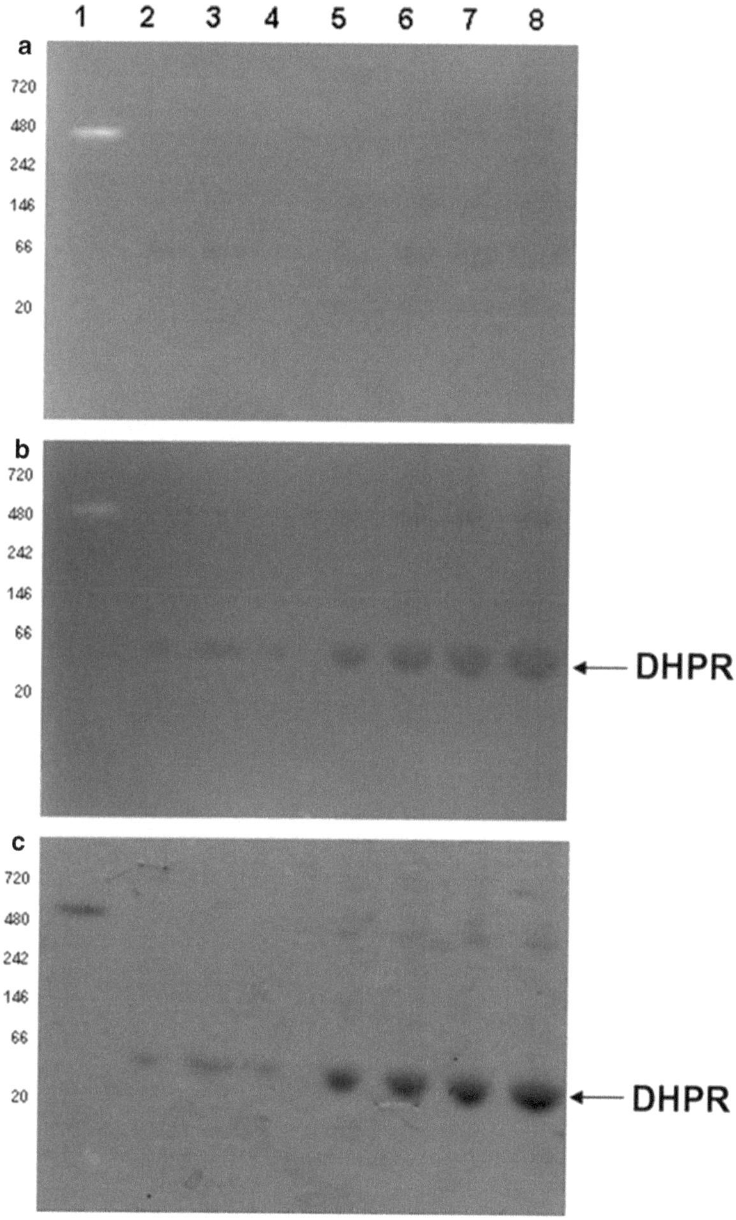

Fig. 2. CRAA staining of a dehydrogenase (DHPR) and fluorescence imaging. (**a**) Two native (10% Tris–glycine) gels were run, and DHPR was stained using 2.0 mM CRAA. *Lane 1*, NativeMark protein standard. *Lanes 2–8*: *E. coli* DHPR (10 μL) at concentrations of 0.22, 0.43, 0.86, 1.29, 1.72, 2.59, and 3.45 μg/μL. (A) Staining with CRAA at pH 6.5 with fluorescence imaging (Excitation at 465 nm, detection at 535 nm). (B) Same as in (A), but stained at pH 8.5. (C) Same as in (B), but imaged using a Canon CanoScan (D1250 U2F) scanner. Data modified from ref. 15. (**b**) In-gel displacement assay. Two native gels were loaded with the same concentrations of *E. coli* DHPR. *Lane 1*: NativeMark MW standard. *Lanes 2–11*: *E. coli* DHPR (10 μL) at concentrations of 0.0072, 0.014, 0.028, 0.072, 0.143, 0.29, 0.72, 1.4, 2.2, and 4.3 μg/μL, respectively. All gels were scanned as in (C). (A) Staining with 0.5 mM NADH and 2.0 mM CRAA (pH 7.8). The gel showed no bands for CRAA bound to DHPR, due to displacement by NADH. (B) Staining as in (A), but in the absence of NADH competitor. Lowest detectable concentration of DHPR was 0.14 μg/μL. (C) The same gel stained with Coomassie blue. The lowest detectable concentration of a DHPR band was 0.072 μg/μL. Data modified from ref. 15.

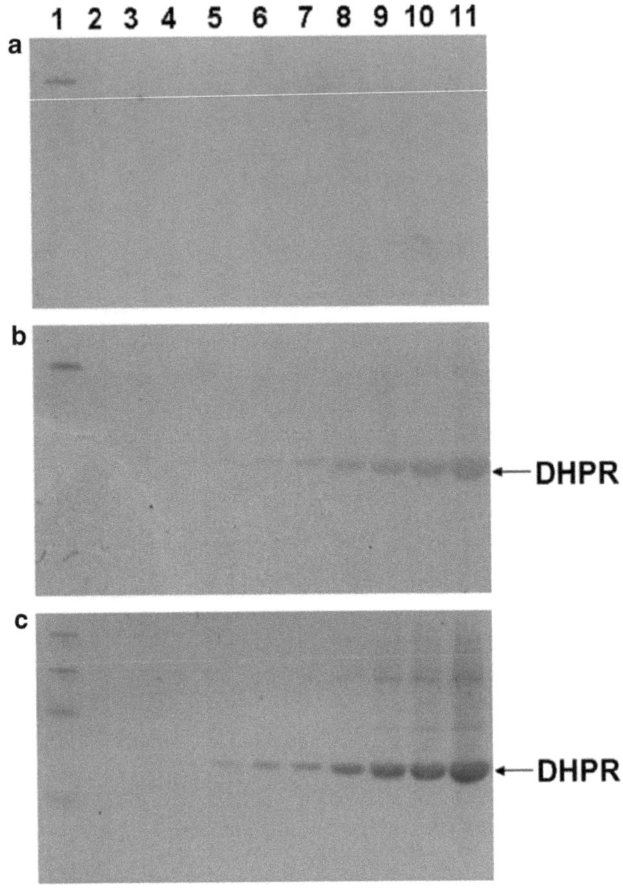

Fig. 2. (continued)

3.3. In-Gel Staining of the Dehydrogenase Subproteome Using CRAA

1. Incubate the first native gel from the previous section with blocking buffer for 20 min, using gentle agitation on an orbital shaker (see Note 4).
2. Rinse the gel with PBS buffer for 20 min before staining.
3. Stain the gel with CRAA staining solution for 20 min (see Note 5).
4. Rinse the gel for about 1 h on an orbital shaker with 25 mM Tris–HCl buffer, pH 8.5, until a clear band is seen as a result of binding of CRAA. The background of the gel turns a pale yellowish color (Fig. 1). The CRAA bands can be imaged with either a fluorescence reader or simply by visible imaging using a document scanner.
5. Fix the gel with fixing solution by incubating it for 15 min (see Note 6). This completes the staining procedure.

6. For comparison, stain the second gel with Coomassie blue staining solution. The gel is incubated on an orbital shaker for 1–3 h.
7. Destain the gel with destaining solution over a period of 1 h.
8. Wash the destained gels with water, then soak in 10% glycerol for 15 min.
9. Wrap the gels in a sheet of cellophane. Add a small amount of 10% glycerol to moisten the cellophane. Be careful to remove any air bubbles which may be trapped between the gel and the cellophane. Fix the wrapped gel between plastic plates and fix it with binder clips (see Note 7).
10. Place the plates with the gels standing upright on a shelf at room temperature and allow them to dry overnight.
11. Remove the gel from the holder. The gel is now ready to be scanned.

3.4. In-Gel Staining in a Competition-Binding Study

1. For the displacement assay, two native gels are run at the same time, and pretreated with blocking buffer (5% milk in pH 7.4 PBS).
2. The first gel is incubated in CRAA competition buffer (contains NADH to compete with CRAA binding).
3. The second gel is incubated in CRAA staining buffer.
4. The gels are rinsed for about 1 h until clear red CRAA-stained bands are observed, clearly distinguished from the pale yellow background (see Note 8).
5. The gel images are recorded.
6. After staining with CRAA and recording the gel images, the gels can be further stained using Coomassie blue to identify all proteins and their positions relative to the molecular weight standards, which do not stain with CRAA (Fig. 3).

3.5. Imaging the Gel

1. The gel can be scanned with a document scanner; the visible red band from CRAA binding to dehydrogenases can be recorded using any common document scanning device.
2. The gel can also be fluorescently imaged; for example, using a Kodak Image Station 2000MM system. CRAA binding to dehydrogenases shows fluorescent bands. Gels are excited at ~465 nm with detection at ~535 nm, using the filters on the Kodak Image Station.
3. Coomassie blue-stained gels are recorded using a Canon document scanner.

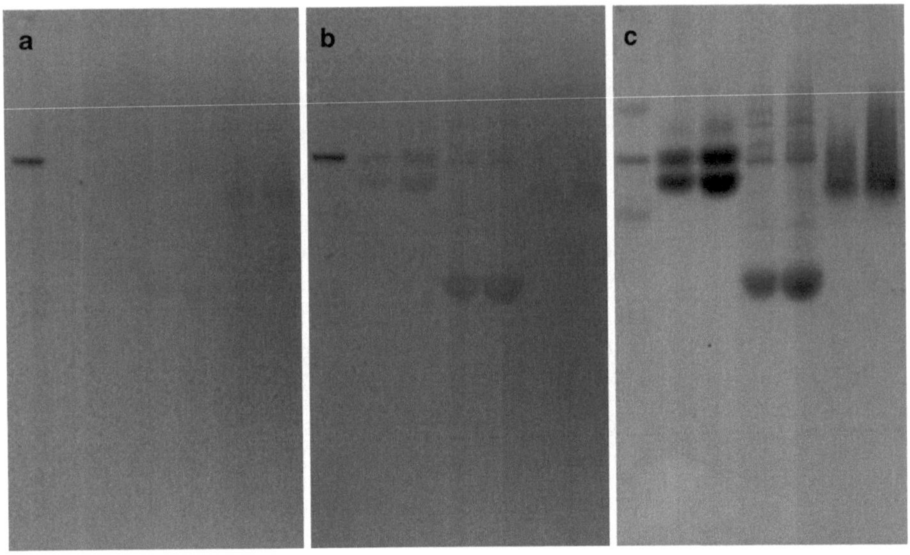

Fig. 3. CRAA staining of multiple dehydrogenases: competition ± NADH. *Lane 1*: NativeMark protein standard. *Lanes 2 and 3*: LDH at concentrations of 5.4 and 10.8 μg/μL. *Lanes 4 and 5*: *E. coli* DHPR at concentrations of 2.2 and 4.4 μg/μL. *Lanes 6 and 7*: *E. coli* DOXPR at concentrations of 0.625 and 1.25 μg/μL. The gels were stained with 2.0 mM CRAA (pH 7.8) in the (**a**) presence of NADH (0.5 mM) or (**b**) absence of NADH. (**c**) The same gel was also stained with Coomassie blue. Data modified from ref. 15.

4. Notes

1. CRAA has low solubility in water, but it can be dissolved to ~4 mM at pH 7.8. CRAA can be oxidized when in solution. For this reason, it is best to prepare CRAA staining buffer fresh, as needed. 2 mM CRAA was chosen to stain dehydrogenases in native gels, to lessen the CRAA yellowish background on the gel.
2. Only native loading buffer (i.e., no SDS) can be used to prepare protein samples. If SDS-PAGE gel loading buffer is erroneously used, protein could be denatured and will therefore not bind to CRAA when staining.
3. The 10% Tris–Glycine Novex® native gel does not have easily visible wells when the comb is removed. It is helpful to label each well with a marker, for correct loading of protein samples in the right position.
4. Gels were soaked in 5% milk in pH 7.4 PBS buffer for 20 min to prevent nonspecific binding of CRAA to the gel itself. If gels were stained with CRAA buffer directly, it usually showed a reddish background and takes much longer to rinse the CRAA background away. However, too extensive washing time may wash away stained dehydrogenases as well.

5. After staining with CRAA for about 20 min, the gel should be washed with 25 mM Tris-HCl buffer for about 1 h to achieve the best contrast. If CRAA staining was too long, it will take a longer time to wash away the background CRAA on the gel.
6. The native gels should be fixed first before staining to avoid any loss of protein from the native gel.
7. When drying the gel with gel drying cellophane film, it is very important to make sure there are no air bubbles between the gel and films, or else the gel will be distorted when dried.
8. If CRAA is used in the competition binding with NADH or other NADH-competitive inhibitors, NADH (or the inhibitor of interest) needs to be present in both the staining and washing process, because CRAA can competitively bind to the dehydrogenases on the native gel at any stage of the process (CRAA washing out of the background can then bind to dehydrogenases in the gel).

Acknowledgments

This work was supported in part by GM085739 (NIH) and shared instrumentation grants S10 RR019012 (NIH) and CHE-0521323 (NSF).

References

1. Fonovic, M., and Bogyo, M. (2007) Activity Based Probes for Proteases: Applications to Biomarker Discovery, Molecular Imaging and Drug Screening. *Curr. Pharm. Des.* 13, 253–261.
2. Sieber, S. A., and Cravatt, B. F. (2006) Analytical Platforms for Activity-Based Protein Profiling – Exploiting the Versatility of Chemistry for Functional Proteomics. *Chem. Commun. (Camb).* 22, 2311–2319.
3. Sadaghiani, A. M., Verhelst, S. H., and Bogyo, M. (2007) Tagging and Detection Strategies for Activity-Based Proteomics. *Curr. Opin. Chem. Biol.* 11, 20–28.
4. Jeffery, D. A., and Bogyo, M. (2003) Chemical Proteomics and Its Application to Drug Discovery. *Curr. Opin. Biotechnol.* 14, 87–95.
5. Kato, D., Boatright, K. M., Berger, A. B., Nazif, T., Blum, G., Ryan, C., Chehade, K. A., Salvesen, G. S., and Bogyo, M. (2005) Activity-Based Probes that Target Diverse Cysteine Protease Families. *Nat. Chem. Biol.* 1, 33–38.
6. Hagenstein, M. C., and Sewald, N. (2006) Chemical Tools for Activity-Based Proteomics. *J. Biotechnol.* 124, 56–73.
7. Kidd, D., Liu, Y., and Cravatt, B. F. (2001) Profiling Serine Hydrolase Activities in Complex Proteomes. *Biochemistry.* 40, 4005–4015.
8. van Swieten, P. F., Maehr, R., van den Nieuwendijk, A. M., Kessler, B. M., Reich, M., Wong, C. S., Kalbacher, H., Leeuwenburgh, M. A., Driessen, C., van der Marel, G. A., Ploegh, H. L., and Overkleeft, H. S. (2004) Development of an Isotope-Coded Activity-Based Probe for the Quantitative Profiling of Cysteine Proteases. *Bioorg. Med. Chem. Lett.* 14, 3131–3134.
9. Ratcliffe, S. J., Yi, T., and Khandekar, S. S. (2007) Synthesis and Characterization of 5′-p-Fluorosulfonylbenzoyl-2′(Or3′)-(Biotinyl) Adenosine as an Activity-Based Probe for Protein Kinases. *J. Biomol. Screen.* 12, 126–132.
10. Chan, E. W., Chattopadhaya, S., Panicker, R. C., Huang, X., and Yao, S. Q. (2004) Developing

Photoactive Affinity Probes for Proteomic Profiling: Hydroxamate-Based Probes for Metalloproteases. *J. Am. Chem. Soc.* 126, 14435–14446.

11. Patricelli, M. P. (2002) Activity-Based Probes for Functional Proteomics. *Brief Funct. Genomic Proteomic.* 1, 151–158.
12. Yee, D. J., Balsanek, V., and Sames, D. (2004) New Tools for Molecular Imaging of Redox Metabolism: Development of a Fluorogenic Probe for 3 Alpha-Hydroxysteroid Dehydrogenases. *J. Am. Chem. Soc.* 126, 2282–2283.
13. Speers, A. E., and Cravatt, B. F. (2004) Chemical Strategies for Activity-Based Proteomics. *Chembiochem.* 5, 41–47.
14. Sem, D. S., Bertolaet, B., Baker, B., Chang, E., Costache, A. D., Coutts, S., Dong, Q., Hansen, M., Hong, V., Huang, X., Jack, R. M., Kho, R., Lang, H., Ma, C. T., Meininger, D., Pellecchia, M., Pierre, F., Villar, H., and Yu, L. (2004) Systems-Based Design of Bi-Ligand Inhibitors of Oxidoreductases: Filling the Chemical Proteomic Toolbox. *Chem. Biol.* 11, 185–194.
15. Ge, X., and Sem, D. S. (2007) Affinity-Based Chemical Proteomic Probe for Dehydrogenases: Fluorescence and Visible Binding Assays in Gels. *Anal. Biochem.* 370, 171–179.
16. Ge, X., Wakim, B., and Sem, D. S. (2008) Chemical Proteomics-based Drug Design: Target and Antitarget Fishing with a Catechol Rhodanine Privileged Scaffold for NAD(P)(H) Binding Proteins. *J. Med. Chem.* 51, 4571–4580.
17. Reddy, S. G., Sacchettini, J. C., and Blanchard, J. S. (1995) Expression, Purification, and Characterization of Escherichia Coli Dihydrodipicolinate Reductase. *Biochemistry.* 34, 3492–3501.
18. Duhamel, R. C., and Johnson, D. A. (1985) Use of Nonfat Dry Milk to Block Nonspecific Nuclear and Membrane Staining by Avidin Conjugates. *J. Histochem. Cytochem.* 33, 711–714.

Chapter 6

Probing Proteomes with Benzophenone Photoprobes

Akira Kawamura and Doina M. Mihai

Abstract

Benzophenone photoprobes represent powerful tools for chemical proteomics. Upon UV irradiation, a benzophenone photoprobe can selectively form a covalent bond with its target protein in complex protein mixtures. Thus, photoprobes can be used to profile a wide variety of proteins in complex proteomes. This chapter describes simple protocols to derivatize fluorenylmethyloxycarbonyl (Fmoc)-protected peptide-nucleic-acid adenine (PNA adenine) into a benzophenone photoprobe and its application in photolabeling its target proteins. The method as described does not require specialized equipment for probe synthesis and photolabeling. In addition, the strategy is applicable to recognition motifs other than PNA adenine, such as peptides, to profile their target proteins in complex proteomes.

Key words: Photoaffinity-labeling, PNA, Adenine, Solid-phase synthesis, Chemical proteomics

1. Introduction

Photoaffinity-labeling has been employed widely to study protein–ligand interactions (1). In this approach, ligands of interest are derivatized with photoactivatable functional groups, such as benzophenone, and used to photocrosslink its target proteins. In theory, photoaffinity-labeling can be employed in chemical proteomics, in which small molecular probes are used to tag functionally or structurally analogous proteins in complex proteomes and analyze their expression and functional states. Yet, compared to the widespread use of isotope-coded affinity tags (ICATs) (2) and activity-based protein profiling (ABPP) (3), the use of photoprobes in chemical proteomics is less widespread (4–11). This is probably because of the chemistry involved in probe synthesis and because photolabeling is perceived difficult. Photoaffinity-labeling as well as synthesis of photoprobes, however, can be carried out by biochemists and

Gerard Drewes and Marcus Bantscheff (eds.), *Chemical Proteomics: Methods and Protocols*, Methods in Molecular Biology, vol. 803, DOI 10.1007/978-1-61779-364-6_6, © Springer Science+Business Media, LLC 2012

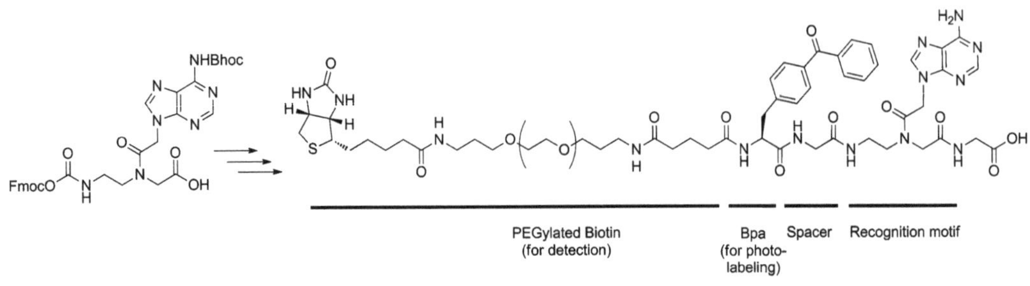

Fig. 1. Structures of Fmoc-protected PNA-adenine monomer and PNA-adenine photoprobe. PNA adenine is used as a recognition motif in our proteomic profiling studies for several reasons. First, adenine is a ubiquitous recognition motif for various proteins, including kinases and transcription factors. Second, the PNA backbone permits the use of simple solid-phase chemistry to append photoactive functional groups and detection tags, such as benzophenone and PEGylated biotin, respectively.

biologists without the use of special equipment for probe synthesis and photolabeling.

The current chapter describes how to design, synthesize, and use photoprobes for photoaffinity-labeling of target proteins in complex proteomes. Specifically, the chapter focuses on peptide-nucleic-acid (PNA) adenine as a recognition motif (molecular bait) for proteins, and explains a simple procedure to prepare a benzophenone photoprobe from fluorenylmethyloxycarbonyl (Fmoc)-protected PNA-adenine monomer (Fig. 1). Photoaffinity-labeling with the resulting photoprobe can be reproducibly carried out with a commercially available UV transilluminator equipped with standard UV lamps. It should be noted that the procedures described in this chapter are applicable to other Fmoc-protected recognition motifs with a carboxylic acid. For example, the same procedure can be used to convert Fmoc-protected peptides into photoprobes and to analyze their target proteins in complex proteomes.

2. Materials

2.1. Solid-Phase Synthesis of the Benzophenone Photoprobe from a PNA-Adenine Monomer

1. Filter funnel, 6 ml, 25–50 μm frit, 5 mm tube fitting with PP screw cap.
2. Anhydrous dimethylformamide (DMF). Used without further purification.
3. *N*-Methyl-pyrolidone. Used without further purification.
4. Fmoc-Gly-Wang resin, Fmoc-glycine (Fmoc-Gly-OH), Fmoc-4-benzoyl-L-phenylaniline (Fmoc-L-Bpa), dicyclohexylcarbodiimide (DCC), and 1-hydroxybenzotriazole (HOBt) (Fluka). Used without further purification.

5. Fmoc-PNA-adenine monomer (ASM Research Chemicals, Burgwedel, Germany). The primary amine on adenine is protected by benzhydryloxycarbonyl (Bhoc) group which is removed at the end of photoprobe synthesis.
6. Fmoc-deprotection solution (20% piperidine in DMF). Prepare fresh before the experiment.
7. PNA-adenine (recognition motif) coupling solution: Dissolve Fmoc-PNA-adenine (151 mg, 0.30 mmol, 2 eq), HOBt (0.041 ml, 0.30 mmol, 2 eq), and DCC (62 mg, 0.30 mmol, 2 eq) in 2 ml of DMF. Prepare fresh before the experiment (see Note 1).
8. Gly (spacer) coupling solution: Dissolve Fmoc-Gly-OH (90 mg, 0.30 mmol, 2 eq), HOBt (0.041 ml, 0.30 mmol, 2 eq), and DCC (62 mg, 0.30 mmol, 2 eq) in 2 ml of DMF. Prepare fresh before the experiment (see Note 1).
9. Bpa (photoactive group) coupling solution: Dissolve Fmoc-L-Bpa-OH (148 mg, 0.30 mmol, 2 eq), HOBt (0.041 ml, 0.30 mmol, 2 eq), and DCC (62 mg, 0.30 mmol, 2 eq) in 2 ml of DMF. Bpa is a photosensitive chemical. Whenever possible, keep the solution in dark and avoid exposure to UV lights. Prepare fresh before the experiment (see Note 1).
10. PEGylated biotin (detection motif) coupling solution: Dissolve *N*-biotinyl-NH-$(PEG)_2$-COOH (210 mg, 0.30 mmol, 2 eq) (Novabiochem, Switzerland; EMD Chemicals), HOBt (0.041 ml, 0.30 mmol, 2 eq), and DCC (62 mg, 0.30 mmol, 2 eq) in 2 ml of DMF. Prepare fresh before the experiment (see Note 1).
11. Resin-cleavage solution (15 ml): Mix 95% trifluoroacetic acid (TFA), 2.5% triisopropylsilane (TIS), and 2.5% water solution. Prepare fresh before the experiment.
12. Petroleum ether, filter paper (Fisherbrand, P2, Qualitative), HPLC grade methanol. Solvents are used without further purification.

2.2. Jurkat Cell Culture and Preparation of Cytosolic Lysate

1. RPMI-10: RPMI-1640 medium supplemented with 10% fetal bovine serum and 1% penicillin-streptomycin-fungizone (Lonza).
2. 1× Hypotonic lysis buffer: 10 mM HEPES, 1.5 mM $MgCl_2$, 10 mM KCl, 0.1 mM EDTA, and 1 mM phenylmethanesulfonylfluoride (PMSF), pH 7.3. 10× Hypotonic lysis buffer without PMSF can be prepared and stored at room temperature. PMSF is added just before the experiment.

2.3. Photolabeling

1. Jurkat cytosol lysates (adjusted to 2 μg/μl with 1× Tris-buffered saline (TBS)).
2. 3.7 mM Photoprobe solution in DMSO. Whenever possible, avoid exposure to light.

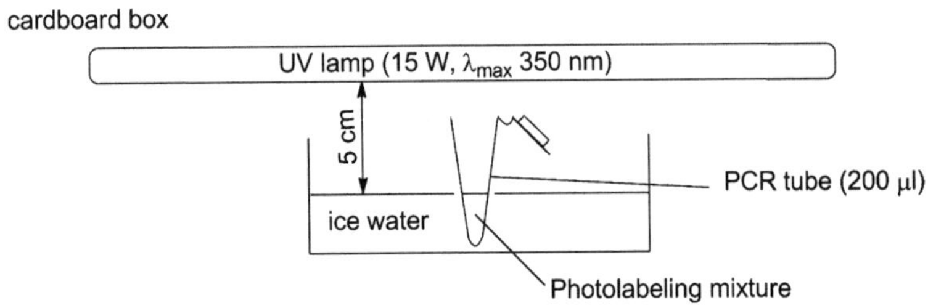

Fig. 2. Schematic diagram of photolabeling apparatus. A commercial UV transilluminator with Sylvania 350 Blacklight lamps (15 W, λ_{max} 350 nm; or equivalent) can be used to set up the photolabeling apparatus for benzophenone photoprobes. Note that safety precautions must be followed to prevent UV exposure and electrical accidents before and during the photolabeling experiment. For efficient photolabeling, the sample should be placed directly underneath the UV lamp and positioned so that the distance between the UV lamp and the photochemical reaction is 5 cm. Since the UV lamp generates heat, ice bath has to be constantly monitored, and ice should be replenished as needed.

3. Crushed ice and ice water.
4. A small UV transilluminator equipped with Sylvania 350 Blacklight lamps (15 W, λ_{max} 350 nm) or equivalent.
5. Power surge protector.
6. A cardboard box, large enough to cover the entire photolabeling apparatus (Fig. 2).

2.4. SDS-Polyacrylamide Gel Electrophoresis and Western Blotting

1. Laemmli sample buffer (BioRad).
2. 10% Polyacrylamide gel 12+2 well, 45 µl, 13.3×8.7 cm (Criterion, BioRad).
3. 1× Tris–Glycine–SDS (TGS) buffer (BioRad).
4. 1× Tris–Glycine (TG) buffer (BioRad).
5. 1× TBS with 0.1% Tween-20 (TBS-T).
6. Millipore Immobilon-P polyvinylidene fluoride (PVDF) membrane (Millipore).
7. Streptavidin–horseradish peroxidase (HRP) conjugate (BioRad).
8. ECL-Plus chemiluminescence reagent (Amersham Biosciences).

3. Methods

3.1. Probe Design Considerations

This section describes the key structural considerations that are important for successful photolabeling experiments with the benzophenone photoprobe (Fig. 1).

1. *Position of the recognition motif.* The position of the adenine recognition motif is an important parameter that significantly affects the target selectivity of photoprobes. If the benzophenone

group is placed at the end of the molecule and the adenine motif is buried in the middle, the resulting probe tends to photocrosslink proteins that recognize the benzophenone moiety instead of adenine (12). The placement of adenine at the end of the photoprobe, on the other hand, results in selective photolabeling of adenine-binding protein in complex proteomes (13). Thus, the recognition motif should be positioned at the end of the photoprobe to maximize its exposure for protein recognition.

2. *Structural requirements for the recognition motif.* Prior to the synthesis, the recognition motif has to be suitably protected such that it is compatible with the Fmoc-chemistry on solid phase (14). The minimum structural requirements are a Fmoc-protected amine and a free carboxylic acid. If a recognition motif of interest contains other reactive functional groups, they have to be capped with orthogonal protecting groups that are compatible with the coupling reactions. For example, the primary amine group on the adenine moiety of PNA-adenine monomer is protected with the Bhoc group, which is removed at the final step of the synthesis (Fig. 1).
3. *The role of spacer and its flexibility.* Our photoprobe contains a glycine residue as a spacer between the adenine and benzophenone motifs. The flexibility of this spacer is important for efficient photolabeling of target proteins (15). A conformationally constrained spacer, such as proline, can dramatically reduce photolabeling efficiency of target proteins because the benzophenone group needs flexibility to attain stereoelectronic requirements for hydrogen abstraction and radical recombination (16, 17). On the other hand, more flexible spacers, such as γ-aminobutyric acid (GABA), can improve photolabeling efficiency (15).
4. *Stereochemistry of Bpa.* Both L- and D-isomers of Fmoc-protected Bpa are commercially available. Although the L-isomer is most commonly used in photoaffinity-labeling studies, it does not necessarily photolabel target proteins of interest more efficiently than the D-isomer. Our previous results suggest that the stereochemistry of Bpa can modulate proteomic profiles of benzophenone photoprobes (12).
5. *Detection tag.* Our photoprobe contains a PEGylated biotin moiety to visualize tagged proteins, for instance, by Western blot. Through this PEGylated biotin moiety, photolabeled proteins can be readily detected with streptavidin–HRP conjugate and chemiluminescence reagent. This moiety can also be used for the affinity purification of photolabeled proteins. Instead of the large PEGylated biotin, however, chemical tags for Click chemistry, such as 5-hexynoic acid, can be attached at this position of

the photoprobe. This would permit modification of photolabeled proteins with various detection tags, such as fluorophores and biotin, after the photocrosslinking reaction (18).

3.2. Solid-Phase Synthesis of the Benzophenone Photoprobe from a PNA-Adenine Monomer

It is important that standard safety measures for organic synthesis, which include wearing protective goggles, gloves, and a lab coat, must be followed during all chemical synthesis procedures.

1. Transfer Fmoc-Gly-Wang resin (200 mg, 0.15 mmol, loading of 0.75 mmol/g, 1 eq.) into a 6-ml filter funnel.
2. To swell the resin, add 5 ml of *N*-methyl-pyrolidone and let the mixture stand at room temperature for 1 h.
3. Drain the *N*-methyl-pyrolidone and wash the resin three times with DMF (5 ml × 3) (see Note 2).
4. To remove the Fmoc-protection group on the resin, soak the resin in the "Fmoc-deprotection solution" (5 ml) for 5 min. Drain the solution. To ensure the complete removal of the Fmoc-group, soak the resin again in the Fmoc-deprotection solution (5 ml) for 20 min, and then drain the solution (see Note 3).
5. Wash the resin three times with 5 ml of DMF.
6. To immobilize the recognition motif, add 2 ml of the "PNA-adenine (recognition motif) coupling solution" to the resin (see Note 2).
7. Shake the mixture at room temperature overnight.
8. Drain the solution. Then, wash the resin three times with 5 ml of DMF. Further wash the resin three times with 5 ml of dichloromethane/methanol (1:1) (see Note 3).
9. To remove the Fmoc-protection group, soak the resin in the "Fmoc-deprotection solution" (5 ml) for 5 min. Drain the solution. To ensure the complete removal of the Fmoc-group, soak the resin again in the "Fmoc-deprotection solution" (5 ml) for 20 min, and then drain the solution (see Note 3).
10. To couple the spacer moiety, add 2 ml of the "Gly (spacer) coupling solution" to the resin (see Note 2).
11. Repeat steps 7–9.
12. To couple the photoaffinity group, add 2 ml of the "Bpa (photoactive group) coupling solution" to the resin (see Note 1).
13. Repeat steps 7–9.
14. To couple the detection motif, add 2 ml of the "PEGylated biotin (detection motif) coupling solution" to the resin (see Note 1).
15. Repeat steps 7 and 8.

16. To cleave the product from the resin and deprotect the primary amine on the adenine moiety, add 5 ml "resin-cleavage solution" and let the mixture stand at room temperature for 1 h.
17. Collect the liquid phase into a round bottomed flask. Add 1 ml of the "resin-cleavage solution" to the resin and collect the rinse into the same round-bottomed flask.
18. To ensure the complete cleavage of the product from the resin, add 5 ml of the "resin-cleavage solution" to the resin and incubate at room temperature for 30 min. Collect the solution into the same round-bottomed flask. Rinse the resin again with 1 ml of the "resin-cleavage solution".
19. Remove the solvent under reduced pressure (e.g., a rotary evaporator). Viscous oil remains at the bottom of the flask.
20. To the viscous oil, add cold petroleum ether (5–10 ml, until a precipitate forms). Store the solution at −20°C overnight (see Note 4).
21. Collect the precipitated product by vacuum filtration using a qualitative filter paper.
22. Dry the precipitate in vacuo for 2–3 h: The sample may be dried using a regular vacuum pump.
23. Purify the crude product with reversed-phase HPLC. For preparative HPLC, an Econosil C-18 10 μ semipreparative column (10 × 250 mm) can be used, with a linear gradient from 20 to 100% methanol aqueous over 30 min (flow rate 3 ml/min). The product can be detected by UV absorption at 220 and 280 nm (see Note 5). To avoid the activation of benzophenone, do not use UV 350 nm for product detection.

3.3. Jurkat Cell Culture and Preparation of Cytosolic Lysate

1. Jurkat cells are maintained in RPMI-10 in a 37°C incubator at 5% CO_2. For each labeling experiment, cells are harvested during the log phase (50–70% confluency) and plated onto a 10-cm tissue culture dish on the day before the experiment and left overnight in the incubator.
2. To obtain cytosolic lysates of Jurkat cells, cells are pelleted by centrifugation and then suspended in 1× hypotonic lysis buffer (1 ml). The cell suspension is incubated at room temperature for 10 min and the swollen cells are lysed by gentle pipetting. The lysates are centrifuged in a microfuge (9,300 × *g*, 4°C) and the supernatant (cytosolic lysate) is carefully transferred into a new Eppendorf tube on ice. The protein concentration in the samples is determined with the Promega Coomassie Plus™ Protein Assay Reagent (Promega) or equivalent (see Note 6). Prior to the photolabeling experiment, the final protein concentration is adjusted to 2 μg/μl with 1× TBS. The best result is obtained when protein solutions are kept on ice

or in a refrigerator (4°C) and used for photolabeling studies on the same day.

3.4. Photolabeling

It is important that suitable safety precautions are followed to prevent UV exposure and electrical accidents during the photolabeling procedures!

1. 20 μl Aliquots of cytosolic lysate (2 μg/μl) are mixed with 3.7 mM photoprobe solution (final probe dilutions are 1:20, 1:40, 1:100, and 1:200 in the lysate) in PCR tubes. Place the tubes on ice and incubate for 30 min. Cover the tubes with aluminum foil to minimize the exposure of the sample to light. After 30 min, remove the aluminum foil, and securely position the tubes in an ice-water bath.
2. Ensure that the UV transilluminator is unplugged. Remove the UV blocking cover from the transilluminator. Carefully turn over the UV transilluminator and securely position it over each photolabeling reaction mixture so that one UV lamp (Sylvania 350 Blacklight lamps, 15 W, λ_{max} 350 nm) is directly above the opening of each PCR tube (Fig. 2). The distance between the lamp and the surface of the photolabeling solution should be kept at 5 cm (see Note 7).
3. Plug the UV transilluminator into a power surge protector. Cover the entire apparatus in a cardboard box to avoid any exposure of skin and eyes to the UV light. Then, turn on the UV lamp through the surge protector. Since the UV lamp generates heat, check the ice-water bath every 10–15 min and when doing so, make sure to turn off the lamp. If needed add ice and adjust the water level. Continue UV irradiation for a total period of 1 h (see Note 8).

3.5. SDS-Polyacrylamide Gel Electrophoresis and Western Blotting

1. Following the photocrosslinking, the sample is mixed with Laemmli sample buffer with 5% (v/v) 2-mercaptoethanol, denatured at 80°C for 5 min, and quickly chilled on ice.
2. The denatured protein samples are separated by SDS-polyacrylamide gel electrophoresis (SDS-PAGE) (10% Tris–HCl gel, 200 V, 1 h) in 1× TGS.
3. The gel is blotted onto PVDF membrane (200 mA, 2 h) in a cold transfer buffer (20% methanol in 1× TG).
4. The blotted membrane is blocked with 5% nonfat milk in TBS-T (50 ml) at room temperature for 1 h. Blocked membrane is rinsed with TBS-T (50 ml) three times (10 min, 3×), treated with streptavidin–HRP conjugate (1:3,000 dilution in 3% nonfat milk in TBS-T, 50 ml) for 30 min, and washed with 50 ml of TBS-T three times for 30 min.
5. The washed membrane is treated with the ECL-Plus chemiluminescence reagent (Amersham Biosciences) for 5 min. Tagged

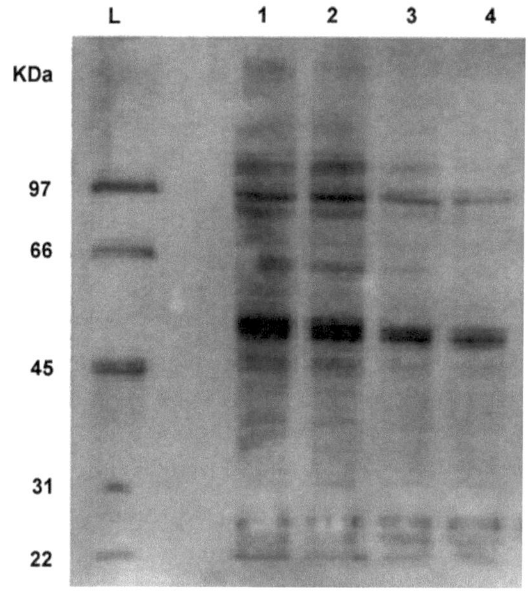

Fig. 3. Photolabeling of Jurkat cytosolic proteins with PNA-adenine photoprobe. Cytosolic lysate of Jurkat cells was photolabeled with the PNA-adenine photoprobe at different concentrations (1:20, 1:40, 1:100, and 1:200 dilutions of 3.7 mM photoprobe in the lysates) and the labeled proteins were visualized by Western blot as described in the text. PNA adenine selectively and dose-dependently photolabeled cytosolic proteins of Jurkat cells. L (Ladder); *Lane 1*, 1:20 dilution; *Lane 2*, 1:40 dilution; *Lane 3*, 1:100 dilution; *Lane 4*, 1:200 dilution.

proteins are visualized with the BioRad ChemiDoc gel documentation system. A typical result of photolabeling experiment is shown in Fig. 3. As demonstrated the PNA-adenine photoprobe selectively and dose-dependently photolabeled cytosolic proteins of Jurkat cells.

4. Notes

1. Before use, the solution is cooled for 1 h (0°C) to precipitate dicyclohexylurea (DCU). The supernatant is then used for the coupling reaction.
2. Solutions in a filter funnel can be simply drained out by loosening the screw cap on top and opening the plastic cap at the bottom. If desired, residual solution can be further removed by attaching a syringe at the bottom and drawing several times, although this step is usually not necessary.
3. The completion of Fmoc-deprotection as well as the coupling reaction can be verified by Kaiser (ninhydrin) test. Prepare the ninhydrin reagent by dissolving 80 mg of ninhydrin in 40 ml

of ethanol. The reagent can be stored at room temperature. Transfer a few pieces of resin beads from the filter funnel to a borosilicate test tube. Add 300 μl of the ninhydrin reagent to the test tube and heat the tube (100–115°C) for 5 min. A boiling water bath can be used for heating. If there is free amine on the resin, the color turns purple (positive test). If there is no free amine, the color stays yellow (negative test). Thus, after Fmoc-deprotection, the ninhydrin reagent should produce a purple color. On the other hand, the ninhydrin test should produce a yellow color after coupling reaction.

4. It appears that the total yield is largely dependent on this precipitation step. Optimization of this step therefore may be critical to obtain the photoprobes in good yields.
5. A typical total yield is 30–50% after HPLC purification.
6. The typical protein concentration of Jurkat cytosolic lysate is 4–5 μg/μl.
7. The distance between the UV lamp and reaction mixture has to be optimized for individual experiments.
8. The UV lamp becomes very hot during the experiment. NEVER touch the UV lamp with your bare hands. In addition, UV lamps should not be in contact with potentially flammable materials, such as paper.

Acknowledgments

The authors would like to thank Dr. Skagit Hindi, who was involved in the initial development of our PNA-adenine photoprobes. This work was supported by PSC CUNY grant (PSCREG-37-813). RR-03037 from NCRR/NIH, which supports the research infrastructure at Hunter College, is also acknowledged. We also thank Mr. Kwang Won Cha, Mr. Laurence James, and Ms. Olga Aminova for technical assistance.

References

1. Dorman, G., and Prestwich, G. D. (2000). Using photolabile ligands in drug discovery and development. *Trends Biotechnol,* **18**, 64–77.
2. Shiio, Y., and Aebersold, R. (2006). Quantitative proteome analysis using isotope-coded affinity tags and mass spectrometry. *Nat Protoc,* **1**, 139–145.
3. Cravatt, B. F., Wright, A. T., and Kozarich, J. W. (2008). Activity-based protein profiling: from enzyme chemistry to proteomic chemistry. *Annu Rev Biochem,* 77, 383–414.
4. Gubbens, J., Ruijter, E., de Fays, L. E., Damen, J. M., de Kruijff, B., Slijper, M., et al. (2009). Photocrosslinking and click chemistry enable the specific detection of proteins interacting with phospholipids at the membrane interface. *Chem Biol,* **16**, 3–14.
5. Tian, R., Li, L., Tang, W., Liu, H., Ye, M., Zhao, Z. K., et al. (2008). Chemical proteomic study of isoprenoid chain interactome with a synthetic photoaffinity probe. *Proteomics,* **8**, 3094–3104.

6. Uttamchandani, M., Li, J., Sun, H., and Yao, S. Q. (2008). Activity-based protein profiling: new developments and directions in functional proteomics. *ChemBioChem,* **9**, 667–675.
7. Salisbury, C. M., & Cravatt, B. F. (2008). Optimization of activity-based probes for proteomic profiling of histone deacetylase complexes. *J. Am. Chem. Soc.,* **130**, 2184–2194.
8. Salisbury, C. M., and Cravatt, B. F. (2007). Activity-based probes for proteomic profiling of histone deacetylase complexes. *Proc Natl Acad Sci USA,* **104**, 1171–1176.
9. Sieber, S. A., Niessen, S., Hoover, H. S., and Cravatt, B. F. (2006). Proteomic profiling of metalloprotease activities with cocktails of active-site probes. *Nat Chem Biol,* **2**, 274–281.
10. Saghatelian, A., Jessani, N., Joseph, A., Humphrey, M., and Cravatt, B. F. (2004). Activity-based probes for the proteomic profiling of metalloproteases. *Proc Natl Acad Sci USA,* **101**, 10000–10005.
11. Chan, E. W., Chattopadhaya, S., Panicker, R. C., Huang, X., and Yao, S. Q. (2004). Developing photoactive affinity probes for proteomic profiling: hydroxamate-based probes for metalloproteases. *J Am Chem Soc,* **126**, 14435–14446.
12. Kawamura, A., and Hindi, S. (2005). Protein fishing with chiral molecular baits. *Chirality,* **17**, 332–337.
13. Hindi, S., Deng, H., James, L., and Kawamura, A. (2006). Selective photolabeling of Lck kinase in complex proteome. *Bioorg Med Chem Lett,* **16**, 5625–5628.
14. Fields, G. B., and Noble, R. L. (1990). Solid phase peptide synthesis utilizing 9-fluorenylmethoxycarbonyl amino acids. *Int J Pept Protein Res,* **35**, 161–214.
15. Kawamura, A., Hindi, S., Mihai, D. M., James, L., and Aminova, O. (2008). Binding is not enough: flexibility is needed for photocrosslinking of Lck kinase by benzophenone photoligands. *Bioorg Med Chem,* **16**, 8824–8829.
16. Winnik, M. A. (1981). Cyclization and the conformation of hydrocarbon chains. *Chem Rev,* **81**, 491–524.
17. Wagner, P. J. (1983). Conformational Flexibility and Photochemistry. *Acc Chem Res,* **16**, 461–467.
18. Speers, A. E., and Cravatt, B. F. (2004). Profiling enzyme activities in vivo using click chemistry methods. *Chem Biol,* 11, 535–546.

Chapter 7

Biotinylated Probes for the Analysis of Protein Modification by Electrophiles

Simona G. Codreanu, Hye-Young H. Kim, Ned A. Porter, and Daniel C. Liebler

Abstract

Formation of covalent protein adducts by lipid electrophiles contributes to diseases and toxicities linked to oxidative stress, but analysis of the adducts presents a challenging analytical problem. We describe selective adduct capture using biotin affinity probes to enrich protein and peptide adducts for analysis by liquid chromatography-tandem mass spectrometry (LC-MS/MS). One approach employs biotinamidohexanoic acid hydrazide to covalently label residual carbonyl groups on adducts. The other employs alkynyl analogs of lipid electrophiles, which form adducts that can be postlabeled with azidobiotin tags by Cu^{+}-catalyzed cycloaddition (Click chemistry). To enhance the selectivity of adduct capture, we use an azidobiotin reagent with a photocleavable linker, which allows recovery of adducted proteins and peptides under mild conditions. This approach allows both the identification of protein targets of lipid electrophiles and sequence mapping of the adducts.

Key words: Electrophile, Click chemistry, Protein adducts, Lipid oxidation, Photocleavable, Biotin, Shotgun proteomics

1. Introduction

Electrophilic lipid oxidation products initiate toxic responses by modifying proteins and triggering specific biochemical and cellular responses (1–4). The evaluation of the protein targets and the biological effects of lipid electrophiles in both their "free" and phospholipid-esterified forms will help identify new candidate biomarkers for oxidative stress at the cellular level (5, 6) as well as in plasma (7, 8). The problem of identifying protein targets of electrophiles is complicated in general by the diversity of targets and the low levels of modification under relevant conditions. It is even harder to map the sites of modifications on proteins for the same reason.

Gerard Drewes and Marcus Bantscheff (eds.), *Chemical Proteomics: Methods and Protocols*, Methods in Molecular Biology, vol. 803, DOI 10.1007/978-1-61779-364-6_7, © Springer Science+Business Media, LLC 2012

To address these problems, we have employed a combination of mass spectrometry (MS)-based shotgun proteomics analysis and novel affinity labeling for selective capture and analysis of the modified proteins. The key to this strategy is the use of probes that enable selective capture of adducts. Two approaches have proven successful in our studies. The first employs biotinamidohexanoic acid hydrazide (biotin hydrazide) for covalent capture of carbonyl-containing adducts. The second employs alkynyl analogs of lipid electrophiles and postlabeling of the resulting adducts using "Click" chemistry with a novel, photocleavable biotinylating reagent during sample workup. Here, we describe biotin probe strategies to capture both adducted proteins and to map adduct sites by selective capture of adducted peptides.

2. Materials

2.1. Cell Culture, HNE Treatment, and Derivatization

1. Dulbecco's modified Eagle's medium (DMEM) (Gibco/BRL, Bethesda, MD) supplemented with 10% fetal bovine serum (FBS, Atlas Biologicals, Fort Collins, CO).
2. Trypsin solution 0.25% (Gibco/BRL, Bethesda, MD).
3. 4-Hydroxy-2-nonenal (HNE, the lipid electrophile) (Cayman Chemical, Ann Arbor, MI), dissolved in ethanol (64 mM), stored in aliquots at −80°C, for addition to tissue culture dishes.
4. M-Per Lysis buffer (Pierce, Rockford, IL), stored at room temperature.
5. Dimethyl sulfoxide (DMSO).
6. Biotin hydrazide (Sigma-Aldrich, St. Louis, MO). A 50-mM stock solution in DMSO is prepared fresh before every experiment. The final concentration of biotin hydrazide per sample is 5 mM.
7. Sodium borohydride. A 500-mM stock solution in distilled H_2O is prepared fresh before every experiment. The final concentration of sodium borohydride is 50 mM.
8. Ice-cold 1× PBS, pH 7.2.
9. Protease inhibitor cocktail: 1.0 mM phenylmethylsulfonylfluoride, 1.0 mM *N*-ethylmaleimide, 10 μg/mL leupeptin, 10 μg/mL aprotinin, and 10 μg/mL pepstatin.
10. Phosphatase inhibitor cocktail: 1.0 mM sodium fluoride, 1.0 mM sodium molybdate, 1.0 mM sodium orthovanadate, and 10.0 mM β-glycerophosphate.
11. BCA Protein Assay Kit (Pierce, Rockford, IL).

2.2. Affinity Capture

1. Amicon Ultra Centrifugal Filter Devices, 10,000 Da molecular weight cutoff (Millipore, Billerica, MA).
2. Streptavidin Sepharose High Performance Beads (GE Healthcare, Uppsala, Sweden).
3. 1% Sodium dodecyl sulfate (SDS).
4. 4 M Urea. Urea must be freshly prepared before every experiment.
5. 1 M Sodium chloride prepared in 1× PBS.
6. Dithiothreitol stock solution (1 M), stored at –20°C. The working concentration of DTT is 50 mM.
7. NuPAGE LDS sample buffer (4×) (Invitrogen, Carlsbad, CA).

2.3. Western Blot

1. NuPage Bis-Tris 10% SDS-PAGE gels (Invitrogen, Carlsbad, CA).
2. Precision Plus Protein Standard Kaleidoscop Molecular Weight Marker (Bio-Rad Laboratories, Hercules, CA).
3. 20×-MES running buffer (Bio-Rad Laboratories, Hercules, CA).
4. 20× Tris–Glycine transfer buffer (Bio-Rad Laboratories, Hercules, CA).
5. Polyvinylidene difluoride (PVDF) membrane (Invitrogen, Carlsbad, CA).
6. Methanol.
7. 1× TBS-Tween (0.05% Tween 20).
8. Blocking buffer for near infra red fluorescent Western blotting (Rockland, Gilbertsville, PA).
9. Anti-HNE-Michel reduced rabbit polyclonal antibody (EMD biosciences, San Diego, CA).
10. Primary antibodies: Antiheat shock protein 90 (HSP90) rabbit polyclonal, antiactin mouse monoclonal, antitubulin rabbit polyclonal, anticofilin rabbit polyclonal, antiglutathione-*S*-transferase P (GSTP) rabbit polyclonal, antiglyceraldehyde-3-phosphate dehydrogenase(GAPDH) mouse monoclonal, antithioredoxin reductase 1 (TrxRd1) mouse monoclonal, and antiperoxiredoxin 6 (Prdx6) rabbit polyclonal antibodies (Abcam, Cambridge, MA). Anticullin3 goat polyclonal and antibax rabbit polyclonal antibodies (Santa Cruz, Santa Cruz, CA).
11. Secondary antibodies: Streptavidin and AlexaFluor®680 conjugated fluorescent antibodies (Molecular Probes, Eugene, OR) and IRDye™800 conjugated fluorescent antibodies (Rockland, Gilbertsville, PA).
12. Odyssey™ Infrared Imaging System and software (Li-Cor, Lincoln, NE).

2.4. LC-MS-MS

1. Colloidal Coomassie Blue (Invitrogen, Carlsbad, CA).
2. 100 mM Ammonium bicarbonate, pH 8.0, 45 mM DTT, 100 mM iodoacetamide. Prepare immediately before use.
3. Trypsin Gold Mass Spectrometry Grade (Promega, Madison, WI) reconstituted in 50 mM acetic acid to a final concentration of 1 mg/mL. Store aliquots at –20°C. Trypsin is diluted in 25 mM ammonium bicarbonate to 0.01 μg/mL and used at a ratio of 1:50 (trypsin:protein).
4. LTQ ion trap mass spectrometer (Thermo Electron, San Jose, CA) equipped with an Eksigent nanoLC (Dublin, CA) and Thermo Surveyor HPLC pump, Nanospray source and Xcalibur 1.4 instrument control. Peptides are resolved on 100 μm × 11 cm fused silica capillary column (Polymicro Technologies, LLC Phoenix, AZ) packed with 5 μm 300 Å Jupiter C18 (Phenomenex, Torrance, CA).
5. Solvent A: 0.1% (v/v) formic acid in water. Solvent B: 0.1% (v/v) formic acid in acetonitrile.
6. Synthetic peptide TpepK (AVAGKAGAR) as standard.

2.5. Plasma Treatment

1. Human plasma, obtained from fresh blood donated by a healthy subject (7). Care must be exerted to proceed in agreement with local protocols and ethical regulations.
2. Vacutainer tubes (BD Worldwide, Franklin Lakes, NJ), EDTA as anticoagulant, 1.5 mL Eppendorf tubes.
3. Antihuman ApoA1 polyclonal antibody (Cayman Chemicals, Ann Arbor, MI), antihuman ApoA1 monoclonal antibody (BioDesign International, Saco, ME), antihuman serum albumin (HSA) (Cortex Biochem™, Concord, MA), and protein G agarose beads.
4. S OASIS HLB® (Waters Corp., Milford, MA) (30 mg) cartridges for desalting of peptides.

3. Methods

The first approach to identify protein targets of lipid oxidation products is to covalently capture carbonyl-containing adducts with biotin hydrazide (Fig. 1a). To generate adducts, we treat RKO cells with the prototypical lipid electrophile HNE at 50 or 100 μM HNE (5, 6). HNE Michael adducts formed contain a residual carbonyl group and can be biotinylated by reaction with biotin hydrazide. Following capture with streptavidin, the biotinylated proteins are eluted and resolved by one-dimensional sodium dodecyl sulfate-polyacrylamide (SDS) gel electrophoresis, digested with trypsin and the resulting peptides are analyzed by liquid chromatography-tandem mass spectrometry (LC-MS/MS).

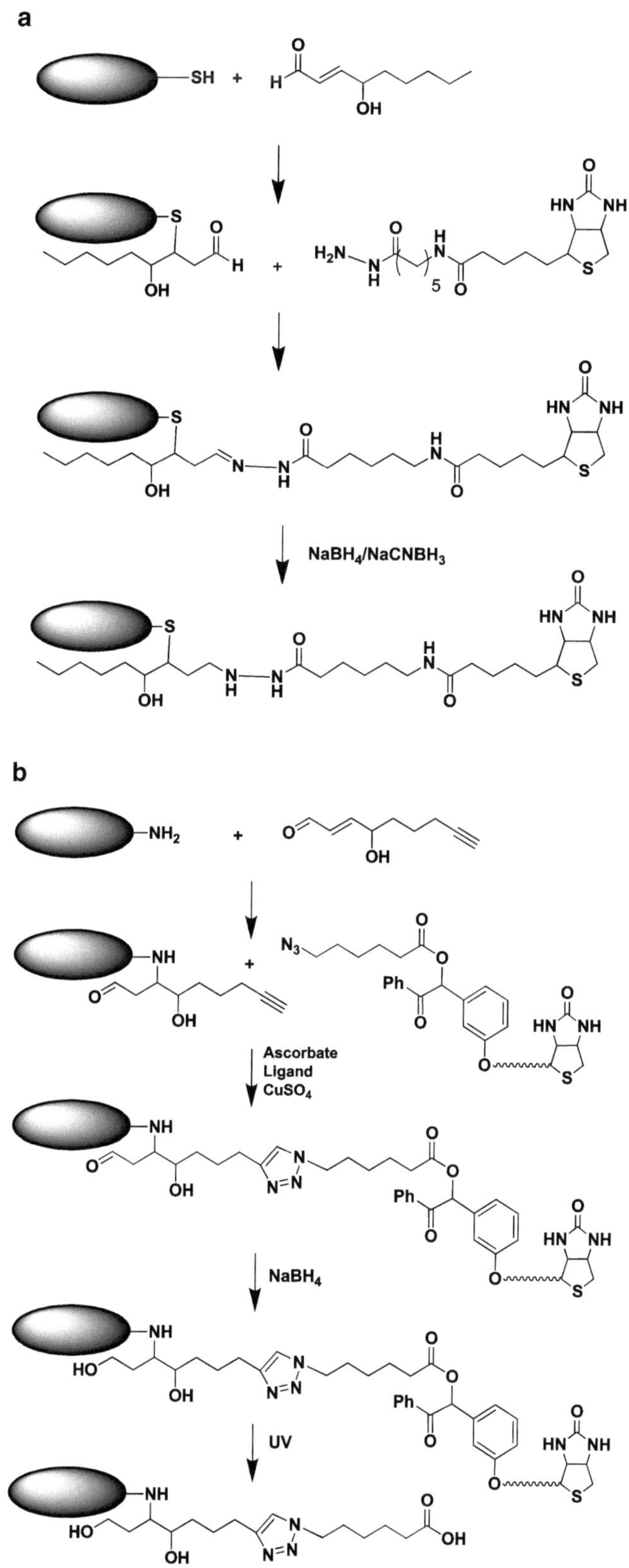

Fig. 1. (**a**) Biotinylation of protein reactive carbonyls with biotin hydrazide and (**b**) biotinylation by click chemistry using a photocleavable probe.

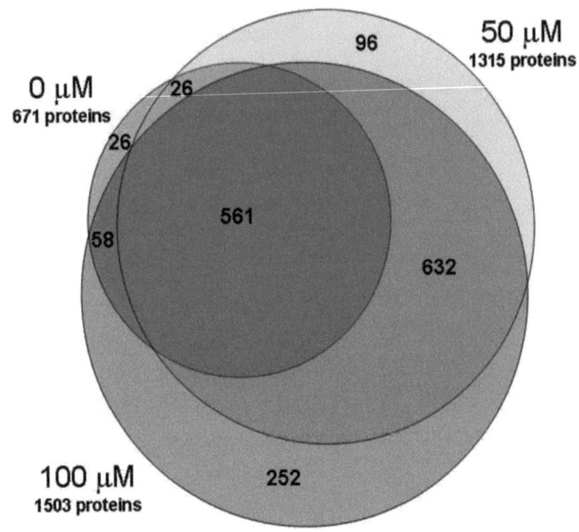

Fig. 2. Overlap of adducted proteins identified from RKO cells treated with 50 or 100 μM HNE or vehicle control. The *numbers represent* total proteins identified in triplicate analyses of each experimental condition. Overlaps between treatments are indicated by the numbers in the corresponding segments. A total of 561 proteins were common to all groups. Reproduced from ref. 5 with permission.

Of the 1,500+ proteins identified, 417 displayed a statistically significant increase in spectral counts with increasing HNE exposure concentration (Fig. 2) (5, 6). This relationship distinguishes true adducts from proteins nonspecifically captured with streptavidin or proteins containing carbonyls not derived from HNE treatment. A subset of the identified HNE protein targets can be verified with a streptavidin capture and immunoblotting approach. The subsequent immunoblotting analysis allows for a rapid screening of individual proteins to determine if these proteins have been adducted. Proteins that have undergone modifications following the exposure to HNE are identified in the eluate fraction.

The second approach to characterize protein and peptide adducts of lipid electrophiles is to use an HNE analog bearing a terminal alkyne (aHNE) (Fig. 1b). The alkynyl-tagged protein or peptide adducts can be biotinylated with an azidobiotin probe using a Cu^{+}-catalyzed cycloaddition (Click reaction). To improve selectivity for the capture of aHNE-labeled proteins and peptides, we use a photocleavable azido-biotin-linker (7, 8). The photocleavable linker allows selective release of adducts under very mild conditions, thus reducing the contamination of protein adducts with other proteins or peptides nonspecifically bound to the beads. The residual tag after photocleavage is small (311 mass units), which minimizes interference with adduct analysis by LC-MS/MS. Human plasma is exposed to the aHNE and then Click reactions are performed with photocleavable azido-biotin-linker to profile both proteins and peptides modified by aHNE. Protein adducts are analyzed by affinity capture of intact proteins prior to digestion (protein catch and release) or capture of peptide adducts after digestion (peptide catch and release). The former approach identified 14 plasma proteins as plasma targets of this lipid-derived electrophile with high confidence (Table 1). The latter approach

Table 1
Plasma proteins modified by *a*HNE as identified by *protein catch and photorelease*

IPI accession	UniProtKB/ SwissProt entry	Protein description	Unique peptide counts	Spectral counts
IPI00021841	P02647	Apolipoprotein A-I precursor, reverse transport of cholesterol	13	41
IPI00021854	P02652	Apolipoprotein A-II precursor, may stabilize HDL (high-density lipoprotein) structure	6	14
IPI00022229	P04114	Apolipoprotein B-100 precursor, major protein constituent of LDL and VLDL	38	44
IPI00021885	P02671	Isoform 1 of Fibrinogen alpha chain precursor, a cofactor in platelet aggregation	7	10
IPI00304273	P06727	Apolipoprotein A-IV precursor, VLDL secretion and catabolism	4	5
IPI00032258	P0C0L4	Complement C4-A precursor, activation of the classical pathway of the complement system	3	4
IPI00553177	P01009	Alpha-1-antitrypsin precursor	5	6
IPI00745872	P02768	Isoform 1 of serum albumin precursor	140	623
IPI00164623	P01024	Complement C3 precursor, activation of the complement system, pyogenic infection	17	25
IPI00022488	P02790	Hemopexin precursor, binds heme and transports it to the liver for breakdown and iron recovery	9	15
IPI00022463	P02787	Serotransferrin precursor, iron binding transport proteins, stimulating cell proliferation	23	34
IPI00022426	P02760	AMBP protein precursor, inhibits trypsin, plasmin, and lysosomal granulocytic elastase, and calcium oxalate crystallization	6	16
IPI00478003	Q9BQ22	Alpha-2-macroglobulin precursor, inhibit all four classes of proteinases by a unique "trapping" mechanism	25	35
IPI00555812	P02774	Vitamin D-binding protein precursor	12	18

dramatically enhances detection of adducted peptides and allowed identification of 50 specific adduction sites in 14 plasma proteins adducted with aHNE (Table 2).

3.1. Biotinylation of HNE Protein Targets in Cellular Lysates

3.1.1. Cell Culture and In Vivo HNE Treatment

1. Grow RKO human colorectal carcinoma cells to 80% confluence in McCoy's 5A medium supplemented with 10% FBS, at 37°C in an atmosphere of 95% air/5% CO_2.
2. Treatments are carried out with varying concentrations of HNE dissolved in ethanol. Wash the confluent cells plated in 150 mm culture dishes with 5 mL cold phosphate-buffered saline. Add HNE in 10 mL McCoy's 5A medium without FBS, at a final concentration of 0, 50, or 100 μM. The total concentration of ethanol per culture should be below 0.1% of the total medium volume.
3. Incubate the cells with the electrophile for 1 h at 37°C in an atmosphere of 95% air/5% CO_2. Use a disposable cell scraper to harvest the cells directly in the treatment medium, and centrifuge at 100 × *g* for 5 min. Remove the treatment medium slowly by aspiration and wash the cell pellets twice with cold PBS.
4. Lyse the cell pellets from each 150 mm treated plate on ice in 2 mL of cold M-PER buffer supplemented with 150 mM NaCl and the protease inhibitor and phosphatase inhibitor cocktails.
5. Clear the lysate by centrifugation at 10,000 × *g* for 10 min to remove cellular debris. Determine the total protein concentration of the supernatant using the BCA protein assay.
6. Adjust the protein concentration to 2 mg/mL for each sample using Lysis buffer. The final volume of each sample should be 1 mL.
7. Transfer a 50 μL aliquot from each sample into a fresh labeled 1.5 mL Eppendorf tube. Add DTT (50 mM) and NuPage Sample buffer (4×) to the samples, heat for 10 min at 95°C, and store at −20°C. These samples are referred to as the *Whole Cell Lysate*.

3.1.2. Derivatization of HNE-Adducted Proteins and Streptavidin Affinity Capture

1. Biotinylation of the reactive carbonyl group in HNE-adducted proteins is achieved with biotin hydrazide (Fig. 1) added to a final concentration of 5 mM to each 1 mL sample (See Note 1). The mixture is incubated with gentle rotation at room temperature for 2 h in the dark.
2. Hydrazone bonds formed during the reaction are reduced with sodium borohydride (100 mM) for an additional of 60 min at room temperature.
3. Excess reagents are removed at the end of the incubation time by filtration using 10,000 Da molecular weight cutoff Amicon

Table 2
Peptides from proteins identified in Table 1 modified by *a*HNE as detected using *peptide catch and photorelease*

IPI accession	Protein	Adducted peptides	Peptide precursor Calculated	Peptide precursor Found	z	Residue Number
IPI00021841	ApoA1	TH*LAPYSDELR	806.9201	806.9237	2	162
			538.2825	538.2843	3	162
		LEALK*ENGGAR	490.2755	490.2772	2	182
		LAEYH*AK	571.8139	571.8154	2	193
		ATEH*LSTLSEK	763.9067	763.9084	2	199
			509.6069	509.6078	3	199
		AK*PALEDLR	441.9259	441.9278	3	208
		K*LNTQ	457.7689	457.7712	2	239
IPI00021854	ApoA2	VK*SPELQAEAK	755.9274	755.9376	2	53
		SPELQAEAK*SYFEK	646.6667	646.666	3	62
		SK*EQLTPLIK	486.9642	486.964	3	69
IPI00022229	ApoB100	VLVDH*FGYTK	497.2732	497.2733	3	737
		DDKH*EQDMVNGIMLSVEK	608.5425	608.5426	4	746
		LLDH*R	482.7424	482.742	2	1214
		H*VGSK	419.7427	419.7425	2	1229
		H*INIDQFVR	484.9368	484.9363	3	2101
		VH*ELIER	402.8995	402.8992	3	2342
		LKQH*IEAIDVR	544.9859	544.983	3	2288
		SFDRH*FEK	459.5701	459.5689	3	3207
IPI00021885	Fibrinogen alpha chain precursor	TVIGPDGH*K	617.8431	617.8444	2	475
		HRH*PDEAAFFDTASTGK	733.359	733.3651	3	513
		TVTK*TVIGPDGHK	555.3174	555.3192	3	467
		ESSSH*HPGIAEFPSR	650.3219	650.326	3	563
		DSH*SLTTNIM^EILR	653.0068	653.0074	2	103
		VQH*IQLLQK	473.289	473.2905	3	151
		SSSYSK*QFTSSTSYNR	714.343	714.3467	3	581

(continued)

Table 2 (continued)

IPI accession	Protein	Adducted peptides	Peptide precursor Calculated	Peptide precursor Found	z	Residue Number
	ApoA4	K*LVPFATELHER	584.3332	584.3328	3	79
		LNH*QLEGLTFQM^K	629.3329	629.3336	3	236
		LGPH*AGDVEGHLSFLEK	706.3724	706.3724	3	332
IPI00032258	Complement C4-A precursor	SH*ALQLNNR	455.2529	455.2526	2	1342
		DK*GQAGLQR	428.5737	428.571	2	749
	Alpha-1-antitrypsin precursor	TDTSHHDQDH*PTFNK	697.6557	697.6569	3	39
		TDTSH*HDQDHPTFNK	697.6557	697.6566	3	34
		LYH*SEAFTVNFGDTEEAK	790.3814	790.3777	3	163
		LVDK*FLEDVK	506.2922	506.2941	3	153
		LGM^FNIQHC*K	506.5916	506.5936	3	255
		QINDYVEK*GTQGK	597.6476	597.6481	3	187
		K*QINDYVEK	483.2646	483.2589	3	179
		LQHLENELTH*DIITK	529.5361	529.5406	4	293
		LQH*LENELTHDIITK	705.7197	705.7221	3	286
		K*LSSWVLLM^K	511.3021	511.3018	3	258
IPI00745872	Isoform 1 of serum albumin precursor	SLH*TLFGDK	443.5785	443.5793	3	90
		RH*PYFYAPELLFFAK	737.3981	737.4041	3	169
		H*PYFYAPELLFFAK	685.3644	685.3738	3	169
		LK*C#ASLQK	629.863	629.8877	2	222
		LK*C#ASLQK	420.2444	420.2453	3	222
		LK~C#ASLQK	414.2409	414.2426	3	222
		RH*PDYSVVLLLR	593.6807	593.681	3	361
		H*PDYSVVLLLR	541.647	541.6522	3	361
		K~QTALVELVK	474.6291	474.6292	3	548
		NLGK*VGSK	557.335	557.3357	2	455
		SEVAH*R	505.2749	505.2757	2	32

IPI00478003	Alpha-2-macroglobulin precursor	TEH*PFTVEEFVLPK	662.0191	662.0232	3	217
		GH*FSISIPVK	465.938	465.937	3	523
IPI00022426	AMBP protein precursor	H*HGPTITAK	424.9068	424.9061	3	141
IPI00164623	Complement C3	AAVYHH*FISDGVR	594.981	594.9813	3	918
		AAVYH*HFISDGVR	594.981	594.9809	3	918
	Hemopexin	GH*GHR	437.7357	437.736	2	238
	Serotransferrin	DGAGDVAFVKH*STIFENLANK	636.8328	636.8326	4	226
IPI00022463		EFQLFSSPH*GK	529.9473	529.9433	3	308
IPI00555812	Vitamin D-binding protein precursor	H*LSLLTTLSNR	522.6385	522.6435	3	208

*Michael adduct

~Imine

#*S*-Carboxyamido

^Methionine sulfoxide

Filter Devices. One filter device per sample is needed and each filter device is pre-equilibrated first with 1 mL 1× PBS.

4. Upon termination of the 60 min incubation, each sample is added to the upper chamber of a pre-equilibrated filter device together with 2.5 mL cold 1× PBS. Samples are centrifuged at 2,500×*g* for 20 min and the filtrate is discarded. Another 2.5 mL of 1× PBS is added to the filter device and the sample is centrifuged at 2,500×*g* for another 20 min. This process is repeated twice for a total of three washes.
5. The washed, biotinylated proteins are resuspended in Lysis buffer to a total volume of 1 mL. Transfer 100 μL of each sample into a new prelabeled 1.5 mL Eppendorf tube, followed by the addition of DTT (50 mM) and NuPage Sample buffer (4×). The samples are heated for 10 min at 95°C and stored at −20°C. This sample is referred to as the *Input.*
6. Equilibrate Streptavidin Sepharose High Performance beads by washing a 50:50 (w/v) bead slurry three times with Lysis buffer. Use 1 mL of bead slurry per sample. Each time the beads are centrifuged at 10,000×*g* for 2 min and the supernatant is carefully discarded.
7. Biotinylated proteins are incubated with streptavidin-agarose beads (~2 mg protein per 1 mL bead slurry previously equilibrated with Lysis buffer) for 2 h with rotation at room temperature.
8. At the end of the incubation period, samples are centrifuged at 10,000×*g* for 2 min and 100 μL of the supernatant is transferred into a new prelabeled 1.5 mL Eppendorf tube. DTT (50 mM) and NuPage Sample buffer (4×) are added to each sample. Samples are heated for 10 min at 95°C and stored at −20°C. This sample is referred to as the flow through. The remainder of the supernatant is transferred into another 1.5 mL Eppendorf tube and stored at −20°C.
9. The bound proteins are washed two times with 1% SDS solution in PBS, two times with 4 M urea solution in PBS (see Note 2), two times with 1 M NaCl solution in PBS, and two times with PBS 1×, pH 7.4. Each first time wash with each of the reagents rotate the beads for 10 min at room temperature. After each wash step, the beads are centrifuged at 10,000×*g* for 1 min and the supernatant is discarded.
10. Proteins are eluted from the beads in 100 μL NuPage Sample buffer (4×) and DTT (50 mM). Heat samples for 10 min at 95°C. This sample is referred to as the eluate.

3.1.3. Western Blotting

1. Aliquots of the protein input, flow through, and eluates (20 μL) are subjected to immunoblot analysis with antibodies directed against individual proteins that were detected by

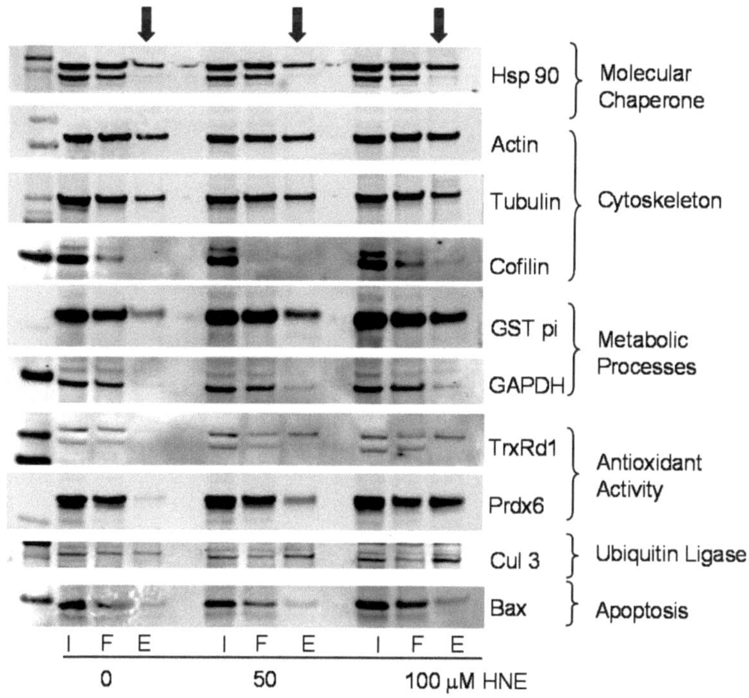

Fig. 3. Immunoblotting validation of ten individual protein targets from RKO cellular extracts treated with increasing concentrations of HNE. The presence of the proteins was confirmed in the input (I), flow through (F), and elution (E) fractions (*red arrows*), which contain adducted proteins. Reproduced from ref. 5 with permission.

LC-MS-MS as putative HNE targets. In the example described, blots are probed for heat shock protein 90 (HSP90), actin, tubulin β, cofilin, glutathione-*S* transferase Pi (GSTP), glyceraldehyde-3-phosphate dehydrogenase (GAPDH), thioredoxin reductase 1 (TrxRd1), peroxiredoxin 6 (Prdx6), cullin3, and bax (Fig. 3).

2. Heat the input, flow through, and eluate samples from each experimental condition for 10 min at 95°C.
3. Set up a NuPage Bis-Tris 10% SDS-PAGE gel and follow the manufacturer's directions to run the gel. Load 5 μL of the MW marker and load 10 μL of each of the samples in the following order: input, flow through, and eluate. This will allow for a more efficient comparison of bands during analysis.
4. Proteins are transferred electrophoretically onto a PVDF membrane by following the manufacturer's directions for the transfer apparatus. Block nonspecific primary antibody binding by placing the membrane into 5 mL Blocking buffer (1:1 1× TBS-Tween:Blocking buffer) for 1 h at room temperature on a rocking platform (see Note 3).
5. Prepare dilutions of the primary antibodies to proteins of interest according to the manufacturer's directions.

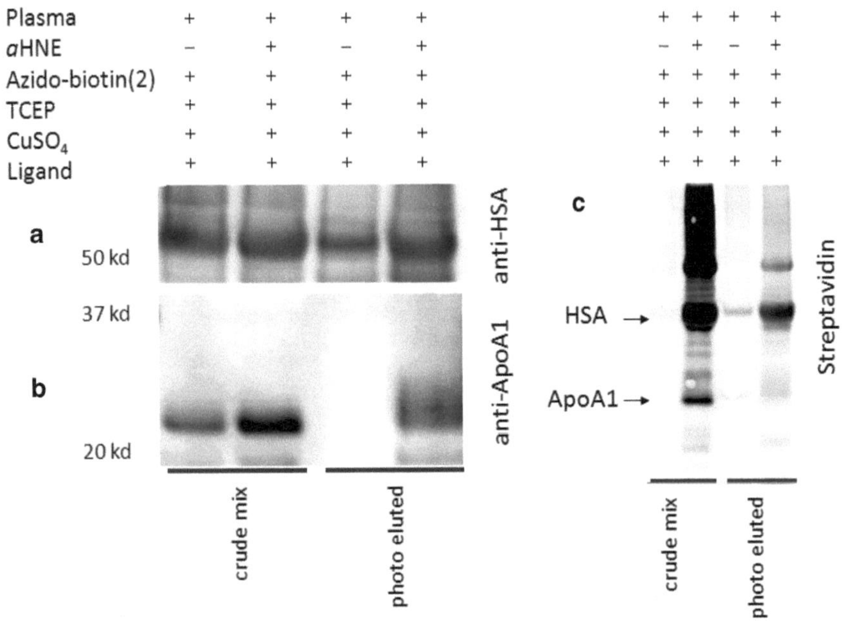

Fig. 4. Immunoblots of plasma supplemented with *a*HNE followed by 1, 3 cycloaddtion with azido-biotin then released by photolysis. The final concentrations of reagents, biotin, TCEP (or ascorbate), $CuSO_4$, and ligand were 1 mM, 2 mM, 2 mM, and 0.2 mM, respectively. (**a**) anti-HSA visualized with Alexa Fluor 680 donkey antigoat, (**b**) anti-ApoA1 visualized with Alexa Fluor 680 goat anti-rabbit, (**c**) visualized by Alexa Fluor 680-conjugated with streptavidin: residual biotin is probed even after the photolysis due to multiple sites of adduction. The whole plasma proteins are shown in the crude mix on the *left* and proteins that are eluted from photolysis of beads on the *right*. Intensities of + or – *a*HNE can be compared in each case; photoeluted fractions are diluted compared with crude mix. Reproduced from ref. 7 with permission.

6. Incubate the membrane with primary antibody overnight at 4°C while shaking on an orbital shaker.
7. Using multiple changes of 1× TBS-Tween, wash membranes for a total of 30 min before adding the secondary antibody.
8. Prepare appropriate dilutions of AlexaFluor®680-labeled secondary antibodies and incubate with the membrane for 1 h at room temperature while shaking. Wash the membrane with multiple changes of 1× TBS-Tween for a total of 30 min before scanning.
9. Immunoreactive proteins are visualized using the Odyssey™ System and software as described by the manufacturer. Examples are shown in Figs. 3 and 4.

3.1.4. In-Gel Trypsin Digestion and MS Analysis

1. Resolve the protein adducts purified by streptavidin capture as described above (eluate) by 10% SDS-PAGE using NuPAGE Bis-Tris gels and stain with Colloidal Coomassie Blue for 1 h, followed destaining in water for 1 h.
2. Desired bands corresponding to different molecular weights are excised from the gel and subjected to in-gel digestion with trypsin. Each excised band is carefully chopped into 1 mm

cubes, placed in 1.5 mL Eppendorf tube containing 100 μL of 100 mM ammonium bicarbonate, pH 8.0, and incubated at room temperature for 15 min.

3. Samples are reduced with 10 μL of 45 mM DTT for 20 min at 55°C and alkylated with 10 μL of 100 mM iodoacetamide for 20 min at room temperature in the dark.
4. The liquid is discarded and 100 μL acetonitrile:50 mM ammonium bicarbonate (50:50, v/v) is added to destain the samples. Incubate at room temperature for 15 min, then discard the liquid. Repeat this step twice.
5. The gel pieces are then dehydrated with 100 μL of acetonitrile, incubated for 15 min at room temperature, and the supernatant liquid is discarded.
6. The rehydrated gel pieces are digested with trypsin (50 μL of 0.01 μg/mL Trypsin Gold in 25 mM ammonium bicarbonate) overnight at 37°C.
7. The peptides are extracted twice with 100 μL of 60% acetonitrile and 0.1% trifluoroacetic acid, each for 15 min at room temperature; the extracts are combined. The extracts are evaporated under vacuum and resuspended in 10–20 μL of H_2O (0.1 %FA) for LC-MS/MS analysis.
8. The resulting peptides are subjected to LC-MS/MS analysis using an LTQ ion trap mass spectrometer. Liquid chromatography is carried out at ambient temperature at a flow rate of 0.6 μL/min using a gradient mixture of 0.1% (v/v) formic acid in water (solvent A) and 0.1% (v/v) formic acid in acetonitrile (solvent B). Centroided MS/MS scans are acquired using an isolation width of 2 *m/z*, an activation time of 30 ms, an activation Q of 0.250, and 30% normalized collision energy using 1 microscan with a max ion time of 100 ms for each MS/MS scan. The mass-spectrometer is tuned prior to analysis using the synthetic peptide TpepK (AVAGKAGAR), so that some parameters may vary slightly from experiment to experiment, but typically the tune parameters are as follows: spray voltage of 2 kV, a capillary temperature of 150°C, a capillary voltage of 50 V, and tube lens of 120 V. The MS/MS spectra of the peptides are collected using data-dependent scanning in which one full MS spectrum is followed by four MS/MS spectra. MS/MS spectra are recorded using dynamic exclusion of previously analyzed precursors for 60 s.

3.1.5. Database Searching

1. The "ScanSifter" algorithm, an in-house developed software, reads MS/MS spectra stored as centroided peak lists from Thermo RAW files and transcoded them to mzData v1.05 files. Spectra that contain fewer than six peaks or that have less than 2e1 measured TIC do not result in mzData files.

Only MS/MS scans are written to the mzData files; MS scans are excluded. If 90% of the intensity of a tandem mass spectrum appears at a lower *m/z* than the precursor ion, a single precursor charge is assumed; otherwise, the spectrum is processed under both double and triple precursor charge assumptions.

2. Tandem mass spectra are assigned to peptides from the IPI Human database version 3.33 (September 2007; 67,837 proteins) by the MyriMatch algorithm (9). To estimate false discovery rates (FDRs), each sequence of the database was reversed and concatenated to the database, for a total of 135,674 entries. Candidate peptides are required to feature trypsin cleavages or protein termini at both ends, though any number of missed cleavages is permitted. All cysteines are expected to undergo carboxamidomethylation and are assigned a mass of 160 kDa. All methionines are allowed to be oxidized. Precursor ions are required to fall within 1.25 *m/z* of the position expected from their average masses, and fragment ions are required to fall within 0.5 *m/z* of their monoisotopic positions. The database searches produced raw identifications in pepXML format.
3. Peptide identification, filtering, and protein assembly are performed with the IDPicker algorithm. Initial filtering takes place in multiple stages (10). First, IDPicker filters raw peptide identification to a target FDR of 5%. The peptide filtering employs reversed sequence database match information to determine thresholds that yield an estimated 5% FDR for the identifications of each charge state by the formula (11) $FDR = (2R)/(R+F)$, where R is the number of passing reversed peptide identifications and F is the number of passing forward (normal orientation) peptide identifications. The second round of filtering removes proteins supported by less than two distinct peptide identifications in the analyses. Indistinguishable proteins are recognized and grouped. Parsimony rules are applied to generate a minimal list of proteins that explain all of the peptides that passes the entry criteria.

3.2. Biotinylation of HNE Protein Targets in Plasma Using Click Chemistry

3.2.1. aHNE Treatment and Biotinylation

1. Draw blood from a healthy human volunteer by venipuncture into 7 mL Vacutainer tubes, using EDTA as anticoagulant (see Note 4). Centrifuge the blood samples at 1500×*g* for 10 min and aliquot the plasma in 1 mL Eppendorf tubes. Store at −80°C until use.
2. Dissolve aHNE in DMSO and add to 0.5 mL human plasma (final concentration 100 μM, <0.3% v). Stir gently for 1.5 h at 37°C. This incubation leads to formation of aHNE plasma protein adduct.
3. Add $NaBH_4$ to a final concentration of 10 mM and rotate the tube for 1 h at room temperature. Adjust the pH to 5.

4. Filter the plasma with a 10,000 Da molecular weight cutoff filter to remove excess reagents. Perform two additional washes with 1 volume each of 1× PBS.
5. Add $CuSO_4$, ligand (TBTA), ascorbic acid (or TCEP), and azido-biotin to the plasma at final concentrations of 1 mM, 0.2 mM, 1 mM, and 0.6 mM respectively. Rotate the reaction mixture for 2 h at room temperature in the dark (see Note 5).
6. Filter the solution with a 10,000 Da molecular weight cutoff filter as described in step 4.
7. Add protein G Agarose bead slurry (50 μL) and rotate the tube for 30 min. Save 2 μL for Western blot analysis with Streptavidin.
8. Half of the collected supernatant is mixed with 0.5 mL of Streptavidin beads and the suspension is then mixed gently overnight at 4°C in the dark. Note that the other half of the supernatant can be processed by "peptide catch and release" to afford a sample enriched in adducted peptides – see below.
9. The beads are washed with 1 mL each of 1% SDS (twice), 4 M urea (twice), 1 M NaCl (twice), H_2O (twice), and ammonium bicarbonate (once) sequentially in the dark.

3.2.2. Photorelease and Analysis of Adducted Plasma Proteins (Protein Catch and Release)

1. The washed beads are resuspended in PBS then irradiated with a hand held UV light (365 nm) for 1 h with stirring. This cleaves the linker and releases the adducted proteins from the Streptavidin beads.
2. The supernatant is collected and concentrated to 50 μL using a 10,000 Da molecular weight cutoff filter.
3. A sample of the supernatant (10 μL) is subjected to Western blot analysis for HSA and ApoA1 using the anti-HSA and anti-ApoA1 antibodies, respectively. These analyses confirm that these abundant plasma proteins were adducted, captured, and then photoreleased (Fig. 4).
4. The remaining supernatant is resolved by SDS-PAGE on a NuPage Bis-Tris® gel. After electrophoresis, gel lanes are cut into 10 MW fractions and subjected to in-gel trypsin digestion as described above.
5. Extracted peptides are redissolved in 0.1% formic acid and analyzed by LC-MS/MS as described above. Typical results are shown in Table 1.

3.2.3. Photorelease of Adducted Plasma Peptides (Peptide Catch and Release)

1. The remaining half of the Protein G Agarose supernatant (step 8 in Subheading 3.2.1) is agitated with 2 volumes of isopropylether/*n*-butanol (6:4, v/v) for 30 min to extract lipid. The aqueous phase reaction mixture is treated with DTT (5 mM) at 50°C for 10 min and then with iodoacetamide

(10 mM) at RT for 10 min in the dark. The reduced, alkylated proteins are then digested in solution with trypsin (1:100, w/w) for 4 h at 37°C with shaking in the dark (see Note 3).

2. The digested peptides are immobilized on a Streptavidin slurry (500 μL) overnight at 4°C in the dark. The beads are then washed sequentially with SDS, urea, NaCl, H_2O, and ammonium bicarbonate as described above. The washed slurry is resuspended in 1 mL of PBS and then exposed to UV light for 1 h as described above. This releases the peptide adducts.
3. The supernatant containing the peptide adducts is evaporated under vacuum and the mixture is redissolved in 0.5 mL of H_2O.
4. The solution containing the peptides is subjected to SPE using OASIS HLB© cartridges (30 mg). The SPE cartridge is activated with 1 mL of 0.1% formic acid in acetonitrile/water (4:1, v/v) and then equilibrated with 1 mL of 0.1% aqueous formic acid. The peptides are loaded onto the cartridge, which is then washed with 1 mL 0.1% aqueous formic acid. The peptides are then eluted with 1 mL of 0.1% formic acid in acetonitrile/water (4:1, v/v). The eluted peptide solution is evaporated under vacuum and the peptides are then redissolved in 0.1% formic acid and analyzed by LC-MS/MS as described above. Typical results are shown in Table 2.

4. Notes

1. Chemical reagents are purchased from commercial sources and used without further purification. Reagents should be prepared fresh before use. Biotin hydrazide stored in solution at −20°C loses reactivity toward carbonyls.
2. Urea should be freshly prepared before each experiment.
3. Incubation with primary antibody overnight at 4°C gives a much stronger signal for Western blotting than 2 h incubation at room temperature.
4. We found that delipidation of plasma significantly increased efficiency of capturing biotinylated peptides using streptavidin.
5. Azido biotin, aHNE, and ligand (TBTA, *Tris[(1-benzyl-1H-1,2,3-triazol-4-yl)methyl] amine*) were synthesized in house (7). TBTA can be purchased from AnaSpec (Fremont, CA).

Acknowledgment

This work was supported by National Institutes of Health Grants ES013125 and ES000267.

References

1. West, J. D., and Marnett, L. J. (2006) Endogenous Reactive Intermediates as Modulators of Cell Signaling and Cell Death. *Chem. Res. Toxicol.* **19**, 173–194.
2. Mashima, R., Witting, P. K., and Stocker, R. (2001) Oxidants and antioxidants in atherosclerosis. *Curr. Opin. Lipidol.* **12**, 411–418.
3. Ames, B. N. (2001) DNA damage from micronutrient deficiencies is likely to be a major cause of cancer. *Mutation Research* **475**, 7–20.
4. Beckman, K. B., and Ames, B. N. (1998) The Free Radical Theory of Aging Matures. *Physiol. Rev.* **78**, 547–581.
5. Codreanu, S. G., Zhang, B., Sobecki, S. M., Billheimer, D. D., and Liebler, D. C. (2009) Global Analysis of Protein Damage by the Lipid Electrophile 4-Hydroxy-2-nonenal. *Mol. Cell. Proteomics* **8**, 670–680.
6. Vila, A., Tallman, K. A., Jacobs, A. T., Liebler, D. C., Porter, N. A., and Marnett, L. J. (2008) Identification of Protein Targets of 4-Hydroxynonenal Using Click Chemistry for ex Vivo Biotinylation of Azido and Alkynyl Derivatives. *Chem. Res. Toxicol.* **21**, 432–444.
7. Kim, H.-Y. H., Tallman, K. A., Liebler, D. C., and Porter, N. A. (2009) An Azido-Biotin Reagent for Use in the Isolation of Protein Adducts of Lipid-derived Electrophiles by Streptavidin Catch and Photorelease. *Mol. Cell. Proteomics* **8**, 2080–2089.
8. Tallman, K. A., Kim, H.-Y. H., Ji, J.-X., Szapacs, M. E., Yin, H., McIntosh, T. J., Liebler, D. C., and Porter, N. A. (2007) Phospholipid-Protein Adducts of Lipid Peroxidation: Synthesis and Study of New Biotinylated Phosphatidylcholines. *Chem. Res. Toxicol.* **20**, 227–234.
9. Tabb, D. L., Fernando, C. G., and Chambers, M. C. (2007) MyriMatch: Highly Accurate Tandem Mass Spectral Peptide Identification by Multivariate Hypergeometric Analysis. *J. Proteome Res.* **6**, 654–661.
10. Zhang, B., Chambers, M. C., and Tabb, D. L. (2007) Proteomic Parsimony through Bipartite Graph Analysis Improves Accuracy and Transparency. *J. Proteome Res.* **6**, 3549–3557.
11. Elias, J. E., and Gygi, S. P. (2007) Target-decoy search strategy for increased confidence in large-scale protein identifications by mass spectrometry. *Nat. Meth.* **4**, 207–214.

Chapter 8

Profiling of Methyltransferases and Other *S*-Adenosyl-L-Homocysteine-Binding Proteins by Capture Compound Mass Spectrometry

Thomas Lenz, Peter Poot, Elmar Weinhold, and Mathias Dreger

Abstract

There is a variety of approaches to reduce the complexity of the proteome on the basis of functional small molecule–protein interactions. We describe a generic approach based on trifunctional Capture Compounds, in which the initial equilibrium-driven interaction between a small molecule probe and target proteins is irreversibly fixed upon photo-crosslinking between an independent photo-activable reactivity function of the Capture Compound and the surface of the target protein(s). Subsequently, Capture Compound - protein conjugates are isolated from complex biological mixtures via the sorting function of the Capture Compound. Here, we describe the application of a trifunctional Capture Compound that carries the methyltransferase product inhibitor *S*-Adenosyl-L-homocysteine as the selectivity function for the isolation of methyltransferases from a complex lysate of *Escherichia coli* DH5α cells. Photo-activated crosslinking enhances yield and sensitivity of the experiment, and the specificity can be readily tested for in competition experiments using an excess of free *S*-Adenosyl-L-homocysteine.

Key words: Capture Compound, Photo-crosslink, Small molecule–protein interaction, Methyltransferase, *S*-Adenosyl-L-homocysteine, SAH-AdoHcy, *S*-Adenosyl-L-methionine, SAM, AdoMet, Functional proteomics, LC-MS/MS

1. Introduction

S-Adenosyl-L-methionine (SAM, Fig. 1a) is probably, second to ATP, the most widely used cofactor in nature (1, 2). It is used as the major methyl group donor in all living organisms with the chemical reaction being catalyzed by SAM-dependent methyltransferases (MTases), which methylate small molecules, DNA, RNA, or proteins. Small molecule MTases participate in a variety of metabolic transformations and regulatory processes (3), among them

Gerard Drewes and Marcus Bantscheff (eds.), *Chemical Proteomics: Methods and Protocols*, Methods in Molecular Biology, vol. 803, DOI 10.1007/978-1-61779-364-6_8,

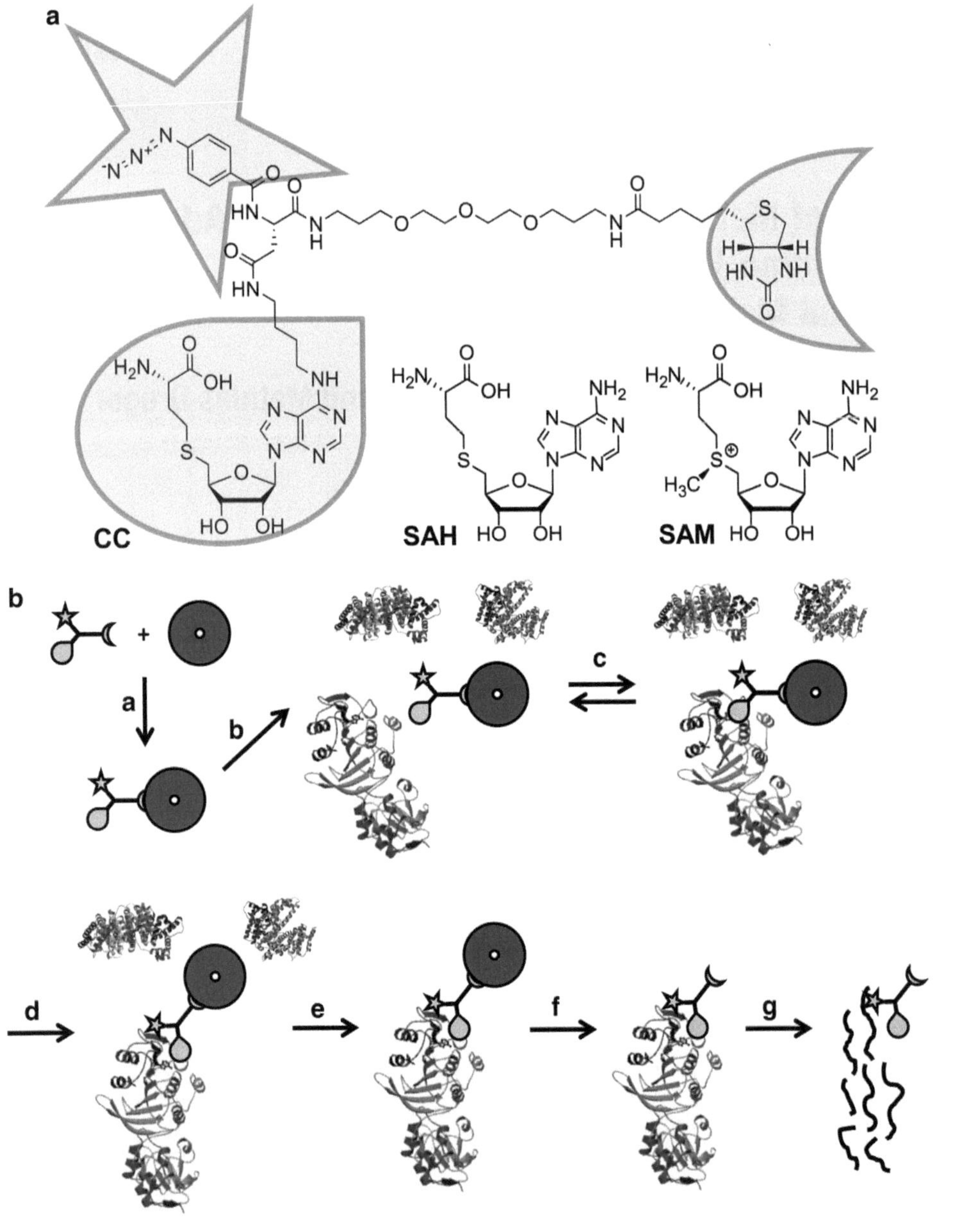

Fig. 1. (**a**) Chemical structure of the trifunctional Capture Compound (CC). The selectivity function is framed with a droplet, the reactivity function with a star, and the sorting function with a half-moon. The chemically stable *S*-adenosyl-L-homocysteine (SAH) is the cofactor product of *S*-adenosyl-L-methionine (SAM) after methyl group transfer by SAM-dependent MTases, for which SAH acts as a product inhibitor. (**b**) CCMS "on-bead" workflow. The CC is bound on the magnetic beads by its sorting function (a), the so formed caproBeads are incubated with the complex protein mixture (b), where a reversible binding equilibrium (c) is established between the selectivity function of the CC and the target proteins. Upon UV irradiation (d), the reactivity function forms a covalent crosslink. After washing the magnetic beads bearing the captured proteins (e), cleavage of the crosslinked CC–protein complexes from the magnetic beads (f), and tryptic digest (g), the captured proteins can be identified by MS analysis of the tryptic peptides.

the catabolism of important transmitter substances such as histamine or dopamine. DNA methylation adds epigenetic information to the genome controlling gene expression. It is the key to genomic imprinting, and hyper- or hypomethylation of DNA in higher eukaryotes is often related to cancer (4). RNA methylation significantly contributes to the more than 100 different posttranscriptional modifications of the standard RNA nucleotides clustered in regions of functional importance and differentiating the interactions of rRNA, tRNA, mRNA, proteins, or ligands such as antibiotics within the ribosome (5). Protein methylation occurs at *N*-atoms of arginine or lysine residues or at *O*-atoms of carboxyl groups of glutamate or isoaspartate or C-terminal (modified) carboxyl groups and plays a role in protein repair, protein sorting, signal transduction, and – as part of the "histone code" – regulation of gene expression (6). Given the crucial role of methylation reactions in diverse physiological scenarios, the profiling of MTases can be expected to become of similar importance in functional proteomics as the profiling of kinases. Analytical tools for their profiling, however, have not been available. We recently introduced a Capture Compound (CC) with *S*-adenosyl-L-homocysteine (SAH) as selectivity group to fill this technological gap (Fig. 1a) (7).

CCs are trifunctional small molecule probes with the generic design of a selectivity function, a photo-activable reactivity function, and a sorting function. The selectivity function is the small molecule of interest attached to the rest of the CC, the so-called scaffold, and interacts with its target proteins under equilibrium conditions. This equilibrium is then fixed through covalent photo-crosslinking of the photo-activable reactivity function. The sorting function (here biotin) serves to isolate the CC-protein conjugates from complex biological mixtures with the help of a solid phase (here streptavidin coated magnetic beads). Two configurations of the experiments are possible: In the "off-bead" configuration, the streptavidin magnetic beads are added to the samples after binding of the CC and photo-crosslinking to the target proteins have been accomplished. In the "on-bead" configuration (Fig. 1b), the CC is prebound to the streptavidin magnetic beads, and the CC-loaded beads are then incubated with the samples to accomplish binding of the CC selectivity group to the target protein(s) and photo-activated crosslinking through the reactivity function. The selectivity group may be virtually any small molecule of interest, e.g., either broadband inhibitors on enzyme classes for functional profiling in proteomics, or drug molecules for drug target discovery or off-target binding and toxicity profiling.

SAH, the product of SAM after methyl transfer, is a known general MTase product inhibitor (8). For this reason and because the natural cofactor SAM is used by further enzymes transferring other parts of the cofactor or initiating radical reactions as well as because of its chemical instability (9), SAH is an ideal selectivity function for a CC to target MTases. Here, we report the utility of the SAH-CC and capture compound mass spectrometry (CCMS) (10) by profiling MTases and other SAH-binding proteins from the strain DH5α of *Escherichia coli* (*E. coli*), one of the best-characterized prokaryotes, which has served as the preferred model organism in countless biochemical, biological, and biotechnological studies.

2. Materials

2.1. Preparation of E. coli Cell Lysate

1. *Escherichia coli* strain DH5α.
2. Autoclaved LB medium: 10 g/L Bacto-tryptone, 5 g/L yeast extract, 10 g/L NaCl, pH 7.5.
3. One 1 L shaker flask with baffles and four 5 L shaker flasks with baffles.
4. Incubator with orbital shaker.
5. UV-vis spectrometer or plate reader, e.g., Varian Cary 50 Bio or Anthos 2010.
6. Temperature-controlled centrifuge, e.g., Hettich centrifuge universal 320 R equipped with a swinging bucket rotor 1494.
7. Cell opening buffer: 6.7 mM MES, 6.7 mM NaOAc, 6.7 mM HEPES, 1 mM EDTA, 10 mM ß-mercaptoethanol, 200 mM NaCl, 10% (w/v) glycerol, and 0.2 mM PMSF, pH 7.5.
8. Sonifier, e.g., Sonopuls HD 2070 (Bandelin electronic GmbH & Co. KG).
9. Ultrafiltration devices, e.g., iCON™ concentrators, 7 mL/9 K (Pierce).
10. Gel filtration for depletion of small molecules like SAM or SAH, e.g., Zeba™ Desalt Spin Columns, 10 mL (Pierce).
11. Mini protease inhibitor cocktail tablets, EDTA free (Roche).
12. Bradford assay materials. Bradford reagent: 100 mg Coomassie Brilliant Blue G-250 in 50 mL 95% ethanol, add 100 mL 85% (w/v) phosphoric acid, dilute to 1 L when the dye has completely dissolved, and filter through Whatman #1 paper just before use.

2.2. Capture Experiments and Controls

1. SAH-CC and nonmodified SAH for competition binding.
2. Streptavidin-coated magnetic beads with 1 μm diameter (Dynabeads® MyOne™ Streptavidin C1, Invitrogen Dynal).
3. 5× Capture Buffer: 5× CB, containing 100 mM HEPES, 250 mM potassium acetate, 50 mM magnesium acetate, and 50% glycerol.

4. 5× Wash Buffer: 5× WB, containing 250 mM Tris–HCl, 5 mM EDTA, 5 M NaCl, and 42.5 μM octyl-ß-D-glucopyranoside, pH 7.5 (see Note 1).
5. caproBox™ (caprotec bioanalytics GmbH, Berlin, Germany). The caproBox is a device for UV-irradiating samples under simultaneous cooling, thereby avoiding denaturation of proteins during photo-crosslinking.
6. A magnet suitable for handling of magnetic beads, e.g., caproMag™ (caprotec bioanalytics) (see Note 2).
7. 200 μL PCR tubes, ideally as 12-well stripes (e.g., 0.2 mL Thermo-Strip, Thermo Scientific).
8. 60% Acetonitrile (ACN) (v/v): Mix 60 mL LC-MS-grade acetonitrile with 40 mL LC-MS-grade water.
9. 60% ACN/0.2% trifluoro acetic acid (TFA) (v/v/v): Mix 60 mL LC-MS-grade acetonitrile with 40 mL LC-MS-grade water and add 0.2 mL LC-MS-grade TFA. Prepare freshly before use.

2.3. SDS-PAGE and Silver Stain

1. Gels for SDS-PAGE: OLS® ProPage 4–20% Tris/glycine precast gels.
2. Gel electrophoresis system: OLS omniPAGE mini.
3. SDS sample buffer: 50 mM Tris–HCl, 320 mM β-mercapto-ethanol, 2.5% SDS, 0.05% bromophenol blue, and 10% glycerol, pH 6.8.
4. SDS running buffer: 25 mM Tris base, 200 mM glycine, and 0.1% SDS, pH 8.3.
5. ProteoSilver™ Silver Stain Kit (Sigma).

2.4. Tryptic Digest of Proteins and Preparation of Peptides for LC-MS/MS

1. Digestion buffer: 50 mM ammonium bicarbonate in LC-MS grade water. Store in aliquots at −20°C, thaw only once.
2. Trypsin (sequencing grade, Roche). Prepare a 0.5 μg/μL solution by dissolving the lyophilized trypsin in 1 mM HCl, store at 4°C for several weeks.
3. In-gel digestion solution: 12.5 ng/μL trypsin in 50 mM ammonium bicarbonate. Prepare by adding 7.5 μL 0.5 μg/μL trypsin solution in 1 mM HCl to 292.5 μL digestion buffer.
4. Centrifugal evaporator, e.g., MiVac DNA concentrator (GeneVac Inc., UK).
5. Ultrasound bath, e.g., Sonorex (Bandelin, Germany).
6. 2–10 μL Stage Tips® (20 μL tip, Proxeon Biosystems A/S, Odense, Denmark).
7. 5% Formic acid (FA) (v/v) is made up of 5 mL FA and 95 mL water, both LC-MS grade.

8. 50% Methanol/5% FA (v/v/v) is made up of 50 mL methanol, 5 mL FA, and 45 mL water, all LC-MS grade.

2.5. NanoLC-MS/MS Analysis

1. For high-performance chromatographic separation and high-accuracy mass spectrometric identification of proteins, a nano LC system and a state-of-the-art mass spectrometer are necessary (see Note 3).
2. LC-MS/MS buffer A: LC-MS grade water with 0.1% FA.
3. LC-MS/MS buffer B: LC-MS grade ACN with 0.1% FA.

2.6. Peptide and Protein Identification via Automated Sequence Database Search

1. Use a recent UniProtKB/Swiss-Prot database release (http://www.expasy.org) of the organism investigated (database used for present study: *E. coli*, strain K12, release 57–11).
2. One or more protein identification algorithms, e.g., SEQUEST implemented in BioworksBrowser 3.3.1 SP1 (Thermo Fisher Scientific, Germany) and X!Tandem (The Global Proteome Machine Organization; version 2007.01.01.1) implemented in the Scaffold 3 software (version Scaffold_3_00_03, Proteome Software Inc., USA).
3. The protein identification software Scaffold 3 for the visualization and validation of complex MS/MS proteomics experiments is recommended (http://www.proteomesoftware.com).

3. Methods

Capturing proteins that specifically interact with SAH in a given biological sample by the SAH-CC in the "on-bead" configuration consists of three consecutive steps (Fig. 1b): first, incubation of the sample with the caproBeads (CC bound by the sorting function (biotin) to the streptavidin-coated magnetic beads) enables equilibrium binding between the CC selectivity function (SAH) and the target proteins. Subsequently, the equilibrium is fixed by irradiation of the sample to induce a covalent photo-crosslink between the CC reactivity function and the bound target protein. CC–protein conjugates located on the streptavidin magnetic beads are then isolated from the protein mixture by using a strong magnet and washed by repeated re-suspension in fresh WB and magnetic collection. To determine any nonspecifically captured proteins, it is essential to perform the capture experiment along with a competition experiment in parallel. In the competition experiment, the binding of the CC is competed for by preincubation of the sample with an excess of the corresponding free small molecule (SAH). The results can be visualized by SDS-PAGE (exemplary data of

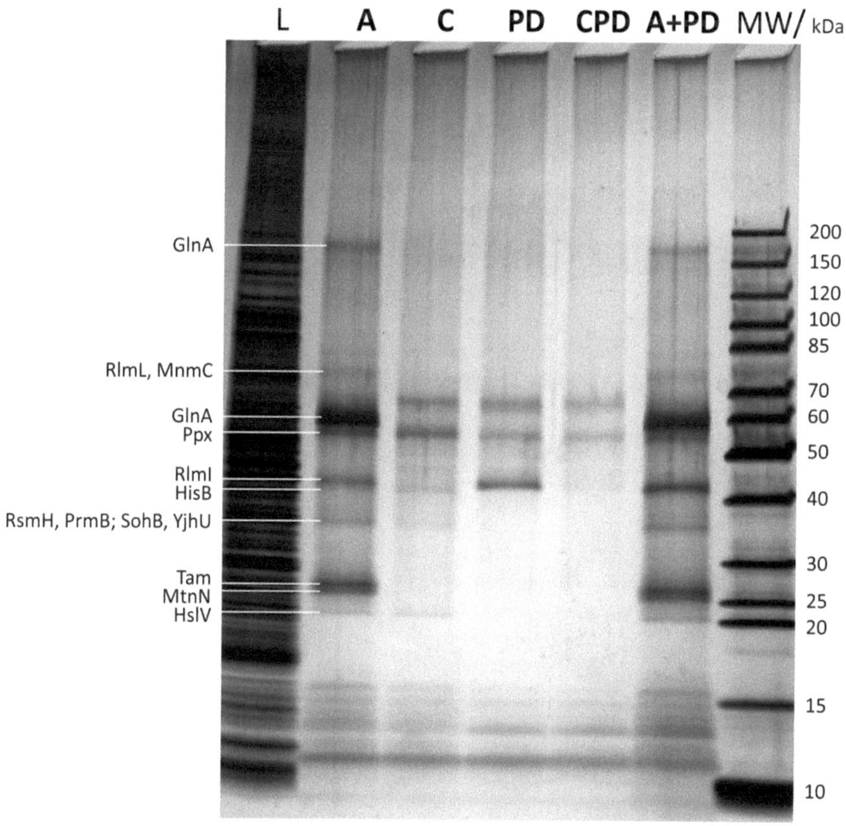

Fig. 2. SDS-PAGE/silver stain analysis of captured proteins (after *step f* in Fig. 1b). The lane description is given on top of the gel (MW: molecular weight marker with the corresponding molecular weights of the marker bands given to the very right; L: 0.25% sample drawn from the *E. coli* DH5a whole cell lysate before adding the caproBeads in *step b* in Fig. 1b; *A*: assay with addition of an excess of free SAH after UV irradiation *step d* in Fig. 1b; *C*: control of assay including an excess of free SAH as competitor during *steps c* and *d* in Fig. 1b; *PD*: pulldown meaning no UV irradiation *step d* in Fig. 1b and no addition of free SAH; *CPD*: control of pulldown using SAH as competitor; *A+PD*: combined assay plus pulldown meaning no addition of free SAH during the workflow). Proteins identified by MS from cut-out protein gel bands after in-gel tryptic digest are given to the very left. See Table 1 for MTases and other selected proteins identified by CCMS experiments of duplicate samples of those shown in the present figure.

captured proteins from cell lysate are shown in Fig. 2). Experiments carried out in parallel without irradiation (pulldown) show less isolated proteins compared to capture experiments. However, instead of analyzing the captured proteins by SDS-PAGE, typically, the captured proteins are submitted to tryptic digest and the resulting peptides are analyzed by LC-MS/MS. A typical result of a CCMS experiment including controls and pulldown experiments from 260 μg *E. coli* lysate, respectively, is given in Tables 1 and 2 as well as in Fig. 3. Much more MTases and other SAH-binding proteins are identified in the CCMS assay (*A*) compared to the pulldown

Table 1
MTases and other selected proteins identified by CCMS experiments

Protein	ORF	MW/kDa	Description	Substrate	*A*	*C*	*PD*	*CPD*	*A+PD*
Dcm	b1961	53.5	DNA-cytosine MTase	DNA (m5C)	1	0	0	0	1
RlmI	b0967	44.4	23S rRNA m5C1962 MTase	rRNA (m5C)	17	0	17	0	20
RlmL	b0948	78.9	23S rRNA m2G2445 MTase	rRNA (m2G)	12	0	0	0	10
TrmB	b2960	27.3	tRNA (guanine-N(7)-)-MTase	tRNA (m7G)	11	0	0	0	13
CmoA	b1870	27.8	tRNA (cmo5U34)-MTase	tRNA (mcmo5U)	7	0	0	0	4
RsmG	b3740	23.4	16S rRNA m7G MTase	rRNA (m7G)	6	0	1	0	5
RsmH	b0082	34.9	16S rRNA m4C1402 MTase	rRNA (m4C)	5	0	0	0	7
RsmD	b3465	21.7	16S rRNA m2G966 MTase	rRNA (m2G)	2	0	0	0	2
RsmB	b3289	48.3	16S rRNA m5C967 MTase	rRNA (m5C)	1	0	0	0	0
MnmC	b2324	74.4	Bifunctional protein includes tRNA (mnm(5)s(2)U34)-MTase	tRNA (mnm5s2U)	1	0	0	0	0
PrmB	b2330	35.0	50S ribosomal protein L3 Gln150 MTase	Protein (Gln)	13	0	0	0	15
CheR	b1884	32.8	Chemotaxis protein MTase	Protein (Glu)	0	0	0	0	1
Cfa	b1661	44.9	Cyclopropane-fatty-acyl-phospholipid synthase	Small molecule	15	0	0	0	14

Tam	b1519	29.0	Trans-aconitate 2-MTase	Small molecule	2	0	0	0	3
CysG	b3368	50.0	Siroheme synthase includes uroporphyrinogen-III C-MTase	Small molecule	1	0	0	0	2
SmtA	b0921	29.8	Protein smtA	(?[a])	7	1	0	0	8
MtnN	b0159	24.4	5′-Methylthioadenosine/ SAH nucleosidase	Small molecule[b]	36	0	0	0	39
GlnA	b3870	51.9	Glutamine synthetase	Small molecule[c]	90	0	0	0	97
RplK	b3983	14.9	50S ribosomal protein L11	Of protein MTase PrmA[d]	2	0	0	0	2

The given numbers denote the unweighted peptide spectral count per protein. Samples are duplicates of those analyzed by SDS-PAGE/silver stain in Fig. 2

[a]Not (fully) characterized

[b]No methylation but cleavage of the glycosidic bond of SAH

[c]No methylation but binding of SAH into the ATP binding site as shown by CCMS experiments with ATP as competitor (data not shown)

[d]Substrate of the 50S ribosomal protein L11 MTase PrmA; reproducible specific identification by CCMS (data not shown)

Table 2
Total number of identified proteins in the CCMS runs and protein overlap between the runs

	A	C	PD	CPD	A+PD
A	111 (64)				
C	65 (41)	107 (46)			
PD	25 (15)	23 (13)	61 (17)		
CPD	23 (13)	22 (12)	20 (14)	47 (14)	
A+PD	87 (61)	64 (41)	23 (14)	22 (12)	124 (67)

The number of proteins identified with at least two peptides are given in parentheses. The high reproducibility of the method can be inferred from the high protein overlap (of mainly unspecific proteins) between comparable experiments (*A* vs. *C* and especially *A* vs. *A+PD* but also *PD* vs. *CPD*) especially with the proteins robustly identified with at least two peptides. See also Fig. 3 for Venn diagrams and *Supplementary Table S1* for a list of all identified proteins

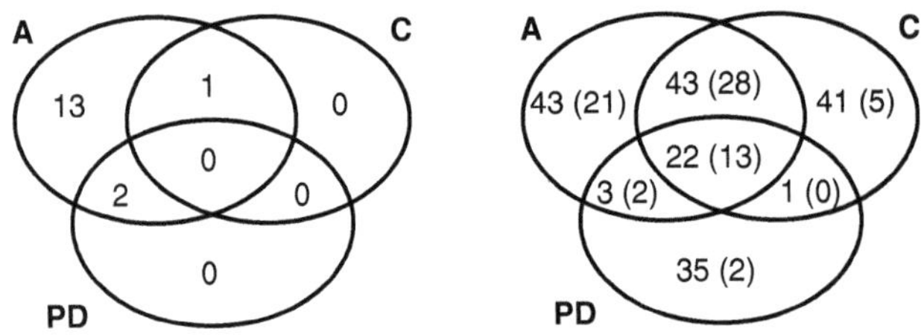

Fig. 3. Overlap of identified proteins in assay (*A*), competition control (*C*), and pulldown (*PD*), represented with Venn diagrams. *Left*: Number of MTases and SAH nucleosidase, only, referring to Table 1. *Right*: Number of all identified proteins referring to Table 2 and *Supplementary Table S1*. The number of proteins identified with at least two peptides is given in parentheses.

(*PD*) and SAH specificity is shown by the almost complete absence of these proteins in the competition control (*C*). In general, the method should also be compatible with any state-of-the-art stable isotope labeling technology or assessment of the capture sample by 2D gel electrophoresis.

3.1. Preparation of E. coli DH5α Cell Lysate

1. Inoculate a 2 mL culture (LB medium in a test-tube) with the *E. coli* DH5α strain directly from a glycerol stock and incubate at 37°C and 250 rpm for 8 h.
2. Inoculate 250 mL LB medium in a 1 L shaker flask with baffles with the 2 mL culture and incubate over night at 37°C and 166 rpm.
3. Inoculate four flasks (5 L) each with 2.5 L LB medium with the 250 mL culture (50 mL for each flask) and incubate the cultures at 37°C and 166 rpm until an OD_{600} of 0.8 is reached.

4. Harvest the cells by centrifugation at 4°C, 3,000×*g* for 20 min.
5. Perform further handling at 0–4°C or on ice.
6. Re-suspend the harvested cell material in Milli-Q water, combine in one centrifuge bucket and centrifuge for a further 30 min at 6,000×*g* and 4°C.
7. Store the resulting 20 g cell material at −20°C or −78°C.
8. Re-suspend the cells in 100 mL ice-cold cell opening buffer and sonicate three times for 1 min in four 25 mL portions on ice using a sonifier (maximal amplitude, continuous output).
9. Centrifuge the lysate over night at 2,370×*g* and 2°C.
10. Concentrate the supernatant to 14 mL by ultrafiltration using 7 mL/9 K concentrators at 2,370×*g* and 2°C.
11. Deplete the viscous concentrate from small molecules by gel filtration at 2°C (four columns; storage buffer removed at 1,000×*g* for 2 min; four times equilibration with 5 mL ice-cold cell opening buffer and buffer removed at 1,000×*g* for 2 min, respectively; 3.5 mL concentrate applied to each column; 45 min centrifugation at 24×*g*, then two times 2 min at 1,000×*g*).
12. Supplement the resulting 13 mL lysate with 13 mL ice-cold glycerol and Roche mini protease inhibitor cocktail tablets, EDTA free. Mix and dissolve the tablets.
13. Store the lysate at −20°C.
14. Determine the total protein concentration by the Bradford assay (21 mg/mL in the present case).

3.2. Capture Assay, Competition Control, Pulldown, Competition Control of Pulldown, and Combined Capture Assay Plus Pulldown

1. For several parallel experiments, it is recommended to prepare a master mix of water, capture buffer and *E. coli* lysate, and to perform reactions within different tubes of one 200 μL-PCR tube strip. Here, quantities for one reaction tube are given. Results for five different experiments will be presented, which are capture assay (*A*), competition control (*C*), pull-down (*PD*), competition control of pulldown (*CPD*), and combined capture assay plus pulldown (*A*+*PD*). For each reaction, prepare 1.5 mL 1× WB by adding 0.3 mL 5× WB to 1.2 mL of Milli-Q water.
2. Prepare SAH-CC loaded streptavidin-coated magnetic beads (caproBeads™) in 200 μL PCR tube strips (0.2 mL Thermo-Strip, Thermo SCIENTIFIC) (see Note 4). Therefore, mix 25 μL of 100 μM SAH-CC with 50 μL of 10 mg/mL streptavidin-coated magnetic beads for each aliquot, vigorously

shake the resulting suspensions at room temperature for 2 min to allow binding of the biotin moiety of the SAH-CC to the streptavidin on the magnetic bead surface, and collect the beads using the strong magnet (in the caps of the PCR tube strips using the caproMag™ magnetic device). Discard the supernatants, re-suspend the resulting caproBeads™ in 200 μL WB, magnetically collect the caproBeads™ (in the caps of the PCR tube strips), and discard the supernating WB. Close tubes to avoid drying of the beads.

3. Prepare aliquots of *E. coli* DH5α whole cell lysate in new PCR tubes at 0–4°C using a master mix consisting of water, capture buffer, and *E. coli* lysate (see Note 5). For one reaction, supplement a volume of Milli-Q water for a final reaction mix volume of 100 μL with 20 μL 5× CB. Mix, add 0.26 mg *E. coli* lysate (protein amount), and gently mix by inversion. Only for *C* and *CPD*, add 20 μL 10 mM SAH competitor solution and gently mix by inversion (add Milli-Q water instead of SAH solution to *A*, *PD*, and *A+PD*). Draw a 1 μL sample from *A* for further analysis (see below).
4. Suspend the caproBeads™ in the respective lysate and incubate for 3 h at 4°C keeping the beads in suspension by rotation (reversible binding of SAH-binding proteins to the SAH selectivity function of the SAH-CC).
5. Place the suspensions *A*, *C*, and *A+PD* in the caproBox™ and irradiate for a total time of 30 min in the closed tubes between 0 and 4°C to form a covalent crosslink between the reactivity function of the SAH-CC and the SAH binding proteins. Therefore, remove the suspensions after irradiation intervals of 2.5 min, respectively, from the caproBox™, cool in ice water for ~15 s (especially the lids), mix several times by inversion, very shortly (~2 s) centrifuge to remove suspension remaining in the lids, and place back into the caproBox™ for the next irradiation interval (see Note 6).
6. Add 20 μL 10 mM SAH solution to *A*, or 20 μL Milli-Q water to *C*, *PD*, *CPD*, and *A+PD* and incubate the suspension for 10 min at 4°C to displace, in *A*, SAH-binding proteins not crosslinked to the SAH-CC. Keep the beads in suspension by rotation or by intermittent manual re-suspension.
7. Collect the caproBeads™ from the suspensions using a strong magnet (e.g., the caproMag™), discard the supernatants, and wash the beads six times – by re-suspension and collection – with 200 μL 1× WB and once with 200 μL Milli-Q water (see Notes 7 and 8).
8. Wash the beads three times with 200 μL 60% ACN and release the captured proteins from the beads by 10 min incubation at

room temperature under vigorous shaking with 200 μL 60% ACN/0.2% TFA, magnetically separate the beads and evaporate the supernatant to dryness (see Note 9).

3.3. SDS-PAGE of Captured Proteins

1. For SDS-PAGE, dissolve the captured proteins released from the magnetic beads (evaporated ACN/TFA solutions from step 8 in Subheading 3.2) in 20 μL SDS sample buffer.
2. Mix the 1 μL sample drawn from the SAH-CC assay with 19 μL SDS sample buffer; use 5 μL of this solution for SDS-PAGE analysis (0.25% of assay).
3. Heat the samples in SDS sample buffer for 10 min at 95°C and analyze by SDS-PAGE (run time 90 min at a constant voltage of 180 V under ice cooling of the SDS running buffer).
4. Silver stain the gel (see Note 10). A representative result is shown in Fig. 2.

3.4. In-Gel Tryptic Digest of Proteins and Peptide Extraction from Gel Bands

1. Wash the silver-stained gel at least three times with 100 mL Milli-Q water for 10 min after the silver stain stop solution was removed.
2. Cut out the gel bands (e.g., using a clean scalpel and transfer into a 0.5 mL Eppendorf tube) and store at –20°C or directly process. Wash the gel bands for 15 min, respectively, with 100 μL water, 100 μL 50% ethanol, 100 μL water, 100 μL 50% ethanol, and for 5 min with pure ethanol. Repeat this washing procedure once again.
3. Re-hydrate the gel band with 10 μL in-gel digestion solution for 45 min at 4°C. Remove the supernatant and replace by 20 μL digestion buffer (without trypsin) followed by incubation at 37°C over night while shaking.
4. Collect the supernatant. For peptide extraction, incubate the gel band with 20 μL 5% FA for 15 min while shaking, add 20 μL ACN and incubate for another 15 min while shaking. Combine the supernatant with the previous supernatant and repeat the peptide extraction procedure once again.
5. Evaporate the combined three supernatants to dryness, dissolve in 10 μL 5% FA while shaking and applying ultrasound (ultrasound bath), and desalt using StageTips as described below.

3.5. Tryptic Digest of Captured Proteins and Preparation of Peptides for LC-MS/MS

1. Dissolve the captured proteins released from the magnetic beads (evaporated ACN/TFA solutions from step 8 in Subheading 3.2) in 10 μL digestion buffer using an ultrasound bath and vortexing, add 1 μL 0.5 μg/μL trypsin in 1 mM HCl, and incubate the solution at 37°C over night.

2. Desalt the solution containing the tryptic peptides of the captured proteins using 2–10 μL StageTips from Proxeon following the manufacturer's procedure (precondition with 10 μL 50% methanol/5% FA, equilibrate with 10 μL 5% FA, load with peptides of captured proteins, wash with 10 μL 5% FA, elute twice with 10 μL 50% methanol/5% FA).
3. Evaporate the desalted peptides in 50% methanol/5% FA to dryness, dissolve in 5.5 μL 0.1% FA while shaking and applying ultrasound.
4. Transfer solution to sample plate for nanoLC-MS/MS analysis.

3.6. NanoLC-MS/MS Analysis

1. Place sample plate into the LC system.
2. Load 5 μL of the peptide solution directly onto the precolumn using 5% ACN/0.1% FA.
3. Use a 80-min linear gradient from 5% ACN/0.1% FA to 40% ACN/0.1% FA followed by an additional 2 min to 100% ACN/0.1% FA and remaining at 100% ACN/0.1% FA for another 8 min with a controlled flow rate of 400 nL/min.
4. Perform the mass spectrometric analysis in the data-dependent mode to automatically switch between orbitrap-MS (profile mode) and LTQ-MS/MS (centroid mode) acquisition (see Notes 11 and 12).
5. Control the mass spectrometer duty cycle by setting the injection time automatic gain control.
6. Acquire survey full-scan MS spectra (from m/z 300 to 2,000) in the orbitrap with resolution $r=60{,}000$ at m/z 400 (after accumulation to a target value of 500,000 charges in the linear ion trap).
7. Set the system to sequentially isolate the five most intense ions for fragmentation in the linear ion trap using collision-induced dissociation (CID) at a target value of 10,000 charges. The resulting fragment ions are recorded in the LTQ.
8. For accurate mass measurements in the MS mode, use the singly charged polydimethylcyclosiloxane background ion $(Si(CH_3)_2O)_6\ H^+$ (m/z 445.120025) generated during the electrospray process from ambient air as the lock mass for real-time internal recalibration.
9. Dynamically exclude target ions already mass-selected for CID for the duration of 60 s.
10. Set charge state screening and rejection of ions for CID with unassigned charge.
11. Further mass spectrometric settings are as follows: set spray voltage to 1.6 kV, set temperature of the heated transfer capillary

to 200°C, and normalized collision energy is 35% for MS^2. The minimal signal required for MS^2 is 500 counts. Apply an activation $q=0.25$ and an activation time of 30 ms for MS^2 acquisitions.

12. To clean the LC system, perform one blank run between two consecutive CCMS measurements.

3.7. Peptide and Protein Identification via Automated Sequence Database Search

1. Use a protein identification algorithm to analyze the MS/MS data (in the present case stored in raw files), e.g., SEQUEST implemented in BioworksBrowser 3.3.1 SP1 and X!Tandem implemented in the Scaffold 3 software.
2. Perform automated database searching against the most recent UniProtKB/Swiss-Prot database release http://www.expasy.org of the organism investigated.
3. Use the following settings for automated database searching within SEQUEST: 5 ppm precursor tolerance, 1 amu fragment ion tolerance, and full trypsin specificity allowing for up to two missed cleavages. Allow as variable modifications phosphorylation at serine, threonine, and tyrosine; oxidation of methionines; deamidation at asparagines and glutamine; acetylation at lysine and serine; formylation at lysine; and methylation at arginine, lysine, serine, threonine, and asparagine. Do not use fixed modifications in the database search.
4. Load srf or dta and out files generated by SEQUEST into the Scaffold 3 software, which performs probability assessment of peptide assignments and protein identifications by combining SEQUEST and X!Tandem database searches. Scaffold is useful for easily comparing and visualizing protein lists from several samples (in the present case *A*, *C*, *PD*, *CPD*, and *A+PD*).
5. Set the parameters within the Scaffold 3 software to consider only peptides with ≥95% probability. Set protein identification probabilities for multiple peptide assignments to ≥95%. For single peptide protein identifications, arbitrarily set protein probability to ≥50% and manually inspect the corresponding peptide MS/MS spectra. Proteins that comprise similar peptides and could not be differentiated based on MS/MS analysis alone are grouped by the software to satisfy the principles of parsimony. The estimated false discovery rate of peptide identifications can be determined using the reversed protein database approach and should be <1%.
6. Representative results of CCMS experiments are given in Tables 1, 2, and Supplementary Table S1 (mind that the protein database is not up-to-date for some proteins, e.g., PrmB (aka YfcB) or RsmH (aka MraW)), as well as in Fig. 3.

Identified Proteins (244)	Accession Number	Molecular Weight	CCMS A	C	PD	C-PD	A+PD
Glutamine synthetase OS=Escherichia coli (strain K12) GN=glnA PE=1 SV=2	GLNA_ECOLI	52kDa	90	0	0	0	97
5'-methylthioadenosine/ S-adenosylhomocysteine nucleosidase OS=Escherichia co	MTNN_ECOU	24 kDa	36	0	0	0	39
UPF0325 protein yaeH OS=Escherichia coli (strain K12) GN=yaeH PE=3 SV=1	YAEH_ECOLI	15kDa	24	15	0	0	22
Ribosomal RNA large subunit methyltransferase I OS=Escherichia coli (strain K12)	RLMI_ECOLI	44 kDa	17	0	17	0	20
Cyclopropane-fatty-acyl-phospholipid synthase OS=Escherichia coli (strain K12) Gl	CFA_ECOLI	44 kDa	15	0	0	0	14
50S ribosomal protein L28 OS=Escherichia coli (strain K12) GN=rpmB PE=1 SV=2	RL28_ECOLI	9 kDa	13	11	5	4	15
30S ribosomal protein S3 OS=Escherichia coli (strain K12) GN=rpsC PE=1 SV=2	RS3_EC0LI	26 kDa	13	5	0	0	16
Uncharacterized adenine-specific methylase yfcB OS=Escherichia coli (strain K12)	YFCB_ECOLI	35 kDa	13	0	0	0	15
50S ribosomal protein L16 OS=Escherichia coli (strain K12) GN=rplP PE=1 SV=1	RL16_EC0U	15 kDa	12	8	4	6	14
Exopolyphosphatase OS=Escherichia coli (strain K12) GN=ppx PE=1 SV=2	PPX_ECOLI	58 kDa	12	6	0	0	15
Ribosomal RNA large subunit methyltransferase L OS=Escherichia coli (strain K12	RLML_ECOLI	79 kDa	12	0	0	0	10
UPF0265 protein yeeX OS=Escherichia coli (strain K12) GN=yeeX PE=3 SV=1	YEEX_ECOU	13 kDa	11	12	0	0	16
tRNA (guanine-N(7)-)-methyltransferase OS= Escherichia coli (strain K12) GN=trmE	TRMB_ECOLI	27 kDa	11	0	0	0	13
50S ribosomal protein L15 OS=Escherichia coli (strain K12) GN=rplO PE=1 SV=1	RL15_EC0U	15 kDa	10	13	8	8	12
30S ribosomal protein S21 OS=Escherichia coli (strain K12) GN=rpsU PE=1 SV=2	RS21_EC0LI	9 kDa	10	12	9	11	11
Uncharacterized protein yjeF OS=Escherichia coli (strain K12) GN=yjeF PE=1 SV=	YJEF_ECOLI	55 kDa	10	11	0	0	11
ATP-dependent protease hsIV OS=Escherichia coli (strain K12) GN=hslV PE=1 SV	HSLV_ECOLI	19 kDa	10	9	0	0	13
Histidine biosynthesis bifunctional protein hisB OS=Escherichia coli (strain K12) GN	HIS7_EC0LI	40 kDa	8	7	0	0	9
DNA protection during starvation protein OS=Escherichia coli (strain K12) GN=dps	DPS_ECOLI	19 kDa	8	6	0	0	7
30S ribosomal protein S18 OS=Escherichia coli (strain K12) GN=rpsR PE=1 SV=2	RS18_EC0LI	9 kDa	8	5	2	3	7
Probable protease sohB OS=Escherichia coli (strain K12) GN=sohB PE=1 SV=1	SOHB_ECOLI	39 kDa	7	6	2	0	8

Integration host factor subunit alpha OS=Escherichia coli (strain K12) GN=ihfA PE=	IHFA_ECOLI	11 kDa	7	5	0	0	6
Protein smtA OS=Escherichia coli (strain K12) GN=smtA PE=1 SV=2	SMTA_ECOLI	30 kDa	7	1	0	0	8
tRNA (cmo5U34)-methyltransferase OS= Escherichia coli (strain K12) GN=cmoA P	CMOA_ECOLI	28 kDa	7	0	0	0	4
Molybdenum cofactor biosynthesis protein B OS=Escherichia coli (strain K12) GN=	MOAB_ECOLI	19 kDa	6	5	0	0	6
Chaperone protein skp OS=Escherichia coli (strain K12) GN=skp PE=1 SV=1	SKP_ECOLI	18 kDa	6	5	0	0	3
Ribosomal RNA small subunit methyltransferase G OS=Escherichia coli (strain K12	RSMG_ECOLI	23 kDa	6	0	1	0	5
50S ribosomal protein L34 OS=Escherichia coli (strain K12) GN=rpmH PE=1 SV=1	RL34_ECOLI	5 kDa	5	8	7	9	4
30S ribosomal protein S14 OS=Escherichia coli (strain K12) GN=rpsN PE=1 SV=2	RS14_EC0LI	12 kDa	5	7	7	5	8
30S ribosomal protein S20 OS=Escherichia coli (strain K12) GN=rpsT PE=1 SV=2	RS20_ECOLI	10 kDa	5	7	4	7	7
50S ribosomal protein L35 OS=Escherichia coli (strain K12) GN=rpml PE=1 SV=2	RL35_ECOU	7 kDa	5	5	10	10	7
50S ribosomal protein L17 OS=Escherichia coli (strain K12) GN=rplQ PE=1 SV=1	RL17_EC0U	14 kDa	5	4	0	0	6
Chorismate synthase OS=Escherichia coli (strain K12) GN=aroC PE=1 SV=4	AROC_ECOU	39 kDa	5	1	0	0	3
S-adenosyl-L-methionine-dependent methyltransferase mraW OS=Escherichia coli	MRAW_ECOLI	35 kDa	5	0	0	0	7
Protein aidB OS=Escherichia coli (strain K12) GN=aidB PE=1 SV=3	AIDB_ECOU	61 kDa	4	6	0	1	5
Putative NAD(P)H nitroreductase ydjA OS= Escherichia coli (strain K12) GN=ydjA F	YDJA_ECOLI	20 kDa	4	6	0	0	6
50S ribosomal protein L22 OS=Escherichia coli (strain K12) GN=rplV PE=1 SV=1	RL22_ECOLI	12 kDa	4	5	0	0	5
DNA-binding protein HU-beta OS=Escherichia coli (strain K12) GN=hupB PE=1 SV	DBHB_ECOU	9 kDa	4	3	0	0	5
Spermidine N(1)-acetyltransferase OS=Escherichia coli (strain K12) GN=speG PE=	ATDA_ECOLI	22 kDa	4	3	0	0	2
Polyphosphate kinase OS=Escherichia coli (strain K12) GN=ppk PE=1 SV=2	PPK_ECOLI	80 kDa	4	2	0	0	6
Integration host factor subunit beta OS=Escherichia coli (strain K12) GN=ihfB PE=	IHFB_ECOU	11 kDa	4	2	0	0	6
30S ribosomal protein S9 OS=Escherichia coli (strain K12) GN=rpsl PE=1 SV=2	RS9_EC0LI	15 kDa	4	2	0	0	2
30S ribosomal protein S5 OS=Escherichia coli (strain K12) GN=rpsE PE=1 SV=2	RS5_EC0LI	18 kDa	4	1	0	0	2
50S ribosomal protein L20 OS=Escherichia coli (strain K12) GN=rplT PE=1 SV=2	RL20_ECOLI	13 kDa	3	3	3	2	2

(continued)

Identified Proteins (244)	Accession Number	Molecular Weight	CCMS A	CCMS C	CCMS PD	CCMS C-PD	CCMS A+PD
50S ribosomal protein L2 OS=Escherichia coli (strain K12) GN=rplB PE=1 SV=2	RL2_EC0LI	30 kDa	3	3	2	2	4
50S ribosomal protein L27 OS=Escherichia coli (strain K12) GN=rpmA PE=1 SV=2	RL27_ECOLI	9 kDa	3	3	0	0	3
30S ribosomal protein S12 OS=Escherichia coli (strain K12) GN=rpsL PE=1 SV=2	RS12_EC0LI	14 kDa	3	2	1	0	2
Uncharacterized transcriptional regulator yjhU OS=Escherichia coli (strain K12) GN	YJHU_ECOLI	36 kDa	3	2	0	0	4
Ribosome modulation factor OS=Escherichia coli (strain K12) GN=rmf PE=1 SV=1	RMF_ECOLI	7 kDa	3	2	0	0	2
GTP cyclohydrolase 1 OS=Escherichia coli (strain K12) GN=folE PE=1 SV=2	GCH1_EC0LI	25 kDa	3	2	0	0	2
30S ribosomal protein S7 OS=Escherichia coli (strain K12) GN=rpsG PE=1 SV=3	RS7_EC0LI	20 kDa	3	0	0	0	2
(3R)-hydroxymyristoyl-[acyl-carrier-protein] dehydratase OS=Escherichia coli (straii	FABZ_ECOLI	17 kDa	2	3	0	0	4
Catabolite gene activator OS=Escherichia coli (strain K12) GN=crp PE=1 SV=1	CRP_ECOLI	24 kDa	2	3	0	0	3
50S ribosomal protein L33 OS=Escherichia coli (strain K12) GN=rpmG PE=1 SV=2	RL33_ECOLI	6 kDa	2	2	3	4	2
Citrate synthase OS=Escherichia coli (strain K12) GN=gltA PE=1 SV=1	CISY_ECOLI	48 kDa	2	2	0	0	2
10 kDa chaperon in OS=Escherichia coli (strain K12) GN=groS PE=1 SV=1	CH10_ECOLI	10 kDa	2	1	7	5	1
Acetyl-coenzyme A carboxylase carboxyl transferase subunit beta OS=Escherichia	ACCD_ECOLI	33 kDa	2	1	0	0	3
Acetyl-coenzyme A carboxylase carboxyl transferase subunit alpha OS=Escherichi	ACCA_ECOLI	35 kDa	2	1	0	0	2
DNA-binding protein HU-alpha OS=Escherichia coli (strain K12) GN=hupA PE=1 S	DBHA_ECOLI	10 kDa	2	1	0	0	1
Trans-aconitate 2-methyltransferase OS=Escherichia coli (strain K12) GN=tam PE	TAM ECOLI	29 kDa	2	0	0	0	3
ATP-dependent protease La OS=Escherichia coli (strain K12) GN=lon PE=1 SV=1	LON_ECOU	87 kDa	2	0	0	0	2
Ribosomal RNA small subunit methyltransferase D OS=Escherichia coli (strain K12	RSMD_ECOLI	22 kDa	2	0	0	0	2
50S ribosomal protein L11 OS=Escherichia coli (strain K12) GN=rplK PE=1 SV=2	RL11_EC0U	15 kDa	2	0	0	0	2
HTH-type transcriptional regulator yiaJ OS=Escherichia coli (strain K12)GN=yiaJ F	YIAJ_ECOLI	31 kDa	2	0	0	0	1
Uncharacterized protein yciY OS=Escherichia coli (strain K12) GN=yciY PE=3 SV=	YCIY_ECOLI	7 kDa	1	2	1	1	0

Elongation factor Tu OS=Escherichia coli (strain K12) GN=tufA PE=1 SV=2	EFTU_ECOLI	43 kDa	1	2	1	0	1
50S ribosomal protein L14 OS=Escherichia coli (strain K12) GN=rplN PE=1 SV=1	RL14_ECOU	14 kDa	1	2	0	0	1
50S ribosomal protein L32 OS=Escherichia coli (strain K12) GN=rpmF PE=1 SV=2	RL32_ECOU	6 kDa	1	1	2	3	1
Putative lipopolysaccharide biosynthesis O-acetyl transferase wbbJ OS=Escherich	WBBJ_ECOLI	22 kDa	1	1	1	1	1
30S ribosomal protein S17 OS=Escherichia coli (strain K12) GN=rpsQ PE=1 SV=2	RS17_ECOLI	10 kDa	1	1	1	1	1
UPF0250 protein ybeD OS=Escherichia coli (strain K12) GN=ybeD PE=1 SV=1	YBED_ECOLI	10 kDa	1	1	1	1	1
Type I restriction enzyme EcoKI R protein OS=Escherichia coli (strain K12) GN=hs	T1RK_EC0U	134 kDa	1	1	1	0	0
Lipopolysaccharide core heptose(ll) kinase rfaY OS=Escherichia coli (strain K12)	RFAY_ECOLI	27 kDa	1	1	0	1	2
50S ribosomal protein L29 OS=Escherichia coli (strain K12) GN=rpmC PE=1 SV=1	RL29_ECOLI	7 kDa	1	1	0	1	1
30S ribosomal protein S15 OS=Escherichia coli (strain K12) GN=rpsO PE=1 SV=2	RS15_ECOLI	10 kDa	1	1	0	1	1
Peptidyl-prolyl cis-trans isomerase C OS=Escherichia coli (strain K12) GN=ppiC PE	PPIC_ECOLI	10 kDa	1	1	0	0	2
Major outer membrane lipoprotein OS=Escherichia coli (strain K12) GN=lpp PE=1	LPP_ECOLI	8 kDa	1	1	0	0	1
30S ribosomal protein S19 OS=Escherichia coli (strain K12) GN=rpsS PE=1 SV=2	RS19_ECOLI	10 kDa	1	1	0	0	1
Virulence factor mviM homolog OS=Escherichia coli (strain K12) GN=mviM PE=1 S	MVIM_ECOLI	34 kDa	1	1	0	0	1
50S ribosomal protein L30 OS=Escherichia coli (strain K12) GN=rpmD PE=1 SV=2	RL30_ECOLI	7 kDa	1	1	0	0	1
Putative transposase insK for insertion sequence element IS150 OS=Escherichia	INSK_ECOU	33 kDa	1	1	0	0	0
50S ribosomal protein L4 OS=Escherichia coli (strain K12) GN=rplD PE=1 SV=1	RL4_ECOLI	22 kDa	1	0	1	1	0
Siroheme synthase OS=Escherichia coli (strain K12) GN=cysG PE=1 SV=1	CYSG_ECOLI	50 kDa	1	0	0	0	2
Ribosomal large subunit pseudouridine synthase C OS=Escherichia coli (strain K12	RLUC_ECOLI	36 kDa	1	0	0	0	1
50S ribosomal protein L6 OS=Escherichia coli (strain K12) GN=rplF PE=1 SV=2	RL6_ECOLI	19 kDa	1	0	0	0	1
Protein translocase subunit secA OS=Escherichia coli (strain K12) GN=secA PE=1	SECA_ECOLI	102 kDa	1	0	0	0	1
DNA-cytosine methyltransferase OS=Escherichia coli (strain K12) GN=dcm PE=1 S	DCM_ECOLI	53 kDa	1	0	0	0	1
Translation initiation factor IF-1 OS=Escherichia coli (strain K12) GN=infA PE=1 SV	IFl_ECOLI	8 kDa	1	0	0	0	1

(continued)

Identified Proteins (244)	Accession Number	Molecular Weight	CCMS A	C	PD	C-PD	A+PD
50S ribosomal protein L13 OS=Escherichia coli (strain K12) GN=rplM PE=1 SV=1	RL13_ECOU	16 kDa	1	0	0	0	1
Universal stress protein D OS=Escherichia coli (strain K12) GN=uspD PE=2 SV=1	USPD_ECOLI	16 kDa	1	0	0	0	1
30S ribosomal protein S16 OS=Escherichia coli (strain K12) GN=rpsP PE=1 SV=1	RS16_ECOLI	9 kDa	1	0	0	0	1
Dihydrolipoyl dehydrogenase OS=Escherichia coli (strain K12) GN=lpdA PE=1 SV=	DLDH_ECOLI	51 kDa	1	0	0	0	0
Ribosomal RNA small subunit methyltransferase B OS=Escherichia coli (strain K12	RSMB_ECOLI	48 kDa	1	0	0	0	0
tRNA 5-methylaminomethyl-2-thiouridine biosynthesis bifunctional protein mnmC C	MNMC_ECOLI	74 kDa	1	0	0	0	0
CTP synthase OS=Escherichia coli (strain K12) GN=pyrG PE=1 SV=2	PYRG_ECOLI	60 kDa	1	0	0	0	0
Uncharacterized protein yneK OS=Escherichia coli (strain K12) GN=yneK PE=1 SV	YNEK_ECOLI	43 kDa	1	0	0	0	0
D-amino acid dehydrogenase small subunit OS=Escherichia coli (strain K12) GN=	DADA_ECOLI	48 kDa	1	0	0	0	0
Maltose/maltodextrin import ATP-binding protein malK OS=Escherichia coli (strain	MALK_ECOLI	41 kDa	1	0	0	0	0
Inosine-guanosine kinase OS=Escherichia coli (strain K12) GN=gsk PE=1 SV=1	INGK_ECOLI	48 kDa	1	0	0	0	0
Nitrate/nitrite sensor protein narQ OS=Escherichia coli (strain K12) GN=narQ PE=3	NARQ_ECOLI	64 kDa	1	0	0	0	0
Ribonuclease P protein component OS=Escherichia coli (strain K12) GN=rnpA PE=	RNPA_ECOLI	14 kDa	1	0	0	0	0
Cyclic di-GMP-binding protein OS=Escherichia coli (strain K12) GN=bcsB PE=1 SV	BCSB_ECOLI	86 kDa	1	0	0	0	0
Anaerobic sulfatase-maturating enzyme homolog asIB OS=Escherichia coli (strain	ASLB_ECOLI	47 kDa	1	0	0	0	0
3-ketoacyl-CoA thiolase OS=Escherichia coli (strain K12) GN=fadl PE=1 SV=1	FADI_ECOLI	47 kDa	1	0	0	0	0
tRNA Delta(2)-isopentenylpyrophosphate transferase OS=Escherichia coli (strain K	MIAA_ECOLI	35 kDa	1	0	0	0	0
Protamine-like protein OS=Escherichia coli (strain K12) GN=tpr PE=4 SV=2	PRTL_ECOU	3 kDa	1	0	0	0	0
Quinone oxidoreductase OS=Escherichia coli (strain K12) GN=qor PE=1 SV=1	QOR_ECOLI	35 kDa	1	0	0	0	0
Rare lipoprotein A OS=Escherichia coli (strain K12) GN=rlpA PE=1 SV=1	RLPA_ECOLI	38 kDa	1	0	0	0	0
Putative beta-xylosidase OS=Escherichia coli (strain K12) GN=yagH PE=3 SV=1	YAGH_ECOU	61 kDa	1	0	0	0	0

Uncharacterized protein yefG OS=Escherichia coli (strain K12) GN=yefG PE=4 SV	YEFG_ECOLI	38 kDa	1	0	0	0	0
Putative uncharacterized protein yghXOS=Escherichia coli (strain K12) GN=yghX I	YGHX_ECOLI	30 kDa	1	0	0	0	0
Ribose-phosphate pyrophosphokinase OS=Escherichia coli (strain K12) GN=prs PI	KPRS_ECOLI	34 kDa	0	5	0	0	0
Uncharacterized protein yghZ OS=Escherichia coli (strain K12) GN=yghZ PE=1 SV	YGHZ_ECOLI	39 kDa	0	2	0	0	1
Protein ydeP OS=Escherichia coli (strain K12) GN=ydeP PE=1 SV=1	YDEP_ECOLI	83 kDa	0	1	1	0	0
Bifunctional aspartokinase/homoserine dehydrogenase 2 OS=Escherichia coli (stra	AK2H_ECOLI	89 kDa	0	1	0	0	1
Glyceraldehyde-3-phosphate dehydrogenase AOS=Escherichia coli (strain K12) G	G3PI_ECOLI	36 kDa	0	1	0	0	0
Flagellin OS=Escherichia coli (strain K12) GN=fliC PE=1 SV=2	FLIC_ECOLI	51 kDa	0	1	0	0	0
Ribosomal large subunit pseudouridine synthase D OS=Escherichia coli (strain K12	RLUD_ECOLI	37 kDa	0	1	0	0	0
Coenzyme A biosynthesis bifunctional protein coaBC OS=Escherichia coli (strain K	COABC_ECOLI	43 kDa	0	1	0	0	0
Sensor-like histidine kinase yojN OS=Escherichia coli (strain K12) GN=yojN PE=1	YOJN_ECOLI	100 kDa	0	1	0	0	0
Mannosyl-D-glycerate transport/metabolism system repressor mngR OS=Eschericl	MNGR_ECOLI	28 kDa	0	1	0	0	0
Aspartate aminotransferase OS=Escherichia coli (strain K12) GN=aspC PE=1 SV=	AAT_ECOLI	44 kDa	0	1	0	0	0
SOS ribosomal protein L36 OS=Escherichia coli (strain K12) GN=rpmJ PE=1 SV=1	RL36_ECOLI	4 kDa	0	1	0	0	0
Uncharacterized protein yfdP OS=Escherichia coli (strain K12) GN=yfdP PE=1 SV=	YFDP_ECOLI	13 kDa	0	1	0	0	0
Probable GTP-binding protein engB OS=Escherichia coli (strain K12) GN=engB PE	ENGB_ECOLI	24 kDa	0	1	0	0	0
Uncharacterized zinc-type alcohol dehydrogenase-like protein ydjJ OS=Escherichia	YDJJ_ECOLI	38 kDa	0	1	0	0	0
Multidrug resistance protein mdtG OS=Escherichia coli (strain K12) GN=mdtG PE=	MDTG_ECOLI	44 kDa	0	1	0	0	0
Ferrichrome-iron receptor OS=Escherichia coli (strain K12) GN=fhuA PE=1 SV=2	FHUA_ECOU	82 kDa	0	1	0	0	0
tRNA modification GTPase mnmE OS=Escherichia coli (strain K12) GN=mnmE PE	MNME_ECOLI	49 kDa	0	1	0	0	0
Multidrug resistance-like ATP-binding protein mdlB OS=Escherichia coli (strain K12	MDLB_ECOU	65 kDa	0	1	0	0	0
Uncharacterized J domain-containing protein djIB OS=Escherichia coli (strain K12)	DJLB_ECOLI	55 kDa	0	1	0	0	0
Uncharacterized protein yagW OS=Escherichia coli (strain K12) GN=yagW PE=4 S	YAGW_ECOLI	60 kDa	0	1	0	0	0

(continued)

Identified Proteins (244)	Accession Number	Molecular Weight	CCMS A	C	PD	C-PD	A+PD
Putative formate acetyltransferase 3 OS=Escherichia coli (strain K12) GN=ybiW PE	PFLF_ECOU	90 kDa	0	1	0	0	0
Uncharacterized protein yqjG OS=Escherichia coli (strain K12) GN=yqjG PE=3 SV=	YQJG_ECOLI	37 kDa	0	1	0	0	0
Putative uncharacterized protein ygaQ OS=Escherichia coli (strain K12) GN=ygaQ	YGAQ_ECOLI	85 kDa	0	1	0	0	0
Putative uncharacterized protein yeeL OS=Escherichia coli (strain K12) GN=yeeL F	YEEL_ECOU	39 kDa	0	1	0	0	0
Protein yihD OS=Escherichia coli (strain K12) GN=yihD PE=1 SV=1	YIHD_ECOU	10 kDa	0	1	0	0	0
Probable general secretion pathway protein C OS=Escherichia coli (strain K12) GN	GSPC_ECOLI	30 kDa	0	1	0	0	0
Putative uncharacterized protein yjiV OS=Escherichia coli (strain K12) GN=yjiV PE	YJIV_ECOU	103 kDa	0	1	0	0	0
Protein hnr OS=Escherichia coli (strain K12) GN=hnr PE=1 SV=1	HNR_ECOLI	37 kDa	0	1	0	0	0
Putative allantoin permease OS=Escherichia coli (strain K12) GN=ybbW PE=1 SV=	ALLP_ECOLI	52 kDa	0	1	0	0	0
Copper homeostasis protein cutC OS=Escherichia coli (strain K12) GN=cutC PE=3	CUTC_ECOLI	27 kDa	0	1	0	0	0
Fructose-6-phosphate aldolase 1 OS=Escherichia coli (strain K12) GN=fsaA PE=1	FSAA_ECOLI	23 kDa	0	1	0	0	0
DNA polymerase III subunit delta' OS=Escherichia coli (strain K12) GN=holB PE=1	HOLB_ECOLI	37 kDa	0	1	0	0	0
Lipoprotein-releasing system ATP-binding protein lolD OS=Escherichia coli (strain	LOLD_ECOLI	25 kDa	0	1	0	0	0
1, 4-dihydroxy-2-naphthoate octaprenyltransferase OS=Escherichia coli (strain K12	MENA_ECOLI	34 kDa	0	1	0	0	0
tRNA-processing ribonuclease BN OS=Escherichia coli (strain K12) GN=rbn PE=1	RBN_ECOLI	33 kDa	0	1	0	0	0
Outer membrane usher protein sfmD OS=Escherichia coli (strain K12) GN=sfmD P	SFMD_ECOLI	96 kDa	0	1	0	0	0
Protein tas OS=Escherichia coli (strain K12) GN=tas PE=1 SV=1	TAS_ECOU	39 kDa	0	1	0	0	0
Uncharacterized HTH-type transcriptional regulator yagl OS=Escherichia coli (strain	YAGI_ECOU	28 kDa	0	1	0	0	0
Uncharacterized HTH-type transcriptional regulator yeaT OS=Escherichia coli (stra	YEAT_ECOLI	35 kDa	0	1	0	0	0
Uncharacterized protein yfjZ OS=Escherichia coli (strain K12) GN=yfjZ PE=1 SV=1	YFJZ_ECOLI	12 kDa	0	1	0	0	0
Inner membranetransporter yhiP OS=Escherichia coli (strain K12) GN=yhiP PE=1	YHIP_ECOU	54 kDa	0	1	0	0	0
Arginyl-tRNA synthetase OS=Escherichia coli (strain K12) GN=argS PE=1 SV=1	SYR_ECOLI	65 kDa	0	0	2	0	0

UvrABC system protein C OS=Escherichia coli (strain K12) GN=uvrC PE=1 SV=1	UVRC_ECOLI	68 kDa	0	0	1	1	0
Naphthoate synthase OS=Escherichia coli (strain K12) GN=menB PE=1 SV=1	MENB_ECOLI	32 kDa	0	0	1	0	1
Chaperone protein dpB OS=Escherichia coli (strain K12) GN=clpB PE=1 SV=1	CLPB_ECOLI	96 kDa	0	0	1	0	0
Respiratory nitrate reductase 1 alpha chain OS=Escherichia coli (strain K12)GN=n	NARG_ECOLI	140 kDa	0	0	1	0	0
Cell division protease ftsH OS=Escherichia coli (strain K12) GN=hflB PE=1 SV=1	FTSH_ECOLI	71 kDa	0	0	1	0	0
Magnesium-transporting ATPase, P-type 1 OS= Escherichia coli (strain K12) GN=rr	ATMA_ECOLI	99 kDa	0	0	1	0	0
Uncharacterized protein yfeX OS=Escherichia coli (strain K12) GN=yfeX PE=3 SV=	YFEX_ECOU	33 kDa	0	0	1	0	0
Multifunctional CCA protein OS=Escherichia coli (strain K12) GN=cca PE=1 SV=1	CCA_ECOLI	46 kDa	0	0	1	0	0
DNA polymerase III subunit beta OS=Escherichia coli (strain K12) GN=dnaN PE=1	DP03B_ECOLI	41 kDa	0	0	1	0	0
Uncharacterized protein yjcC OS=Escherichia coli (strain K12) GN=yjcC PE=4 SV=	YJCC_ECOLI	61 kDa	0	0	1	0	0
Protein flmC OS=Escherichia coli (strain K12) GN=flmC PE=4 SV=1	FLMCl_ECOLI	8 kDa	0	0	1	0	0
Uncharacterized protein yjhG OS=Escherichia coli (strain K12) GN=yjhG PE=1 SV=	YJHG_ECOLI	70 kDa	0	0	1	0	0
Putative potassium channel protein OS=Escherichia coli (strain K12) GN=kch PE=	KCH_ECOLI	46 kDa	0	0	1	0	0
HTH-type transcriptional regulator malT OS=Escherichia coli (strain K12) GN=mal	MALT_ECOLI	103 kDa	0	0	1	0	0
Probable transport protein yifK OS=Escherichia coli (strain K12) GN=yifK PE=1 SV	YIFK_ECOU	50 kDa	0	0	1	0	0
Transcriptional regulator kdgR OS=Escherichia coli (strain K12) GN=kdgR PE=1 S'	KDGR_ECOLI	30 kDa	0	0	1	0	0
Putative acyl-CoA thioester hydrolase ybhC OS=Escherichia coli (strain K12) GN=	YBHC_ECOLI	46 kDa	0	0	1	0	0
Putative uncharacterized protein cysX OS= Escherichia coli (strain K12) GN=cysX F	CYSX_ECOLI	15 kDa	0	0	1	0	0
Glutamate-ammonia-ligase adenylyltransferase OS=Escherichia coli (strain K12) G	GLNE_ECOLI	108 kDa	0	0	1	0	0
Uncharacterized protein ilvX OS=Escherichia coli (strain K12) GN=ilvX PE=1 SV=1	ILVX_ECOLI	2 kDa	0	0	1	0	0
Putative lambdoid prophage defective integrase OS=Escherichia coli (strain K12) G	INTG_ECOLI	11 kDa	0	0	1	0	0
Lipopolysaccharide core biosynthesis protein rfaG OS=Escherichia coli (strain K12	RFAG_ECOLI	42 kDa	0	0	1	0	0
L-rhamnonate dehydratase OS=Escherichia coli (strain K12) GN=yfaW PE=1 SV=2	RHAMD_ECOU	44 kDa	0	0	1	0	0

(continued)

Identified Proteins (244)	Accession Number	Molecular Weight	CCMS				
			A	C	PD	C-PD	A+PD
Ribonuclease PH OS=Escherichia coli (strain K12) GN=rph PE=1 SV=2	RNPH_ECOU	25 kDa	0	0	1	0	0
Regulatory protein soxS OS=Escherichia coli (strain K12) GN=soxS PE=1 SV=2	SOXS_ECOLI	13 kDa	0	0	1	0	0
UPF0174 protein yaaW OS=Escherichia coli (strain K12) GN=yaaW PE=3 SV=1	YAAW_ECOLI	27 kDa	0	0	1	0	0
Uncharacterized protein ydaM OS=Escherichia coli (strain K12) GN=ydaM PE=4 S'	YDAM_ECOLI	46 kDa	0	0	1	0	0
Uncharacterized protein ydhT OS=Escherichia coli (strain K12) GN=ydhT PE=4 SV	YDHT_ECOLI	29 kDa	0	0	1	0	0
Uncharacterized protein yfgJ OS=Escherichia coli (strain K12) GN=yfgJ PE=1 SV=	YFGJ_ECOLI	8 kDa	0	0	1	0	0
Uncharacterized protein ygaF OS=Escherichia coli (strain K12) GN=ygaF PE=1 SV	YGAF_ECOLI	46 kDa	0	0	1	0	0
Uncharacterized protein yhcM OS=Escherichia coli (strain K12) GN=yhcM PE=1 SV	YHCM_ECOLI	43 kDa	0	0	1	0	0
2,3-diketo-L-gulonate TRAP transporter large permease protein yiaN OS=Eschericl	YIAN_ECOLI	45 kDa	0	0	1	0	0
Uncharacterized protein yjel OS=Escherichia coli (strain K12) GN=yjel PE=4 SV=1	YJEI_ECOU	12 kDa	0	0	1	0	0
GpG OS=Enterobacteria phage St-1 GN=G PE=4 SV=1	C6K2G8_BPST1	20 kDa	0	0	1	0	0
DNA topoisomerase 4 subunit A OS=Escherichia coli (strain K12) GN=parC PE=1	PARC_ECOLI	84 kDa	0	0	0	1	1
GTPase obgE/cgtA OS=Escherichia coli (strain K12) GN=obgE PE=1 SV=1	OBG_ECOU	43 kDa	0	0	0	1	0
3-dehydroquinate dehydratase OS=Escherichia coli (strain K12) GN=aroD PE=1 S'	AROD_ECOLI	27 kDa	0	0	0	1	0
ATP-dependent DNA helicase rep OS=Escherichia coli (strain K12) GN=rep PE=1	REP_ECOLI	77 kDa	0	0	0	1	0
Galactokinase OS=Escherichia coli (strain K12) GN=galK PE=1 SV=2	GALl_ECOLI	41 kDa	0	0	0	1	0
Uncharacterized ABC transporter ATP-binding protein ybiT OS=Escherichia coli (st	YBIT_ECOLI	60 kDa	0	0	0	1	0
Sensor kinase protein rcsC OS=Escherichia coli (strain K12) GN=rcsC PE=1 SV=3	RCSC_ECOLI	107 kDa	0	0	0	1	0
Flagellar motor switch protein fliM OS=Escherichia coli (strain K12) GN=fliM PE=1	FLIM_ECOU	38 kDa	0	0	0	1	0
RNA polymerase sigma factor rpoS OS=Escherichia coli (strain K12) GN=rpoS PE	RPOS_ECOLI	38 kDa	0	0	0	1	0
Uncharacterized lipoprotein yfhM OS=Escherichia coli (strain K12) GN=yfhM PE=4	YFHM_ECOU	182 kDa	0	0	0	1	0
Uncharacterized protein yeiR OS=Escherichia coli (strain K12) GN=yeiR PE=3 SV=	YEIR_ECOLI	36 kDa	0	0	0	1	0

TorCAD operon transcriptional regulatory protein torR OS=Escherichia coli (strain	TORR_ECOLI	26 kDa	0	0	0	1	0
Uncharacterized protein ygcL OS=Escherichia coli (strain K12) GN=ygcL PE=4 SV	YGCL_ECOLI	56 kDa	0	0	0	1	0
Protein aegA OS=Escherichia coli (strain K12) GN=aegA PE=3 SV=2	AEGA_ECOLI	72 kDa	0	0	0	1	0
Anaerobic C4-dicarboxylate transporter dcuB OS=Escherichia coli (strain K12) GN=	DCUB_ECOLI	48 kDa	0	0	0	1	0
Protein inaA OS=Escherichia coli (strain K12) GN=inaA PE=1 SV=3	INAA_ECOLI	25 kDa	0	0	0	1	0
Phosphomannomutase OS=Escherichia coli (strain K12) GN=manB PE=3 SV=1	MANB_ECOLI	50 kDa	0	0	0	1	0
Manganese transport protein mntH OS=Escherichia coli (strain K12) GN=mntH PE	MNTH_ECOLI	44 kDa	0	0	0	1	0
Uncharacterized protein ydiY OS=Escherichia coli (strain K12) GN=ydiY PE=4 SV=	YDIY_ECOLI	28 kDa	0	0	0	1	0
Uncharacterized HTH-type transcriptional regulator yhjC OS=Escherichia coli (strai	YHJC_ECOU	33 kDa	0	0	0	1	0
Uncharacterized membrane protein yliF OS=Escherichia coli (strain K12) GN=yliF I	YLIF_ECOLI	50 kDa	0	0	0	1	0
Uncharacterized oxidoreductase yghA OS=Escherichia coli (strain K12) GN=yghA I	YGHA_ECOU	31 kDa	0	0	0	1	0
Protein fdhE OS=Escherichia coli (strain K12) GN=fdhE PE=3 SV=2	FDHE_ECOLI	35 kDa	0	0	0	1	0
ATP-dependent Clp protease ATP-binding subunit clpA OS=Escherichia coli (strair	CLPA_ECOLI	84 kDa	0	0	0	0	2
Mrr restriction system protein OS=Escherichia coli (strain K12) GN=mrr PE=1 SV=	MRR_ECOLI	34 kDa	0	0	0	0	2
Dihydrofolate reductase folM OS=Escherichia coli (strain K12) GN=folM PE=1 SV=	FOLM_ECOLI	26 kDa	0	0	0	0	2
Dihydrolipoyllysine-residue acetyltransferase component of pyruvate dehydrogenas	ODP2_ECOLI	66 kDa	0	0	0	0	1
Beta-galactosidase OS=Escherichia coli (strain K12) PE=4 SV=1	B8LFD6_ECOLI	116 kDa	0	0	0	0	1
Cysteine desulfurase OS=Escherichia coli (strain K12) GN=iscS PE=1 SV=1	ISCS_ECOLI	45 kDa	0	0	0	0	1
NAD-dependent deacetylase OS=Escherichia coli (strain K12) GN=npdA PE=1 SV	NPD_ECOU	31 kDa	0	0	0	0	1
ATP-binding protein phnN OS=Escherichia coli (strain K12) GN=phnN PE=4 SV=1	PHNN_ECOLI	21 kDa	0	0	0	0	1
Curved DNA-binding protein OS=Escherichia coli (strain K12) GN=cbpA PE=1 SV=	CBPA_ECOLI	34 kDa	0	0	0	0	1
Nitrogen regulatory protein P-ll 1 OS=Escherichia coli (strain K12)GN=glnB PE=1	GLNB_ECOLI	12 kDa	0	0	0	0	1
Uncharacterized protein yfiQ OS=Escherichia coli (strain K12) GN=yfiQ PE=4 SV=	YFICLECOLI	98 kDa	0	0	0	0	1

(continued)

Identified Proteins (244)	Accession Number	Molecular Weight	CCMS				
			A	C	PD	C-PD	A+PD
Side tail fiber protein homolog from lambdoid prophage Rac OS=Escherichia coli (s	STFR_ECOLI	114 kDa	0	0	0	0	1
Transposase insH for insertion sequence element IS5Y OS=Escherichia coli (strair	INSH5_ECOU	38 kDa	0	0	0	0	1
Tail fiber assembly protein homolog from lambdoid prophage e14 OS=Escherichia	TFAE_ECOLI	22 kDa	0	0	0	0	1
Putative uncharacterized protein yjiT OS=Escherichia coli (strain K12) GN=yjiT PE=	YJIT_ECOLI	59 kDa	0	0	0	0	1
Carbamate kinase-like protein yah I OS=Escherichia coli (strain K12) GN=yahl PE=	ARCM_ECOLI	34 kDa	0	0	0	0	1
Uncharacterized protein yjbL OS=Escherichia coli (strain K12) GN=yjbL PE=4 SV=	YJBL_ECOLI	10 kDa	0	0	0	0	1
Chemotaxis protein methyltransferase OS=Escherichia coli (strain K12) GN=cheR	CHER_ECOLI	33 kDa	0	0	0	0	1
Protein yibA OS=Escherichia coli (strain K12) GN=yibA PE=1 SV=1	YIBA_ECOLI	32 kDa	0	0	0	0	1
Heat shock protein 15 OS=Escherichia coli (strain K12) GN=hslR PE=1 SV=1	HSLR_ECOLI	15 kDa	0	0	0	0	1
Uncharacterized protein ydbD OS=Escherichia coli (strain K12) GN=ydbD PE=4 SV	YDBD_ECOLI	87 kDa	0	0	0	0	1
Penicillin-binding protein 2 OS=Escherichia coli (strain K12) GN=mrdA PE=3 SV=1	PBP2_ECOLI	71 kDa	0	0	0	0	1
Uncharacterized protein yfhG OS=Escherichia coli (strain K12) GN=yfhG PE=4 SV	YFHG_ECOLI	27 kDa	0	0	0	0	1
HTH-type transcriptional repressor cytR OS=Escherichia coli (strain K12) GN=cytR	CYTR_ECOLI	38 kDa	0	0	0	0	1
Dihydrofolate reductase OS=Escherichia coli (strain K12) GN=folA PE=1 SV=1	DYR_ECOLI	18 kDa	0	0	0	0	1
Fimbriae Z protein OS=Escherichia coli (strain K12) GN=fimZ PE=3 SV=1	FIMZ_ECOLI	24 kDa	0	0	0	0	1
Glutamyl-tRNA reductase OS=Escherichia coli (strain K12) GN=hemA PE=1 SV=1	HEMl_ECOLI	46 kDa	0	0	0	0	1
Lipoprotein signal peptidase OS=Escherichia coli (strain K12) GN=lspA PE=1 SV=	LSPA_ECOLI	18 kDa	0	0	0	0	1
Psp operon transcriptional activator OS=Escherichia coli (strain K12) GN=pspF PE	PSPF_ECOU	37 kDa	0	0	0	0	1
Trimethylamine-N-oxide reductase 2 OS=Escherichia coli (strain K12) GN=torZ PE	TORZ_ECOLI	89 kDa	0	0	0	0	1
Lipopolysaccharide biosynthesis protein wzzE OS=Escherichia coli (strain K12) GN	WZZE_ECOLI	39 kDa	0	0	0	0	1
Esterase ydil OS=Escherichia coli (strain K12) GN=ydil PE=1 SV=1	YDII_ECOU	15 kDa	0	0	0	0	1
Uncharacterized zinc-type alcohol dehydrogenase-like protein ydjL OS=Escherichia	YDJL_ECOLI	39 kDa	0	0	0	0	1

4. Notes

1. CCs containing a variety of selectivity functions are available from caprotec bioanalytics GmbH (http://www.caprotec.com) or may be typically commissioned as custom synthesis.
2. It is possible to separate the beads from the reaction and wash solutions by holding a strong magnet to the side of the tubes or using any other commercially available rack for magnetic beads. However, the caproMag is recommended since this system is adjusted to tube strips and allows the collection of beads in the lids, so that tubes can be emptied without any pipetting steps. It has been designed especially for capture experiments, saves time, and reduces the risk of contaminations.
3. We use the high-accuracy LTQ Orbitrap XL mass spectrometer system connected to a split-less Easy-nLCTM liquid chromatography system (Proxeon Biosystems A/S) equipped with a Biosphere C18 precolumn (5 μm, 120 Å, 20 × 0.1 mm) and a Biosphere C18 analytical column (5 μm, 120 Å, 105 × 0.075 mm).
4. A major advantage of CCs lies in the formation of a covalent bond between the CC and the MTase, as this permits subsequent stringent washing conditions. The covalent crosslink is achieved by a photoreaction at 310 nm. Normal overhead light contains only a small fraction of UV, however, care should be exerted to protect the SAH-CC from longer exposure to overhead or even sun light prior to the controlled and cooled irradiation in the caproBox.
5. The biological samples from which the MTases are to be isolated may contain proteins prone to denaturation. Consequently it is mandatory to keep the samples cool and to avoid frothing at all times.
6. The caproBox cools the samples to 4°C, the lamps emitting the UV, however, also emit heat. Therefore, it is necessary to briefly centrifuge the vials before irradiation, so proteins adhering to the lids or vial walls cannot form precipitation seeds. After irradiation, avoid contact of the reaction mixture with the warm lids. Cool the vials and lids in ice water prior to mixing.
7. If manual re-suspension is not possible, briefly apply ultrasound by holding the samples into an ultrasound bath.
8. The final analysis of captured proteins is carried out by LC-MS/MS. Mass spectrometry is a highly sensitive method. It is necessary to use exclusively LC-MS grade reagents in the final steps (starting with the water for the final bead wash). Avoid contamination of the experiments by external protein sources,

e.g., keratin originating from dust or from the experimenter. Particularly during the final digestion steps, it is recommended to pay attention to a clean work space, to wear a lab coat, and possibly a hair net or ideally perform the final steps under a clean bench.

9. Alternatively to releasing the proteins from the beads by 60% ACN/0.2% TFA, the proteins can be directly tryptically digested within a bead suspension or, for SDS-PAGE, the proteins can be released by suspending and heating the beads at 95°C for 10 min in SDS sample buffer (both, the whole suspension or only the supernatant can be loaded in the gel pocket). These alternative procedures are described in ref. 11. Before the 60% ACN wash step, the beads can be stored in Milli-Q water for several weeks at 4°C.
10. Western blots using streptavidin-horseraddish peroxidase can also be used to visualize successful crosslinking of the biotin containing CC to the proteins.
11. To obtain reliable mass spectra, it is essential to have a stable spray in ESI-MS/MS analysis.
12. For other LC-MS/MS systems, measurement parameters and peptide identification algorithms must be adjusted individually.

Acknowledgments

This work was supported by the Human Frontier Science Program Organization (HFSP Award 2007, RGP0058/2007-C). We thank Prof. Richard Roberts for initiating the project and for fruitful discussions.

References

1. Lu, S.C. (2000) S-Adenosylmethionine. *Int. J. Biochem. Cell Biol. 32*, 391–3952.
2. Cantoni, G.L. (1975) Biological methylation, selected aspects. *Annu. Rev. Biochem. 44*, 435–451.
3. Fujioka, M. (1992) Mammalian small molecule methyltransferases, their structural and functional features. *Int. J. Biochem. 24*, 1917–1924.
4. Jeltsch, A. (2002) Beyond Watson and Crick, DNA methylation and molecular enzymology of DNA methyltransferases. *ChemBioChem 3*, 274–293.
5. Chow, C.S., Lamichhane, T.N., and Mahto, S.K. (2007) Expanding the nucleotide repertoire of the ribosome with post-transcriptional modifications. *ACS Chem. Biol. 2*, 610–619.
6. Grillo, M.A., and Colombatto, S. (2005) S-adenosylmethionine and protein methylation. *Amino Acids 28*, 357–362.
7. Dalhoff, C., Hueben, M., Lenz, T., Poot, P., Nordhoff, E., Koster, H., and Weinhold, E. (2010) Synthesis of *S*-Adenosyl-L-homocysteine Capture Compounds for Selective Photoinduced Isolation of Methyltransferases. *ChemBioChem 11*, 256–265.
8. Borchardt, R.T. (1977) In, Salvatore F, Borek E, Zappia V, Williams-Ashman HG, and Schlenk F (eds) The Biochemistry of

S-Adenosylmethionine, pp 151–171, Columbia University Press, New York, NY.

9. Hoffman, J.L. (1986) Chromatographic analysis of the chiral and covalent instability of S-adenosyl-L-methionine. *Biochemistry 25*, 4444–4449.

10. Koster, H., Little, D.P., Luan, P., Muller, R., Siddiqi, S,M,, Marappan, S., and Yip, P. (2007) Capture compound mass spectrometry, a technology for the investigation of small molecule protein interactions. *Assay Drug Dev. Technol. 5*, 381–390.

11. Fischer, J.J., Graebner Nee Baessler, O.Y., and Dreger, M. (2012) Proteome-wide identification of staurosporine-binding kinases using capture compound mass spectrometry. *Methods Mol. Biol. 795*, 135–147.

Part III

Target Discovery and Target Validation

Chapter 9

Identifying Cellular Targets of Small-Molecule Probes and Drugs with Biochemical Enrichment and SILAC

Shao-En Ong, Xiaoyu Li, Monica Schenone, Stuart L. Schreiber, and Steven A. Carr

Abstract

Sequencing of the human genome in the last decade has not yet led to a concomitant increase in the numbers of novel drug targets. While the pharmaceutical industry has invested heavily in improving drugs for existing protein targets, it has not tended toward a similar investment in experimental approaches to identify cellular targets of drugs. It is striking that the targets of numerous widely used FDA-approved drugs remain unknown. The development of robust, unbiased methods for target identification would greatly enhance our understanding the mechanisms-of-action of small molecules. Cell-based phenotypic screens followed by unbiased target identification have the potential to identify novel combinations of small molecules and their protein targets, shed light on drug polypharmacology, and enable unbiased screening approaches to drug discovery. Classical biochemical enrichment with immobilized small molecules has been used for over four decades but has been limited by issues concerning specificity and sensitivity. The application of mass spectrometry-based quantitative proteomics in combination with these affinity reagents has proven to be especially useful in addressing these common issues in affinity purification experiments. We describe the use of SILAC in identifying proteins that bind small-molecule probes and drugs in a cellular context.

Key words: Drug-target identification, Proteomics, SILAC, Quantitation, Small molecule, Affinity chromatography

1. Introduction

Applying small-molecule libraries to cell-based phenotypic screens is an increasingly popular approach to identify bioactive small molecules (SMs) (1–3). These chemical probes undergo re-synthesis, re-testing, and structural optimization to yield panels of related compounds for downstream drug development. Such approaches are yielding new biologically active small molecule

Gerard Drewes and Marcus Bantscheff (eds.), *Chemical Proteomics: Methods and Protocols*, Methods in Molecular Biology, vol. 803, DOI 10.1007/978-1-61779-364-6_9, © Springer Science+Business Media, LLC 2012

leads, in some cases with protein targets that were previously thought to be undruggable (4). Small-molecule probes are important tools to study biology, often providing new insights implicating specific biological pathways in systems where little is known beyond the observed phenotype. Researchers are applying small molecules like the HDAC inhibitor valproic acid and a TGF-beta kinase inhibitor RepSox in reprogramming cells to induced pluripotent stem cells (5, 6). Identifying biologically active SM probes and their phenotypic effects is just the first step in gaining new insights into biology and developing these probes into drugs.

As cell-based screening has the potential to provide large numbers of SM probes, identifying the cellular targets of these probes (7, 8) will enable downstream studies like structure–activity relationship (SAR) studies, biochemical assays and pharmacokinetics. In this protocol, we describe the combination of a quantitative proteomics approach (SILAC (9)) with biochemical enrichment using SM affinity matrices to identify their interacting protein targets (10).

2. Materials

2.1. Preparation of Affinity Reagents

1. Organic solvents:acetonitrile (ACN), dimethylformamide (DMF), dimethylsulfoxide (DMSO).
2. Disuccinimidyl carbonate (DSC, >95.0%, Fluka, Cat. No. 43720).
3. Triethylamine (TEA).
4. Affigel 102 (Bio-Rad, Hercules, CA).
5. Phosphate-buffered saline (PBS).

2.2. SILAC Cell Culture and Preparation of Lysates

1. Custom SILAC cell culture medium lacking arginine and lysine but otherwise formulated as normal culture medium like Dulbecco's modified Eagle's medium (DMEM, Caisson Laboratories Inc., North Logan, UT) supplemented with 10% dialyzed fetal bovine serum (Caisson), antibiotics, and glutamine (Invitrogen, Carlsbad, CA). It is possible to fully customize medium formulations for use in SILAC. The primary requirements are the use of SILAC amino acids and dialyzed serum.
2. Stable isotope labeled, "Heavy," arginine (Arg-13C615N4, Arg10) and lysine (Lys-13C615N2, Lys8) (Sigma-Isotec, St. Louis, MO) are prepared as concentrated (1,000× of DMEM concentration) stocks in PBS. Stock solutions of natural isotope abundance, "light," amino acids (Sigma, cell culture grade) are prepared accordingly.

3. 0.22 μm Sterile filter bottles to filter SILAC media, where needed.
4. Cells, like the human adenocarcinoma cell line HeLa (ATCC, Manassas, VA), cultured in SILAC media for at least five population doublings or until full incorporation of SILAC amino acids is achieved (11) (see Notes 1–4).
5. Cell lysis and wash (LWB) buffer containing 1% NP-40, 0.25% sodium deoxycholate, 150 mM NaCl, 50 mM Tris, Complete™ protease inhibitor cocktail tablets (Roche, Indianapolis, IN), pH 7.8.

2.3. Proteomic Analyses of Protein Interactors

1. Protein gel electrophoresis equipment and materials.
2. Dithiothreitol (DTT (Sigma), 1 M stock in water) to reduce disulfide bonds of cysteine and iodoacetamide (IAA (Sigma), 600 mM freshly prepared in water) to alkylate the free sulfhydryl groups.
3. Trypsin, TPCK-modified sequencing grade (Promega, Madison, WI) to generate tryptic peptides terminated in arginine or lysine residues. Trypsin is the enzyme of choice when labeling with arginine and lysine, to maximize quantitative coverage. 50 mM Ammonium bicarbonate (ABC) is the buffer used for trypsin digests.
4. Mass spectrometer (MS): A high-performance MS instrument (12, 13) capable of resolving power 10,000 FWHM at mass 400 m/z and better than 25 ppm average mass accuracy is desirable for good quantitative accuracy. Online nano-flow reversed-phase liquid chromatography for separation of peptide mixtures will aid in analysis of complex samples.
5. Mass spectrometric peptide identification software (e.g., Mascot; Matrix Science Ltd., London, UK) and quantification software from commercial vendors or academic labs (e.g., MaxQuant (14) or MSQuant (15)).

3. Methods

Applying SILAC to identify small molecule/protein interactions using small molecules as affinity matrices (16) combines traditional biochemical enrichment of proteins with quantitative proteomics to compare multiple pull-down experiments in a single MS experiment. The salient feature is that peptides and proteins from separate pull-down experiments with SILAC-labeled cell lysates are distinguishable by the mass differences introduced by SILAC amino acids. Combining biochemical purification workflows and comparing relative intensities of MS signals from light and heavy proteins allows proteins that bind specifically to one bait to be discriminated from proteins that bind nonspecifically to both baits (Fig. 2). Because this approach uses quantitative comparisons to evaluate each quantified protein's specific binding to soluble SM, it is possible to use relatively gentle wash conditions that may favor the identification of weaker SM–protein interactions. It is not uncommon to find a small percentage (<10%) of total identified proteins to be specific interactions. The need for ad hoc optimization for individual SM baits is substantially reduced and consequently the method is amenable to higher throughput and automation.

The protocol that follows describes a single SILAC experiment. To take full advantage of the quantitative ratios provided in these experiments, it is best to perform a replicate experiment that reverses the order of the lysate-bait combination (the "reverse" experiment). Therefore, when a specifically bound protein shows a high heavy/light (H/L) ratio in the forward experiment, the reverse replicate should yield the same protein with a low H/L ratio. Nonspecifically bound proteins will have ratios close to 1 in both experiments. As seen in Fig. 3a, the requirement that quantitative ratios reverse in forward–reverse experiments provides a very stringent criterion for bona fide interacting proteins with the small molecule bait.

Small-molecule probes are functionalized to allow covalent attachment to a solid-phase matrix at sites unrelated to the biological activity of the SM. If SAR data are unavailable, functionalization of the SM followed by re-testing of the modified-SM in the phenotypic screen is necessary to ensure that its original phenotype-inducing properties are not compromised. Functional groups such as primary amine, primary alcohol, carboxylic acid, azide, etc. can be used, depending on the structure of the SM and the feasibility of synthesis (Fig. 1). Affigel 102 (Bio-Rad) is an example of a

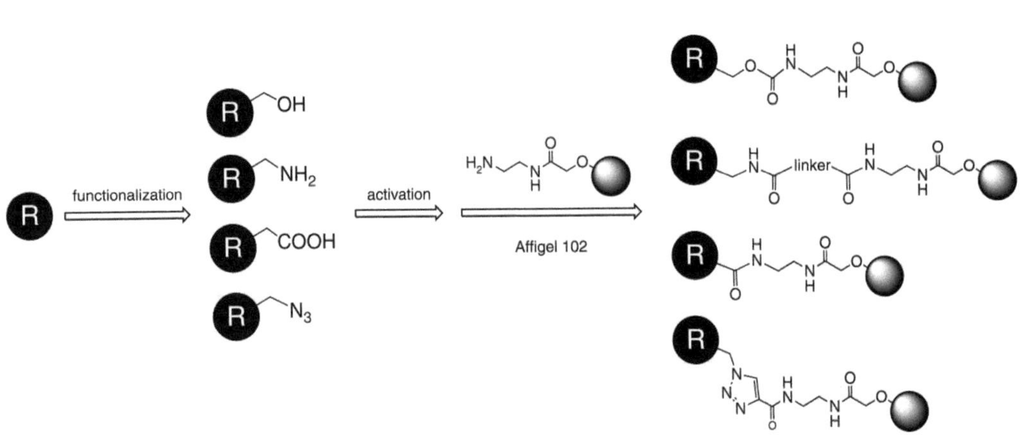

Fig. 1. Generation of small-molecule affinity matrices with covalent coupling to agarose beads. The small molecule of interest (labeled R) is functionalized with common functional groups like primary alcohols, amines, carboxylic acid, or azides to facilitate coupling to a solid-phase matrix like the agarose bead Affigel 102.

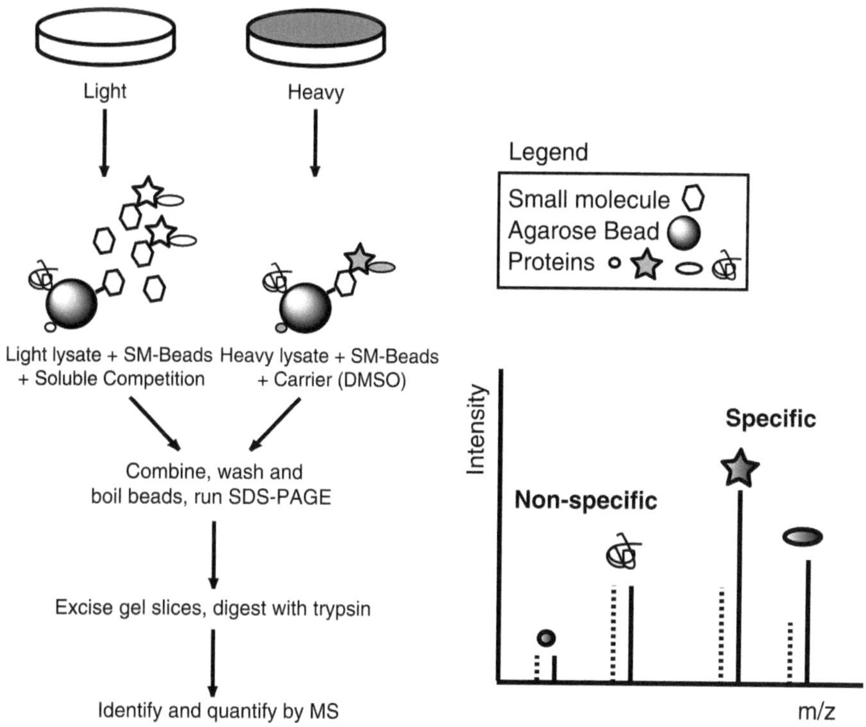

Fig. 2. Identifying specific SM–protein interactions with quantitative proteomics. SILAC identifies specific protein interactions with SM baits. Cell populations are fully labeled with light (*white*) and heavy amino acids (*gray*) and lysates incubated either with SM-loaded beads (*SM-beads*) and soluble SM competitor or SM-Beads with DMSO. Proteins interacting directly with the SM or via secondary and/or higher-order interactions are enriched in the heavy state over the light and are identified with differential ratios. Nonspecifically bound proteins are enriched equally in both states and have ratios close to one.

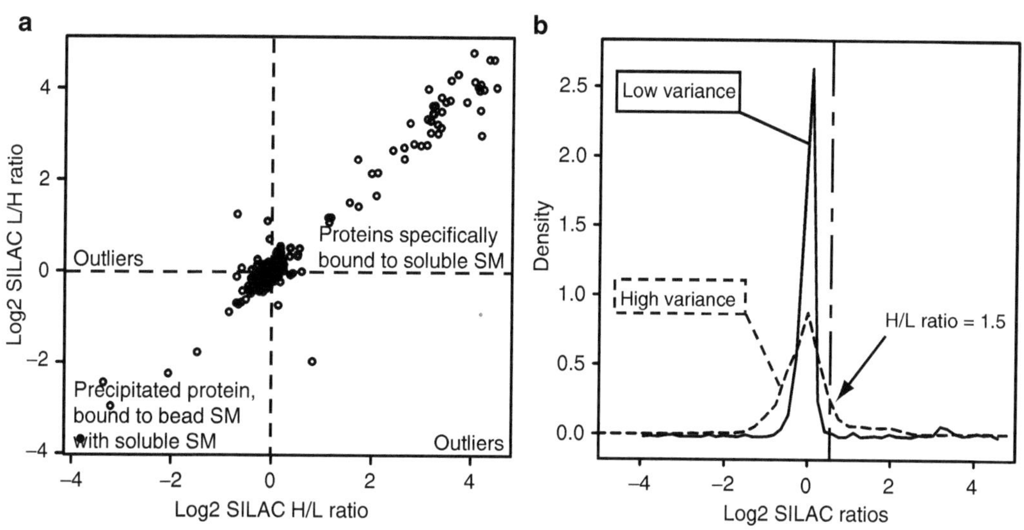

Fig. 3. Analysis of quantitative proteomics data in small molecule target identification. (**a**) Scatter plot of Log2 SILAC ratios forward and reverse replicate experiments. Note that in this representation, the ratios in the reverse experiment are inverted (L/H) with respect to the forward (H/L) experiment. With the experimental design as shown in Fig. 2, target proteins with ratios that invert in the two bait-swap experiments have ratios with the same sign and are found in the top right quadrant. Ratios of outlier proteins do not invert and proteins that precipitate with SM treatment co-precipitate in the affinity matrix and are likely to be found in the bottom left quadrant.

suitable solid phase for immobilization of SM baits. Affigel 102 is an agarose-based gel with a six-atom hydrophilic amino-terminated arm with a typical loading of 12 μmol/mL. In this protocol, we describe the generation of an affinity matrix with a SM modified with a primary alcohol as an example.

3.1. Preparing the Small-Molecule Bait/Affinity Matrix

1. Dissolve the SM bait molecule in anhydrous acetonitrile (or DMF).
2. Dissolve 3 equivalents (eq.) of DSC in anhydrous acetonitrile (or DMF) and add to the SM bait solution before TEA (4 eq.) is added. The reaction solution is stirred at 50°C for 1 h and the activation efficiency is monitored by LC-MS.
3. Transfer bead suspension to a microcentrifuge tube and wash with 6× bead volumes of H_2O, followed by DMF. Resuspend the beads in anhydrous DMF before adding the solution of activated SM bait molecule.
4. Vortex the activated bead suspension at room temperature for 1 h and monitor the depletion of free activated bait molecule by LC-MS to assess the completion of the compound-loading step.
5. Centrifuge the reaction vial at 1,000 ×*g* at room temperature to collect the beads, remove supernatant, and wash beads with 6× bead volumes of DMF and 3× bead volumes of H_2O. The beads are subsequently suspended in PBS and stored at 4°C before use.
6. Set aside a 50% (v/v) PBS/SM-bead slurry (SM-bead) for use in biochemical enrichments (Subheading 3.3).
7. A 30 mM stock solution of unbound small molecule in DMSO is set aside for use as a soluble competitor (SC) (Subheading 3.3).

3.2. Preparing SILAC Cell Lysates for Pull-Down Experiments

1. Fully SILAC-labeled cells (11) are washed twice with PBS to remove serum proteins before lysing of cells in LWB buffer. Add enough LWB buffer to obtain protein concentrations of around 1.5–2 mg/mL. A higher protein concentration may lead to excessive nonspecific binding.
2. Light- and heavy-labeled cell lysates are spun down in a centrifuge at 4°C at 14,000 ×*g* for 10 min to clarify lysates.
3. Determine protein concentrations for clarified light and heavy lysates using a compatible protein concentration assay like Bradford or BCA. If necessary, equalize concentrations of SILAC lysates by diluting with LWB buffer. Set aside lysates on ice until ready for use.
4. Use between 1.5 and 2.0 mg of protein per pull-down with 15 μL of SM-bead. (see Notes 5 and 6).

3.3. Biochemical Enrichment of Small-Molecule Interactors

1. Estimate the amount of loaded small molecule on 15 μL of SM-bead and add a fivefold excess of SC (stock solution in Subheading 3.1 step 7) to light cell lysate. For example, it is common to have ~200 nmol of small molecule on bead and to add 1 mmol of SC in 2 mL of lysate to yield a 0.5 mM solution of SC (see Note 7).
2. Incubate light (with SC) and heavy (with DMSO) tubes on an end-over-end rotator for 1 h at 4°C. At the end of an hour, inspect tubes for any precipitation, spinning tubes down if necessary.
3. Add 30 μL of SM-beads (50% slurry in step 6 of Subheading 3.1) to both light and heavy lysates tubes using a wide-bore pipette tip, taking care to add equal amounts of beads to both tubes. This is critical to allow good quantitative comparisons of the two pull-down experiments.
4. Rotate on end-over-end rotator for 4–16 h. It may be necessary to adjust the incubation period for specific experiments.
5. After incubation, spin tubes at 1,000 × *g* for 1 min to collect SM-beads, allowing beads to settle in tubes for an additional minute on the bench. Carefully aspirate supernatants and wash beads with 1.5 mL of LWB buffer by gently inverting tubes a few times. Do not vortex or agitate, avoid foaming if possible. Spin beads down in tubes and aspirate washes.
6. After the first wash is complete, use 1.5 mL of LWB buffer to resuspend and combine SM-beads from light and heavy tubes into a single tube. Spin tube again and repeat washing again.
7. After the second wash, collect SM-beads and leave a small volume (~15 μL) of LWB buffer in headspace above beads.

3.4. Gel separation and LC-MS Analysis

1. Add DTT to a final concentration of 1 mM and incubate at 56°C for 30 min with shaking. Let tubes cool to room temperature, add IAA to a final concentration of 6 mM, and incubate in the dark with intermittent vortexing for 30 min. Include volume of SM-beads when calculating amounts of DTT and IAA to add.
2. Add SDS-PAGE loading buffer, boil and spin down tubes, and prepare samples for separation on a SDS-PAGE gel.
3. When gel separation is complete, stain gel with MS-compatible stains. Colloidal Coomassie stain is sufficiently sensitive for almost all applications. The primary objective is the visualization of the lane to allow excision of gel slices.
4. Divide and excise the entire gel lane into 6–10 pieces rather than cutting out specific resolved bands. It is preferable to isolate heavily stained, therefore highly abundant, protein bands to their own slices. This increases the dynamic range of the gel separation and LC-MS analysis (GeLC-MS).

5. Perform typical proteolytic digestion of gel-separated proteins (17).
6. Perform off-line reversed-phase cleanup of peptides using a microcolumn (e.g., StageTip (18)). This step is optional but recommended after gel digestion to remove salts or tiny gel pieces that can otherwise block nano-LC columns.
7. Separate peptides in nanoscale reversed-phase LC-MS. Typical conditions for LC-MS are flowrates of 200 nL/min, 1 h gradient 0.5%/min, 7–37% acetonitrile in 0.1% formic acid/water.
8. Collect MS full scan data to acquire accurate peptide mass and intensity information at a scan rate where multiple scans are obtained across the eluting peptide peak. The chromatographic peak width of a typical peptide in nano-LC-MS is between 10 and 25 s. Ideally, quantitative accuracy in measuring peak intensity is good when more than nine MS scans are available for each peptide peak. MS experimental methods should be specifically designed to balance acquisition of peptide MS/MS sequencing information with MS full scans for quantitation.

3.5. Data Analysis

The principles of peptide quantification are similar in most software packages, though the implementation may vary considerably depending on instrument vendor and data analysis software. Analysis of SILAC data collected on the FTMS/Orbitrap platform (Thermo Fisher) can be performed with the freely available software MSQuant (15) or MaxQuant (14, 19). We expect quantitative proteomics data analysis software to continue to evolve and thus encourage readers of this protocol to evaluate and apply the latest versions of available software when analyzing their data.

1. Raw data files from MS runs are collected, processed for peptide identification and quantification. The resulting data file should contain a list of identified proteins and their quantified ratios. Forward and reverse replicates should be separated so that ratios from each experiment are calculated.
2. Density plots of SILAC heavy/light (H/L) ratios for different experiments provide an overview of the experiment. Typically, most identified proteins are not specifically competed by soluble small molecule (Schematic in Fig. 2) and should form the main distribution with log2 ratios centered around zero (Fig. 3a, b). The spread of this main distribution should be small with log2 ratios within ±0.37. As light and heavy lysates were combined in equal proportions, the theoretical protein ratios should be around one. As quantitative errors in MS range between 10 and 20%, a fold change greater than 1.3 (log2 ratio ± 0.37) should be detectable from the main distribution in typical SILAC experiments.

3. If no changes were introduced to either light or heavy cells, mixing the two lysates would ideally result in all proteins with log2 ratios of 0. A number of factors affect this; the main issue is not taking care to treat light and heavy cells/lysates identically. A large spread in the main distribution may indicate that individual protein abundances in light and heavy lysates are different. For instance, the amount of total protein in light and heavy lysates may be equal yet levels of individual proteins can vary considerably. An example of this is illustrated in Fig. 3b, the spread of ratios in null distribution centered about log2 SILAC ratio of 0 is different in the two density plots. This may also arise when comparing SILAC datasets prepared on different days or by different operators. If a fixed ratio cut-off (e.g., 1.5-fold) is used to determine the number of significant SM–protein interactions, the number of targets identified in these two experiments would be very different. Several software packages address such inter-experimental variability and calculate probability or FDR values for target proteins (14, 20, 21).

4. Notes

1. Although SILAC was originally developed for use with mammalian cells, it has been adapted in yeast (22) and even higher organisms like mice (23). While SILAC is especially convenient in mammalian cell culture, it is possible to use other quantitative proteomics approaches like ^{15}N-labeling or chemical labeling instead.
2. Incorporating technical replicates in quantitative proteomics experiments is strongly encouraged and one set of replicates should be performed with the order of lysate-bait incubations reversed from the second set of experiments ("bait-swap" experiment). In bait-swap experiments, target proteins have differential ratios and should invert between the two replicates (Fig. 3a). The requirement that protein ratios invert is a stringent filter for bait binding specificity and helps eliminate false positives arising from systematic errors.
3. Triple encoding SILAC can be applied to label three cell populations instead of two (24), thus allowing the comparison of protein enrichments with three small-molecule baits in a single experiment. This additional state can be used to test the binding of the small molecule under a different experimental condition, e.g., pretreatment of cells with a drug. It is also especially useful to apply SILAC-labeled lysates for analytical evaluations of affinity reagents. For example, by using SILAC triple encoded cell lysates (Arg0-Lys0, Arg6-Lys4, Arg10-Lys8)

in pull-downs with beads loaded with different levels of compound, we can quantitatively compare the ability of these affinity matrices to bind target proteins in a single experiment.

4. Certain cell lines retain the ability to metabolize arginine to proline and can lead to quantitative inaccuracies when using arginine labeling (25). This issue can be circumvented by either titrating the amount of arginine used (11) or supplementing with additional proline (26) when performing a labeling test to measure levels of heavy arginine or proline present.
5. The protocol herein may be modified in a number of ways, to enrich for proteins in a specific cellular compartment, like a nuclei preparation. This will not affect the ability to identify targets as long as modifications are made to both light and heavy cell states.
6. In some cases, pull-downs with SM affinity matrices yield very little bound protein. This may be an issue with compound loading or the nature of the SM itself. It is often useful to test the affinity matrix in a separate pull-down experiment using only the stained SDS-PAGE gel as readout. This provides a quick survey of the pull-down experiment. If the SM-beads do not bind much protein, it may be useful to increase the amount of protein applied or try subcellular fractionation.
7. This is a high concentration of small molecule and may not be possible if the compound has poor solubility in aqueous solution. Look for any precipitation upon addition of SC. If there is any precipitation, it is necessary to use lower amounts of SC. In general, it is desirable to have as high a concentration of SC as solubility limits allow (10). Add an equal volume of DMSO to the heavy lysate.

Acknowledgments

The authors gratefully acknowledge the helpful discussions and useful insights of our colleagues in the Chemical Biology Program, Chemical Biology Platform, and Proteomics Platform at the Broad Institute. This research has been funded in part with Federal funds from the National Cancer Institute's Initiative for Chemical Genetics (Contract No. N01-CO-12400) as well as the NIH grant for Genomics Based Drug Discovery – Target ID project (NIH RL1HG004671), administratively linked to RL1CA133834, RL1GM084437, and UL1RR024924.

References

1. Gamo, F. J., Sanz, L. M., Vidal, J., de Cozar, C., Alvarez, E., Lavandera, J. L., Vanderwall, D. E., Green, D. V., Kumar, V., Hasan, S., Brown, J. R., Peishoff, C. E., Cardon, L. R., and Garcia-Bustos, J. F. Thousands of chemical starting points for antimalarial lead identification, *Nature* **465**, 305–310.
2. Guiguemde, W. A., Shelat, A. A., Bouck, D., Duffy, S., Crowther, G. J., Davis, P. H., Smithson, D. C., Connelly, M., Clark, J., Zhu, F., Jimenez-Diaz, M. B., Martinez, M. S., Wilson, E. B., Tripathi, A. K., Gut, J., Sharlow, E. R., Bathurst, I., El Mazouni, F., Fowble, J. W., Forquer, I., McGinley, P. L., Castro, S., Angulo-Barturen, I., Ferrer, S., Rosenthal, P. J., Derisi, J. L., Sullivan, D. J., Lazo, J. S., Roos, D. S., Riscoe, M. K., Phillips, M. A., Rathod, P. K., Van Voorhis, W. C., Avery, V. M., and Guy, R. K. Chemical genetics of Plasmodium falciparum, *Nature* **465**, 311–315.
3. Stockwell, B. R. (2004) Exploring biology with small organic molecules, *Nature* **432**, 846–854.
4. Stanton, B. Z., Peng, L. F., Maloof, N., Nakai, K., Wang, X., Duffner, J. L., Taveras, K. M., Hyman, J. M., Lee, S. W., Koehler, A. N., Chen, J. K., Fox, J. L., Mandinova, A., and Schreiber, S. L. (2009) A small molecule that binds Hedgehog and blocks its signaling in human cells, *Nat Chem Biol* **5**, 154–156.
5. Huangfu, D., Maehr, R., Guo, W., Eijkelenboom, A., Snitow, M., Chen, A. E., and Melton, D. A. (2008) Induction of pluripotent stem cells by defined factors is greatly improved by small-molecule compounds, *Nat Biotechnol* **26**, 795–797.
6. Ichida, J. K., Blanchard, J., Lam, K., Son, E. Y., Chung, J. E., Egli, D., Loh, K. M., Carter, A. C., Di Giorgio, F. P., Koszka, K., Huangfu, D., Akutsu, H., Liu, D. R., Rubin, L. L., and Eggan, K. (2009) A small-molecule inhibitor of tgf-Beta signaling replaces sox2 in reprogramming by inducing nanog, *Cell Stem Cell* **5**, 491–503.
7. Rix, U., and Superti-Furga, G. (2009) Target profiling of small molecules by chemical proteomics, *Nat Chem Biol* **5**, 616–624.
8. Terstappen, G. C., Schlupen, C., Raggiaschi, R., and Gaviraghi, G. (2007) Target deconvolution strategies in drug discovery, *Nat Rev Drug Discov* **6**, 891–903.
9. Ong, S. E., Blagoev, B., Kratchmarova, I., Kristensen, D. B., Steen, H., Pandey, A., and Mann, M. (2002) Stable isotope labeling by amino acids in cell culture, SILAC, as a simple and accurate approach to expression proteomics, *Mol Cell Proteomics* **1**, 376–386.
10. Ong, S. E., Schenone, M., Margolin, A. A., Li, X., Do, K., Doud, M. K., Mani, D. R., Kuai, L., Wang, X., Wood, J. L., Tolliday, N. J., Koehler, A. N., Marcaurelle, L. A., Golub, T. R., Gould, R. J., Schreiber, S. L., and Carr, S. A. (2009) Identifying the proteins to which small-molecule probes and drugs bind in cells, *Proc Natl Acad Sci USA* **106**, 4617–4622.
11. Ong, S. E., and Mann, M. (2006) A practical recipe for stable isotope labeling by amino acids in cell culture (SILAC), *Nat Protoc* **1**, 2650–2660.
12. Ong, S. E., and Mann, M. (2005) Mass spectrometry-based proteomics turns quantitative, *Nat Chem Biol* **1**, 252–262.
13. Domon, B., and Aebersold, R. (2006) Mass spectrometry and protein analysis, *Science* **312**, 212–217.
14. Cox, J., and Mann, M. (2008) MaxQuant enables high peptide identification rates, individualized p.p.b.-range mass accuracies and proteome-wide protein quantification, *Nat Biotechnol* **26**, 1367–1372.
15. Mortensen, P., Gouw, J. W., Olsen, J. V., Ong, S. E., Rigbolt, K. T., Bunkenborg, J., Cox, J., Foster, L., Heck, A. J., Blagoev, B., Andersen, J. S., and Mann, M. (2009) MSQuant, an open source platform for mass spectrometry-based quantitative proteomics, *J* Proteome Res. 9(1): 393–403.
16. Cuatrecasas, P. (1970) Protein purification by affinity chromatography. Derivatizations of agarose and polyacrylamide beads, *J Biol Chem* **245**, 3059–3065.
17. Shevchenko, A., Tomas, H., Havlis, J., Olsen, J. V., and Mann, M. (2006) In-gel digestion for mass spectrometric characterization of proteins and proteomes, *Nat Protoc* **1**, 2856–2860.
18. Rappsilber, J., Mann, M., and Ishihama, Y. (2007) Protocol for micro-purification, enrichment, pre-fractionation and storage of peptides for proteomics using StageTips, *Nat Protoc* **2**, 1896–1906.
19. Cox, J., Matic, I., Hilger, M., Nagaraj, N., Selbach, M., Olsen, J. V., and Mann, M. (2009) A practical guide to the MaxQuant computational platform for SILAC-based quantitative proteomics, *Nat Protoc* **4**, 698–705.
20. Margolin, A. A., Ong, S. E., Schenone, M., Gould, R., Schreiber, S. L., Carr, S. A., and Golub, T. R. (2009) Empirical bayes analysis of quantitative proteomics experiments, *PLoS One* **4**, e7454.
21. Ting, L., Cowley, M. J., Hoon, S. L., Guilhaus, M., Raftery, M. J., and Cavicchioli, R. (2009)

Normalization and statistical analysis of quantitative proteomics data generated by metabolic labeling, *Mol Cell Proteomics* **8**, 2227–2242.

22. Jiang, H., and English, A. M. (2002) Quantitative analysis of the yeast proteome by incorporation of isotopically labeled leucine, *J Proteome Res* **1**, 345–350.
23. Kruger, M., Moser, M., Ussar, S., Thievessen, I., Luber, C. A., Forner, F., Schmidt, S., Zanivan, S., Fassler, R., and Mann, M. (2008) SILAC mouse for quantitative proteomics uncovers kindlin-3 as an essential factor for red blood cell function, *Cell* **134**, 353–364.
24. Blagoev, B., Ong, S. E., Kratchmarova, I., and Mann, M. (2004) Temporal analysis of phosphotyrosine-dependent signaling networks by quantitative proteomics, *Nat Biotechnol* **22**, 1139–1145.
25. Ong, S. E., Kratchmarova, I., and Mann, M. (2003) Properties of 13 C-substituted arginine in stable isotope labeling by amino acids in cell culture (SILAC), *J Proteome Res* **2**, 173–181.
26. Bendall, S. C., Hughes, C., Stewart, M. H., Doble, B., Bhatia, M., and Lajoie, G. A. (2008) Prevention of amino acid conversion in SILAC experiments with embryonic stem cells, *Mol Cell Proteomics* 7, 1587–1597.

Chapter 10

Determination of Kinase Inhibitor Potencies in Cell Extracts by Competition Binding Assays and Isobaric Mass Tags

Carsten Hopf, Dirk Eberhard, Markus Boesche, Sonja Bastuck, Birgit Dümpelfeld, and Marcus Bantscheff

Abstract

Chemical proteomics offers a unique approach for target identification of small molecule inhibitors directly from cell extracts, thus enabling characterization of target proteins under close to physiological conditions. Here, we describe a competition binding procedure that is based on affinity enrichment of potential target proteins on a probe matrix in the presence of increasing amounts of free test compound in solution. Reduced binding of target proteins to the probe matrix as a function of test compound concentration can be measured and thus, enables calculation of IC_{50} values. The method employs quantitative mass spectrometry using isobaric mass tags which enables determination of potency for a large number of target proteins in a single analysis.

Key words: IC_{50}, Drug potency, Competition binding, Isobaric mass tags, Mass spectrometry, Selectivity profiling

1. Introduction

In recent years, chemical proteomics methods have developed into commonly used tools for deciphering a drug's or a tool compound's mode-of-action. In a typical experiment, linkable analogs are synthesized and immobilized on a solid support. The resulting probe matrix is subsequently used to affinity capture target proteins directly from cell or tissue extracts. The efficiency of this approach, however, depends on several factors including but not limited to the affinity of the immobilized ligand to target proteins,

Gerard Drewes and Marcus Bantscheff (eds.), *Chemical Proteomics: Methods and Protocols*, Methods in Molecular Biology, vol. 803, DOI 10.1007/978-1-61779-364-6_10, © Springer Science+Business Media, LLC 2012

expression levels and thus concentrations of target proteins in available extracts, and the ability to distinguish between specific and unspecific binding to the probe matrix. For example, immobilization of compounds on a solid support requires the presence of functional groups such as amines, carboxylic acids, or hydroxyls. If such groups are not available in the probe compound of interest or if these groups are required for interaction with the target proteins, analogs need to be synthesized. Special care must be taken that modifications to the molecule do not impede its bioactivity or change its selectivity (1). Furthermore, it has been observed that the linker between the resin and the compound can also significantly affect the binding of affinity-enriched proteins (2). Consequently, the target profile of an immobilized ligand can only approximate a test compound's interactome. Further complication in the analysis of such experiments is attributed to the fact that the amounts of individual proteins captured by the probe matrix do not correlate with the affinities of these proteins for the immobilized compound, as highly abundant low-affinity binders may be more prominent in the eluate of the probe matrix than high-affinity interactors that are of very low abundance. As a consequence, additional experiments are required to distinguish between these two cases (3–7).

Recent advances in chemical proteomics have addressed these issues by (a) performing target class specific enrichment of proteins using a universal (set) of probe compounds, e.g., unspecific kinase inhibitors and (b) by performing competition binding experiments using free/unmodified compounds. Pretreatment of cells or cell lysates with test compounds at varying concentrations before incubation with the generic probe matrix leads to reduced binding of target and off-target proteins to the matrix. Using quantitative mass spectrometry, cellular targets of inhibitor molecules and their associated binding potencies (IC_{50} values) can be determined (8, 9).

Here, we describe a generic protocol for performing competition binding experiments by spiking cell lysates with increasing concentrations of test compound and affinity capturing target proteins on a probe matrix sharing the same binding mode as the test compound. Protein quantification is then performed using mass spectrometry and isobaric mass tags and the resulting data are then fitted to inhibition curves to yield potencies of the test compounds against all targets captured on the beads (Fig. 1). Example data is provided for characterization of the target spectrum of the ATP competitive kinase inhibitor staurosporine by using a bead matrix consisting of a set of unspecific kinase inhibitors (8) sharing the same binding mode.

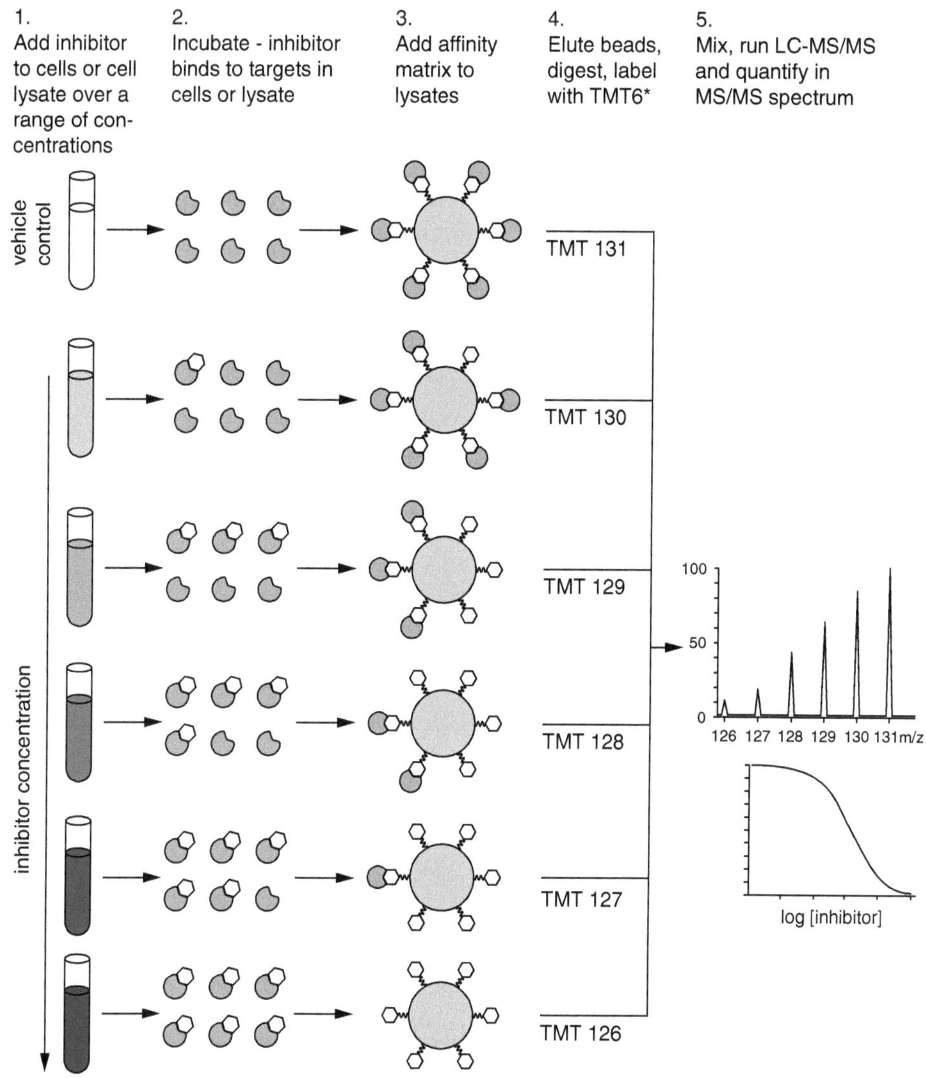

Fig. 1. Schematic representation of the chemoproteomics workflow.

2. Materials

Reagents and equipment listed in the following paragraphs were primarily sourced from Europe.

Unless otherwise stated, the suppliers of laboratory chemicals are not critical but reagent purity should be at least analytical grade. Likewise, most materials required might be purchased from alternative suppliers.

2.1. Cell Lysis

1. Dounce Tissue Grinder (Wheaton, Millville, NJ).
2. Lysis buffer: 50 mM Tris–HCl pH 7.5, 5% (w/v) glycerol, 1.5 mM $MgCl_2$, 150 mM NaCl, 25 mM NaF, 1 mM sodium

vanadate. Add detergent (e.g., Nonidet NP-40 or IGEPAL CA-630) to a concentration of 0.8% (w/v). Add one complete EDTA-free tablet (protease inhibitor cocktail, Roche Applied Science) per 25 mL of lysis buffer. Filter lysis buffer through a 0.22 μm filter (Millipore, SLGV013SL).

3. A standard cell homogenizer is used during cell lysis (e.g., Potter S from B. Braun, Biotech International).
4. Ultracentrifugation: any ultracentrifuge and rotor may be used provided it can process the appropriate volumes and delivers the required *g*-force (e.g., 20,000 × *g* step: Sorvall SLA600; 100,000 × *g* step: Ti50.2).
5. Bradford protein assay (Bio-Rad).

2.2. Preparation of Affinity Matrix

1. NHS-activated Sepharose 4 Fast Flow in isopropanol (GE Healthcare Life Sciences).
2. DMSO, ethanol, isopropanol, methanol.
3. Triethylamine (TEA, SIGMA, 99% pure).
4. End-over-end shaker (Roto Shake Genie, Scientific Industries, Inc.).
5. 15 mL Conical polypropylene tubes (Falcon, BD Biosciences, USA).
6. Laboratory centrifuge with rotor suitable for 15 mL conical tubes (e.g., Multifuge 3S-R Heraeus).
7. 2-Aminoethanol (Sigma-Aldrich).
8. HPLC system with UV detector.

2.3. Competition Binding Experiment

1. DP buffer: 50 mM Tris–HCl, pH 7.5, 5% (w/v) glycerol, 1.5 mM $MgCl_2$, 150 mM NaCl, 1 mM sodium vanadate filtered through 0.22 μm filter (Millipore, USA).
2. Protease inhibitor tablet (EDTA-free tablet, protease inhibitor cocktail, Roche Applied Science, Germany).
3. Detergent, e.g., Nonidet NP-40 or IGEPAL CA-630.
4. Ultracentrifuge with matching rotor (e.g., Ti50.2).
5. Mini columns with filter (2.5 mL column, with 90 μm filter pore size, MoBiTec GmbH, Germany).
6. Laboratory centrifuge as in Subheading 2.2.
7. Sample buffer NuPAGE LDS Sample Buffer (Invitrogen, USA) containing 50 mM DTT.
8. Shaker for polypropylene microfuge tubes (e.g., Eppendorf Thermomixer).
9. 200 mg/mL iodoacetamide in water (always prepare freshly).
10. NuPAGE 4–12% Bis-Tris gel (Invitrogen, USA), gel electrophoresis equipment.

2.4. In-Gel Digestion

1. Low-binding v-shaped 96-well plates, polypropylene (Nunc, Germany).
2. Tetraethylammonium bicarbonate: TEAB, 1 M aqueous solution (Sigma-Aldrich, USA).
3. Destaining solution: 60% 5 mM TEAB, 40% ethanol.
4. Acetonitrile, ethanol.
5. 250–1,250 μL Multichannel Pipette.
6. Trypsin (Promega, USA).
7. HPLC grade water.
8. 1 M aqueous HCL solution.
9. Centrifuge (e.g., Eppendorf type 5810, Germany) with rotor for 96-well plates.
10. Laminar flow hood.
11. Speed Vac.
12. Lids for 96-well plates (Costar, Sigma-Aldrich, USA).

2.5. Labeling of Tryptic Digests with Isobaric Mass Tags

1. Glycine, hydroxylamine, acetonitrile.
2. Thermoshaker (Eppendorf Thermomix comfort).
3. D1 solution: 200 mM TEAB 90%/10% acetonitrile, pH 8.5.
4. D2 solution: 200 mM TEAB 60%/40% acetonitrile.
5. Wash solution: 40% acetonitrile/H_2O (HPLC).
6. Stop solution: 10 mM TEAB/100 mM glycine, pH 8.5.
7. Hydroxylamine solution 2.5%/H_2O (HPLC).
8. 20% Formic acid solution: 20% formic acid/H_2O.
9. 10% Formic acid solution: 10% formic acid/H_2O.
10. TMT reagents, e.g., TMT sixplex Label Reagent Set (Thermo Fisher Scientific, USA).

2.6. Mass Spectrometry

1. Solution A: 0.1% formic acid in water.
2. Solution B: 70% acetonitrile/29.9% water/0.1% formic acid (v/v/v).
3. Ultrasonic bath.
4. Nano-flow HPLC system.
5. Mass spectrometer.

3. Methods

3.1. Cell Lysis

This protocol describes a generic cell lysis protocol for frozen cell pellets using 0.8% (w/v) IGEPAL CA-630 as detergent. Depending on the nature and subcellular localization of the target proteins (e.g., cytosolic, plasma membrane, nucleus, and other intracellular membranous compartments), the cell lysis protocols will need to be adapted and optimized.

1. Mix cell pellet and cold lysis buffer 1/1 (v/v) and thaw cells at 4°C.
2. Transfer the suspension to a precooled Dounce Tissue Grinder.
3. Dounce 20× using mechanized POTTER S (B. Braun, Biotech International), speed setting 300/min.
4. Transfer the homogenate into precooled 50 mL conical tubes.
5. Incubate for 30 min rotating in cold room.
6. Spin cells for 10 min at 20,000 × *g* at 4°C (10,000 rpm in Sorvall SLA600).
7. Transfer supernatant to a UZ-polycarbonate tube.
8. Spin supernatant for 1 h at 145,000 × *g* at 4°C (40,000 rpm in Ti50.2).
9. Save supernatant (remove and discard most of the lipid layer if possible), transfer supernatant into fresh 50 mL conical tube.
10. Determine protein concentration by Bradford assay (Bio-Rad). Typical protein concentrations range from 5 to 20 mg/mL.
11. Prepare several aliquots according to the requirements of the project which has requested the lysate.
12. Freeze aliquots in liquid nitrogen and store at −80°C.

3.2. Preparation of Affinity Matrix

Sepharose beads are available with different functional groups enabling immobilization of probe compounds with a variety of different chemical procedures. Likewise, availability of suitable residues on the probe compound itself is an important parameter for selection of the appropriate immobilization chemistry and thus, the functionalization, of the Sepharose resin. The generic protocol described below uses NHS-activated Sepharose beads and requires a primary or secondary amine function on the probe compound for immobilization.

1. Wash 1 mL (settled volume) of NHS-Sepharose beads three times with 10 mL of DMSO using a 15 mL conical tube. Gently spin down beads after each washing step using a suitable centrifuge and discard supernatant.
2. After the last washing step, resuspend beads in 1 mL of DMSO.
3. Add 10 μL of a 100 mM probe compound solution in DMSO (see Note 1).

4. Add 15 μL of TEA, shake, and settle the beads using a centrifuge.
5. Pipette 20 μL of supernatant into a HPLC autosampler vial and dilute with 20 μL methanol.
6. Incubate beads at room temperature in darkness on an end-over-end shaker for 16–20 h.
7. Pipette 20 μL of supernatant into a HPLC autosampler vial and dilute with 20 μL methanol.
8. Add 50 μL of 2-aminoethanol to beads and incubate on an end-over-end shaker for 4 h to block remaining NHS groups of Sepharose beads.
9. Wash beads once with 10 mL DMSO and three times with ethanol. Gently spin down beads after each washing step using a suitable centrifuge and discard supernatant.
10. After the last washing step resuspend beads in 1 mL isopropanol and keep beads in the refrigerator (or at −20°C in a tightly sealed conical tube until further use (see Note 2)).

3.3. Competition Binding Experiment

1. Wash 500 μL probe matrix twice with 10 mL DP buffer and once with 10 mL DP buffer containing 0.4% (w/v) IGEPAL CA-630, allow beads to settle after each wash step and discard supernatant.
2. After the last washing step, prepare a 1:1 slurry with DP buffer containing 0.4% NP40.
3. Dilute cell lysate with one volume of DP buffer containing protease inhibitors (1 tablet per 25 mL buffer).
4. Dilute lysate further with DP buffer containing 0.4% (w/v) IGEPAL CA-630 and protease inhibitors to a final protein concentration of 5 mg/mL.
5. Ultracentrifuge for 20 min at 100,000 × *g* at 4°C.
6. Discard pellet and keep supernatant on ice.
7. Dissolve test compound in DMSO and prepare a 30 mM stock solutions. Dilute stock in DMSO to the following concentrations: 500, 125, 31, 8, and 2 μM.
8. For each data point, mix 1 mL of cleared lysate (step 6) with 5 μL compound solution in a microfuge tube. Final inhibitor concentrations in lysate are: 2.5, 0.625, 0.156, 0.039, and 0.010 μM (see Note 3). In addition, a vehicle control is performed by adding 5 μL of DMSO into 1 mL of lysate.
9. Incubate samples for 45 min at 4°C using an end-over-end shaker.
10. Wash beads as prepared in Subheading 3.2 three times with 1 mL of DP buffer, after each step let beads settle and discard supernatant.

11. After the last washing step, add 250 μL of DP buffer and create a homogenous suspension by turning the vial several times.
12. Add 70 μL of bead suspension to each lysate compound mix prepared in step 8.
13. Incubate for 60 min at 4°C using an end-over-end shaker.
14. Gently settle beads for 3 min in a centrifuge (1,500 × *g*, in Multifuge 3 S) and carefully remove supernatant.
15. For each of the six samples, resuspend beads in 600 μL DP buffer containing 0.4% (w/v) IGEPAL CA-630 and transfer into 2.5 mL minicolumn.
16. Connect a 10 mL syringe with the upper luer lock of the mini-columns and remove the bottom lid.
17. Wash beads with 10 mL DP buffer containing 0.4% (w/v) IGEPAL CA-630 and 5 mL DP buffer containing 0.2% (w/v) IGEPAL CA-630.
18. After the last washing step, the columns should run dry. They are then placed in microfuge tubes and centrifuged for 1 min at 4°C (e.g., 800 rpm, Multifuge 3S).
19. Close bottom lids of columns and add 50 μL sample buffer, then close top lids and place into microfuge tubes.
20. Incubate for 30 min at 50°C in a shaker.
21. Remove first top then bottom lid of columns and transfer into new siliconized tubes.
22. Spin down for 2 min at 2,000 rpm in a table top centrifuge at room temperature.
23. Transfer 35 μL of eluate into a new tube, add 3.5 μL iodoacetamide solution and incubate for 30 min at room temperature.
24. Spin samples for 5 min at maximum speed in a table top centrifuge.
25. Apply 35 μL of each sample onto a precast gel.
26. Perform SDS gel electrophoresis according to the manufacturer's instructions but stop electrophoresis after the sample has migrated approximately 1.5 cm into the gel (see Note 4).
27. Wash gel in water and either directly proceed with Subheading 3.4 or stain proteins in gel using a mass spectrometry compatible stain (see Note 5).

3.4. In-Gel Digestion

This in-gel digestion protocol follows established routes with two modifications: (1) Primary and secondary amine containing buffers are commonly used for tryptic digestion of proteins but are not compatible with NHS-ester containing isobaric mass tags. Triethylammonium bicarbonate containing buffers are adequate alternatives. (2) This protocol is compatible with relatively high throughput, since all steps are performed in 96-well plates.

1. Drill holes into the center of 96-well plates using a 0.4 mm drill.
2. Wash plate extensively using water and ethanol.
3. Cut the protein containing part of each of the six-gel lanes obtained in Subheading 3.3 into three pieces and place each piece into a separate well of the 96-well plate.
4. If gel pieces were stained with colloidal Coomassie, incubate twice with 100 μL destaining solution at 55°C for 60 min and discard solutions.
5. Spin plate for 1 min at 1,000 rpm in a centrifuge (it is best to stack the drilled plate on top of a normal plate to collect the waste).
6. Add 100 μL ethanol per well and incubate at room temperature for 10 min (the gel pieces should be white) and remove supernatant in a centrifuge.
7. Add 100 μL 5 mM TEAB solution and incubate for 10 min.
8. Discard solution and repeat steps 6 and 7 and then again step 6.
9. Dissolve trypsin at 1 mg/mL in 1 mM HCl and dilute 20 μL trypsin solution in 1,980 μL 5 mM TEAB.
10. Add 10 μL of diluted trypsin solution per well (step 8) and incubate at 4°C for 10 min.
11. Remove remaining solution (e.g., using a centrifuge) and add 20 μL 5 mM TEAB per well.
12. Place a collector plate underneath the sample plate and pub both plates into a plastic bag.
13. Incubate for 4 h at 37°C.
14. Stop the digestion reaction by adding 5 μL 5% formic acid.
15. Extract gel pieces twice by adding 20 μL 0.1% formic acid, once with 20 μL 60% acetonitrile, 0.04% formic acid and once with 20 μL pure acetonitrile and by incubating for 20 min at room temperature. After each step, centrifuge supernatant into the collector plate.
16. Dry extracted peptides in collector plate in vacuo using a Speed Vac.

3.5. Labeling of Tryptic Digests with Isobaric Mass Tags

1. Dissolve TMT 6-plex reagents in acetonitrile to a final concentration of 24 mM (see Note 6).
2. Dissolve each sample in step 16 of Subheading 3.4 in 10 μL D1 solution, shake 15 min at room temperature on a thermo shaker.

3. Add 10 μL TMT reagent to each sample: apply different labels to samples corresponding to different compound concentrations (see Subheading 3.3 step 8).
4. Incubate 60 min at room temperature on a thermo shaker.
5. Add 5 μL stop solution to each sample and incubate for 15 min on a shaker.
6. Combine all 18 samples into one reaction tube.
7. Freeze in liquid nitrogen and dry in vacuo.
8. Dissolve sample in 20 μL D2 solution and vortex.
9. Add 2 μL 2.5% hydroxylamine solution.
10. Incubate on thermo shaker at room temperature for 15 min.
11. Add 10 μL 10% formic acid and vortex.
12. Dry sample in a Speed Vac.
13. Store at −20°C.

3.6. LC-MS/MS Analysis

Appropriate instrumentation for mass spectrometric analysis and quantification of the samples obtained in Subheading 3.5 is available at many institutions and core facilities. Since many different instrument platforms are similarly well suited for this purpose, the following section gives guidelines for appropriate analyses. Furthermore, instrument settings will be described in some detail for the currently very common LTQ-Orbitrap Velos mass spectrometers.

1. For complex samples, chromatographic separation is required to achieve best results. This is typically achieved by online coupling of a nano-LC reversed-phase chromatography system to an electrospray ionization mass spectrometer. Alternatively, off-line separation (e.g., in conjunction with MALDI mass spectrometers) can be performed. In some cases, it might be desirable to perform multiple chromatographic separation steps.
2. Choice of mass spectrometer.
   (a) Isobaric mass tags based quantification is based on measuring relative signal abundances of reporter ions in the low mass range of tandem-MS spectra. Hence the mass spectrometer of choice must have the ability to perform tandem-MS spectra.
   (b) High-resolution Q-TOF and Fourier Transform mass spectrometers (FT-ICR, Orbitrap) typically give most accurate results for peptide identification.
   (c) Older ion trap instruments are typically less suitable, since reporter ion detection suffers from the inability of these instruments to stabilize and detect fragment ions with m/z values below approximately one-third of the precursor ion.

Newer instrumentation such as linear ion traps offer the ability to perform pulsed-Q-dissociation (PQD; different names apply for different instrument manufacturers) instead of the conventional resonance activation collision induced dissociation which circumvents this problem. However, careful optimization of instrument parameters is required to achieve good results (see refs. 10, 11).

(d) Instruments performing dissociation in a collision cell rather than an ion trap tend to give higher transmission of low mass fragments which is generally advantageous for precision of reporter ion-based quantification. Nonetheless, instrument parameters often require careful optimization in order to achieve a good compromise for ion transmission of low and high *m/z* range ions.

3. Typical instrument settings for an Orbitrap Velos mass spectrometer coupled online to a nano-flow HPLC system.

 (a) In-depth analysis of samples requires online coupling of a nano-flow liquid chromatography system to the mass spectrometer. For this purpose, reversed-phase C18 separation columns (50–100 μm I.D.) can be connected to gold coated nano-spray emitters or alternatively separation column and emitter are manufactured in one piece and voltage is applied via a liquid junction set-up. Typically, 1.5–2 kV are applied to induce the electrospray process. Peptide separation is achieved by a linear gradient from 100% solution A to 70% solution B within 2–3 h at a flow rate of 100–250 nL/min (see Note 7).

 (b) As for most Fourier transform mass spectrometers, the instrument should be operated using automated gain control in order to achieve the highest mass accuracy and precision. Typical ion target values for the FT cell are one million ions in MS mode and 20,000–40,000 ions in MS/MS mode; and for the ion trap 30,000 ions in MS mode and 3,000–5,000 ions in MS/MS mode. Switch on lock mass correction (12).

 (c) In order to achieve high sampling rates, the maximum ion accumulation times for each scan event should be kept at <200 ms and only one microscan is acquired.

 (d) One survey scan is followed by six MS/MS scans of the most abundant multiply charged ion species detected.

 (e) Higher energy collision induced dissociation (HCD) should be used for precise reporter ion-based quantification and might be combined with ion trap CID for peptide identification in an independent MS/MS scan (10, 13, 14).

3.7. Data Analysis

At this stage, the objectives are to (a) identify the proteins bound to the probe matrix, (b) accurately quantify the captured amounts of these proteins as a function of free inhibitor concentration used in the experiment, and (c) calculate IC_{50} values of the free inhibitor for each identified protein.

1. For protein identification, tandem mass spectra are typically converted to peak lists and submitted to a data base search against a nonredundant sequence data base of the species of interest. A variety of free and commercial search engines such as Mascot and Sequest are available for this purpose (15, 16).
2. Relative quantification is then performed for each identified protein by comparing relative reporter ion intensities or peak

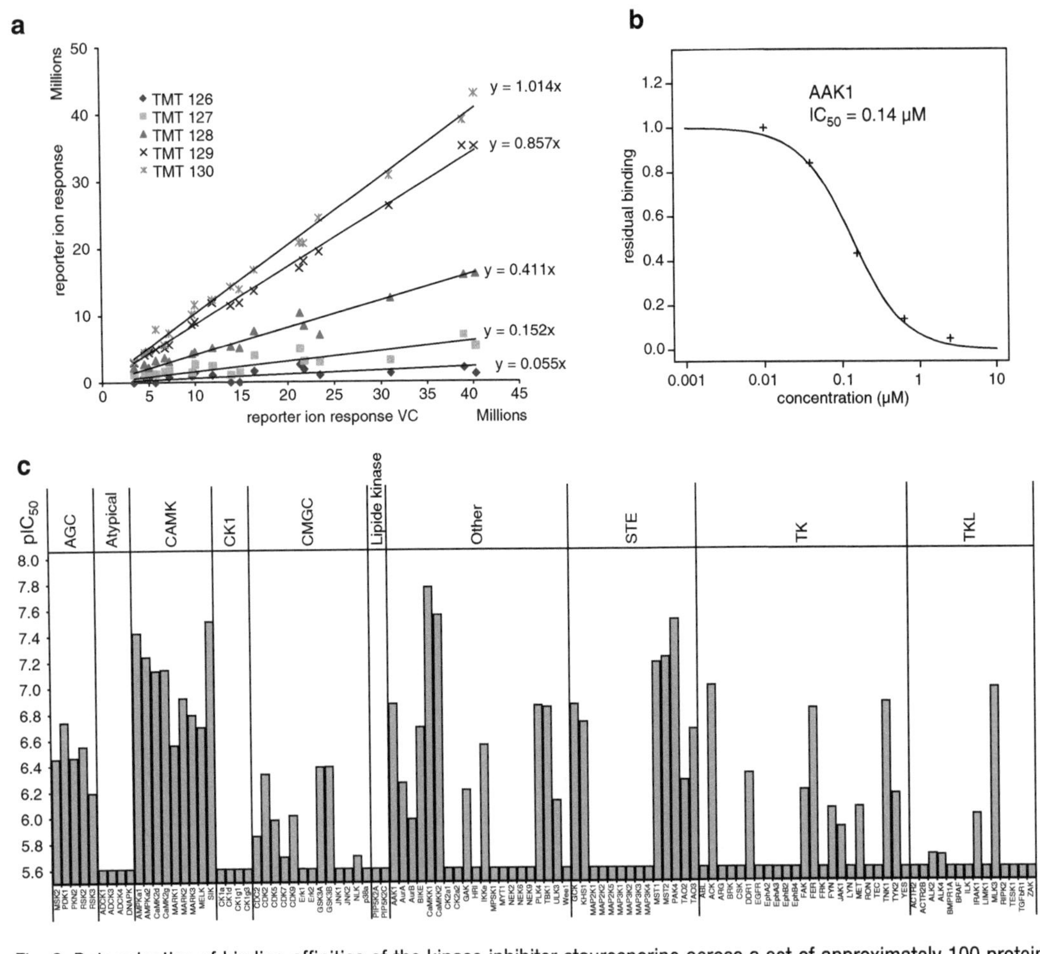

Fig. 2. Determination of binding affinities of the kinase inhibitor staurosporine across a set of approximately 100 protein kinases directly from HeLa lysate. (**a**) Relative quantification of binding of the kinase AAK1 to the mixed kinase inhibitor matrix as a function of staurosporine concentration in the lysate. For each peptide-to-spectrum match, reporter ion abundance at a given inhibitor concentration is plotted against the vehicle control treated sample. Protein fold-changes are then calculated using a linear model. (**b**) Dose–response curve and IC_{50} value as calculated from the data points shown in (**a**). (**c**) Bar plot representing potencies of staurosporine against the captured kinome. Data have been acquired in a single LC-MS/MS run. Size of the bars represents pIC_{50} values.

areas of matching high-confidence spectra (Fig. 2). A variety of software tools is available for this purpose. For very simple samples, this can also be done by manual inspection of reporter ion intensities. Relative protein abundance in compound treated samples should be expressed as fold change to the vehicle control treated sample (see Note 8).

3. IC_{50} values are then calculated by performing a sigmoidal curve fit using standard statistics software packages (Fig. 2).

4. Notes

1. 100 mM is quite a high concentration for some probe compounds. Alternatively, DMF might be used for increased solubility. Turbid solutions indicate incomplete dissolution of the test compound that might cause problems during subsequent steps in the experiment.
2. This protocol yields a concentration of 1 mM for the probe compound on the matrix. This corresponds to derivatization of approximately 5% of the available functional groups. A higher coupling density might be desirable for compounds exhibiting low-affinity binding to their target proteins. However, in our hands, too high a compound concentration on beads often leads to increased unspecific binding of highly abundant proteins. Furthermore, high-affinity probe matrices and high-ligand densities can shift the measured IC_{50}s toward higher values. At low-ligand densities, this effect is minimized and by measuring the relative amounts of target proteins captured in consecutive incubation steps with the probe matrix, matrix independent K_d values can be determined using the Cheng-Prusoff equation (9).
3. The fourfold dilution steps applied in the example are suitable for high-quality dose–response curves. However, the concentration range that can be tested with a 6-plex isobaric labeling experiment is limited. Hence, if compound potency is entirely unclear, it is often advised to first perform a range-finding experiment for which larger dilution steps are applied.
4. Gel-separation is not strictly required, since proteins could also be eluted with, e.g., acid and digested in solution. However, we find this step quite useful, since it allows very efficiently removal of detergents used in the experiment.
5. Albeit not absolutely necessary, staining the gels with a mass spectrometry compatible stain offers the advantage of being able to compare the results of the six independent pull-down experiments. Strong variation in staining often indicates

experimental variation. In such cases, it might be better to repeat the experiment rather than proceeding with the protocol.

6. iTRAQ 8-plex reagents might be used for experiments using seven different compound concentrations and a vehicle control.
7. The optimal flow rate depends on the inner diameter of the separation column used. In most cases, a 75 μm I.D. column packed with 3 μm C18 material and a flow rate of 200 nL/min will produce good results. Solvents should always be degassed in an ultrasonic bath prior to use.
8. Best results are obtained if only significant peptide-to-spectrum matches are considered for quantification.

Acknowledgments

We would like to thank Frank Fischer, Sonja Ghidelli-Disse, and Thilo Werner for contributions to the development of the protocol, Frank Weisbrodt for help with the figures and Gerard Drewes for discussions and advice.

References

1. Scholten, A., Poh, M. K., van Veen, T. A., van Breukelen, B., Vos, M. A., and Heck, A. J. (2006) Analysis of the cGMP/cAMP interactome using a chemical proteomics approach in mammalian heart tissue validates sphingosine kinase type 1-interacting protein as a genuine and highly abundant AKAP, *J Proteome Res* **5**, 1435–1447.
2. Shiyama, T., Furuya, M., Yamazaki, A., Terada, T., and Tanaka, A. (2004) Design and synthesis of novel hydrophilic spacers for the reduction of nonspecific binding proteins on affinity resins, *Bioorg Med Chem* **12**, 2831–2841.
3. Brehmer, D., Godl, K., Zech, B., Wissing, J., and Daub, H. (2004) Proteome-wide identification of cellular targets affected by bisindolylmaleimide-type protein kinase C inhibitors, *Mol Cell Proteomics* **3**, 490–500.
4. Brehmer, D., Greff, Z., Godl, K., Blencke, S., Kurtenbach, A., Weber, M., Muller, S., Klebl, B., Cotten, M., Keri, G., Wissing, J., and Daub, H. (2005) Cellular targets of gefitinib, *Cancer Res* **65**, 379–382.
5. Godl, K., Gruss, O. J., Eickhoff, J., Wissing, J., Blencke, S., Weber, M., Degen, H., Brehmer, D., Orfi, L., Horvath, Z., Keri, G., Muller, S., Cotten, M., Ullrich, A., and Daub, H. (2005) Proteomic characterization of the angiogenesis inhibitor SU6668 reveals multiple impacts on cellular kinase signaling, *Cancer Res* **65**, 6919–6926.
6. Remsing Rix, L. L., Rix, U., Colinge, J., Hantschel, O., Bennett, K. L., Stranzl, T., Muller, A., Baumgartner, C., Valent, P., Augustin, M., Till, J. H., and Superti-Furga, G. (2009) Global target profile of the kinase inhibitor bosutinib in primary chronic myeloid leukemia cells, *Leukemia* **23**, 477–485.
7. Rix, U., Hantschel, O., Durnberger, G., Remsing Rix, L. L., Planyavsky, M., Fernbach, N. V., Kaupe, I., Bennett, K. L., Valent, P., Colinge, J., Kocher, T., and Superti-Furga, G. (2007) Chemical proteomic profiles of the BCR-ABL inhibitors imatinib, nilotinib, and dasatinib reveal novel kinase and nonkinase targets, *Blood* **110**, 4055–4063.
8. Bantscheff, M., Eberhard, D., Abraham, Y., Bastuck, S., Boesche, M., Hobson, S., Mathieson, T., Perrin, J., Raida, M., Rau, C., Reader, V., Sweetman, G., Bauer, A., Bouwmeester, T., Hopf, C., Kruse, U., Neubauer, G., Ramsden, N., Rick, J., Kuster,

B., and Drewes, G. (2007) Quantitative chemical proteomics reveals mechanisms of action of clinical ABL kinase inhibitors, *Nat Biotechnol* *25*, 1035–1044.

9. Sharma, K., Weber, C., Bairlein, M., Greff, Z., Keri, G., Cox, J., Olsen, J. V., and Daub, H. (2009) Proteomics strategy for quantitative protein interaction profiling in cell extracts, *Nat Methods 6*, 741–744.
10. Bantscheff, M., Boesche, M., Eberhard, D., Matthieson, T., Sweetman, G., and Kuster, B. (2008) Robust and sensitive iTRAQ quantification on an LTQ Orbitrap mass spectrometer, *Mol Cell Proteomics 7*, 1702–1713.
11. Griffin, T. J., Xie, H., Bandhakavi, S., Popko, J., Mohan, A., Carlis, J. V., and Higgins, L. (2007) iTRAQ reagent-based quantitative proteomic analysis on a linear ion trap mass spectrometer, *J Proteome Res 6*, 4200–4209.
12. Olsen, J. V., de Godoy, L. M., Li, G., Macek, B., Mortensen, P., Pesch, R., Makarov, A., Lange, O., Horning, S., and Mann, M. (2005) Parts per million mass accuracy on an Orbitrap mass spectrometer via lock mass injection into a C-trap, *Mol Cell Proteomics 4*, 2010–2021.
13. Kocher, T., Pichler, P., Schutzbier, M., Stingl, C., Kaul, A., Teucher, N., Hasenfuss, G., Penninger, J. M., and Mechtler, K. (2009) High precision quantitative proteomics using iTRAQ on an LTQ Orbitrap: a new mass spectrometric method combining the benefits of all, *J Proteome Res 8*, 4743–4752.
14. Savitski, M. M., Fischer, F., Mathieson, T., Sweetman, G., Lang, M., and Bantscheff, M. (2010) Targeted data acquisition for improved reproducibility and robustness of proteomic mass spectrometry assays, *J Am Soc Mass Spectrom 21*, 1668–1679.
15. Eng, J. K., McCormack, A. L., and Yates Iii, J. R. (1994) An approach to correlate tandem mass spectral data of peptides with amino acid sequences in a protein database, *J Am Soc Mass Spectrom 5*, 976–989.
16. Perkins, D. N., Pappin, D. J., Creasy, D. M., and Cottrell, J. S. (1999) Probability-based protein identification by searching sequence databases using mass spectrometry data, *Electrophoresis 20*, 3551–3567.

Chapter 11

Affinity-Based Profiling of Dehydrogenase Subproteomes

Xia Ge and Daniel S. Sem

Abstract

The high cost of drug discovery and development requires more efficient approaches to the identification and inhibition of tractable protein targets. One strategy is to pursue families of proteins that already possess affinity for a drug lead scaffold, where that scaffold plays the dual role of serving (a) when tethered to a resin, as a ligand to purify a subproteome of interest, and (b) as a lead molecule that has the potential for optimization for a given member of the subproteome. Here, we describe an example of the purification of a subproteome using a scaffold tailored to the dehydrogenase family of enzymes. Combined with modern LC-MS/MS methods and subsequent searching of proteome databases, such affinity chromatography strategies can be used to purify and identify any proteins with affinity for the scaffold molecule. The method is exemplified using the CRAA (catechol rhodanine acetic acid) privileged scaffold, which is tailored to dehydrogenases. CRAA affinity column chromatography, combined with LC-MS/MS, is described as a method for profiling dehydrogenase subproteomes.

Key words: Dehydrogenase, Oxidoreductase, Catechol rhodanine, Chemical proteomics, Subproteome, Affinity chromatography, Tandem mass spectrometry, Drug discovery

1. Introduction

Chemical proteomics aims to develop and apply technologies for the characterization of protein function on a global, proteome-wide scale (1). The completion of the human genome-sequencing project has provided significant information on complex biological systems, and laid the foundation for a comprehensive analysis of protein function via chemical proteomics (2). Chemical proteomics can be used to distill this flood of genomic information, to provide useful information about basic cell function, as well as new approaches to disease treatment (3). This approach makes use of small-molecule protein ligands that can be used to identify

Gerard Drewes and Marcus Bantscheff (eds.), *Chemical Proteomics: Methods and Protocols*, Methods in Molecular Biology, vol. 803, DOI 10.1007/978-1-61779-364-6_11, © Springer Science+Business Media, LLC 2012

proteins, which might be pursued as drug targets. The strategy of profiling drug targets, by using affinity chromatography coupled to subsequent high-resolution MS and bioinformatic analyses, is becoming increasingly popular as a postgenomic application of chemical proteomics (4). This method allows rapid biochemical analysis and small-molecule screening of drug targets and off-targets (undesired targets), thereby accelerating the target validation process in drug discovery (5–9).

Dehydrogenases comprise ~5% of most proteomes (10, 11), many of which could be important tractable ("druggable") targets. For example, Isoniazid (INH) binds to multiple dehydrogenases in *Mycobacterium tuberculosis* (12), Epalrestat targets aldose reductase for the treatment of diabetic neuropathy (13, 14), and the statin drugs inhibit HMG-CoA reductase. Using an NAD(P)-INH affinity column, Argyrou et al. found that Isoniazid, a widely used drug for treating tuberculosis, does not bind to only one enzyme target, but rather binds to multiple dehydrogenases. In fact, this may well by why isoniazid is effective at killing *M. tuberculosis* (12). This represents effectively a chemical proteomic approach to profiling the isoniazid drug, as a covalent adduct with $NADP^{+}$. While the general strategy of profiling dehydrogenases using cofactor-based affinity chromatography has already been pursued 30 years ago (15), the ability to readily identify eluted proteins using tandem mass spectrometry, and the application to profiling drugs, is relatively recent. However, the use of NAD(P) as a ligand, either as a scaffold for building a drug or as part of an affinity matrix, is not ideal because of its instability and poor bioavailability, which is why we developed the use of the catechol rhodanine ligand (9, 16, 17). Most recently, Kim et al. used a Cibacron Blue F3GA dye affinity column to ligand-specifically elute and identify aldehyde dehydrogenases from *M. tuberculosis* (18). These approaches demonstrate that dehydrogenase subproteomes can be purified and analyzed using affinity chromatography (19) (Fig. 1).

Affinity column chromatography combined tandem mass spectroscopy (MS) provides an especially useful approach to characterizing subproteomes, based on the affinity of the purified proteins for the ligand that is covalently attached to the resin (12, 20). We have developed this method for dehydrogenase subproteome studies using the recently reported catechol rhodanine acetic acid (CRAA) ligand (17). CRAA was designed to be a privileged scaffold for dehydrogenases (20). Using CRAA affinity chromatography, dehydrogenase protein targets can be purified from the larger proteome, based on affinity for the CRAA probe which binds in the NAD(P)(H) binding site. Then, higher affinity and specificity bi-ligand variants of the CRAA scaffold can be constructed, which selectively bind to the desired dehydrogenase drug target(s) (17).

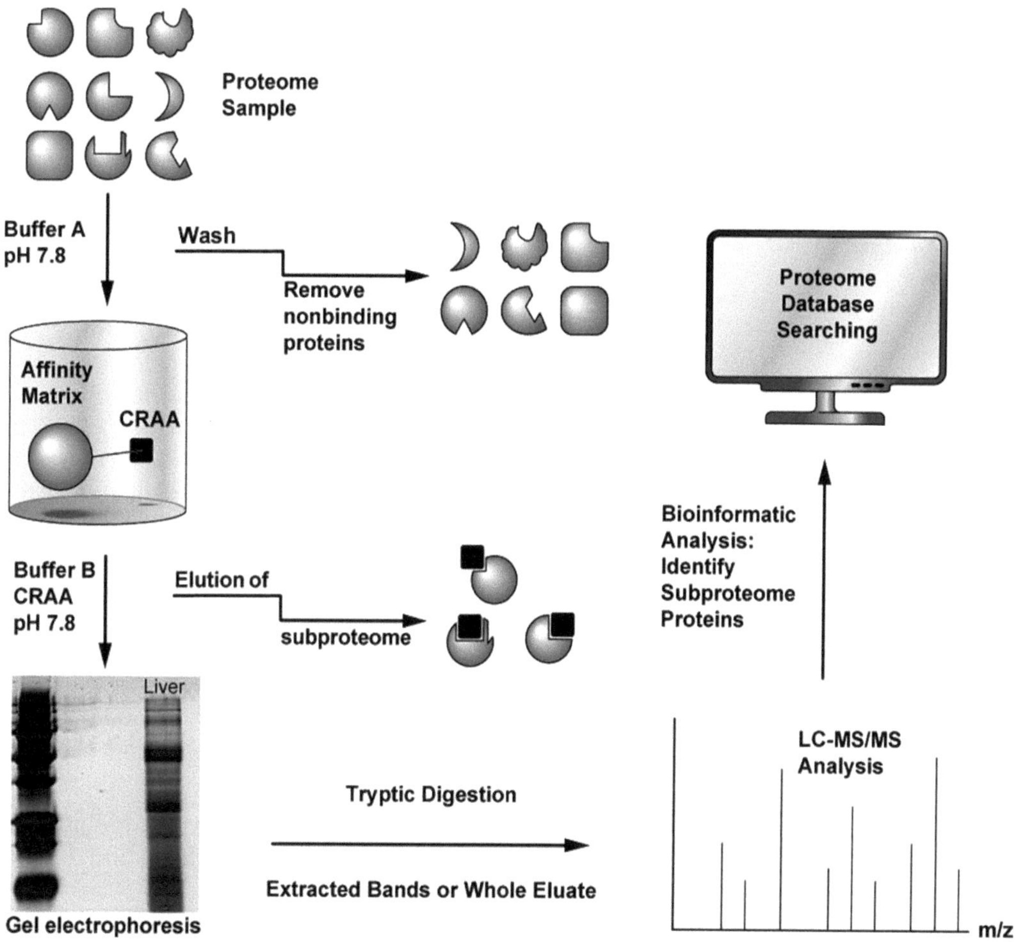

Fig. 1. Schematic representation of the process whereby a dehydrogenase subproteome is purified using a CRAA-based resin, then characterized using tandem mass spectrometry.

2. Materials

2.1. Preparation of Catechol Rhodanine Acetic Acid

1. Mixture of 9.6 g 3-rhodanine acetic acid and 7.6 g 3,4-dihydroxybenzaldehyde (1.0:1.1).
2. 8.2 g Sodium acetate.
3. 150-mL Acetic acid.

2.2. Preparation of the N-hydroxy-succinimide Active Ester of CRAA (21)

1. 6.22 g CRAA (from the previous step).
2. 5.75 g *N*-Hydroxysuccinimide (NHS).
3. 20.6 g *N*,*N*′-Dicyclohexylcarbodiimide (DCC).
4. DMSO.
5. 0.2 g DMAP (4-dimethylaminopyridine) as catalyst.

6. EMD silica gel $60F_{254}$ plates for thin layer chromatography (TLC).
7. Chloroform/methanol/acetic acid, 12:3:1 v/v/v, for TLC.
8. TLC developing tanks/chambers and a 254-nm UV light.

2.3. Preparation of the CRAA–Agarose Matrix (22)

1. Coupling reaction buffer: 600 mL of 100-mM phosphate buffer, pH 10.0 (at 7°C).
2. Quenching buffer: 1 M Tris–HCl buffer, pH 6.5.

2.4. Affinity Chromatography of the Dehydrogenase Subproteome (23)

1. Human liver proteins (Sigma-Aldrich).
2. *Mycobacterium tuberculosis* H37Rv cell lysate in PBS with 8 mM EDTA and protease inhibitors (a gift from Colorado State University). The lysate was prepared from cells grown in glycerol–alanine stocks for 14 days, then washed with PBS. After inactivation by γ-irradiation cells were disrupted using a French Press and the lysate was centrifuged to remove cell debris.
3. Buffer A: 25 mM Tris–HCl, 50 mM NaCl, and 0.1% NaN_3, pH 7.8.
4. Buffer B: Buffer A with 4 mM CRAA, pH 7.8.
5. Novex gels and buffers, and SilverQuest® staining kit for SDS-PAGE (Invitrogen).

2.5. Tandem MS Analysis for the Identification of Dehydrogenases

1. Polyacrylamide gel mixture: 100 μL of acrylamide/bis (30% T/2.67% C), 2 μL of 10% ammonium persulfate, and 2 μL of TEMED.
2. Proteolysis mix: 20 mM ammonium bicarbonate containing 1 μg of trypsin (Promega), pH 8.0.
3. Extraction buffer: 6 M guanidine·HCl in 5 mM potassium phosphate and 1 mM DTT, pH 6.5.
4. Centricon filters with 10 kDa cutoff (Millipore).
5. Speed Vac apparatus (Savant).
6. C_{18} ZipTips (Millipore) and vial inserts (Waters).
7. LTQ mass spectrometer (Thermo-Fisher) coupled to a Surveyor HPLC system (Thermo-Fisher) equipped with a Finnigan Micro AS autosampler, interfaced with an Aquasil C18 PicoFrit capillary column (75 μm × 10 cm) (New Objective).

3. Methods

3.1. Preparation of CRAA

The first intermediate in the synthesis is catechol rhodanine acetic acid (5-[(3,4-dihydroxyphenyl)methylene]-4-oxo-2-thioxo-3-thiazolidineacetic acid). Synthesis of CRAA is a typical aldol condensation and it is necessary to use flame dried glassware to run the reaction (Fig. 2).

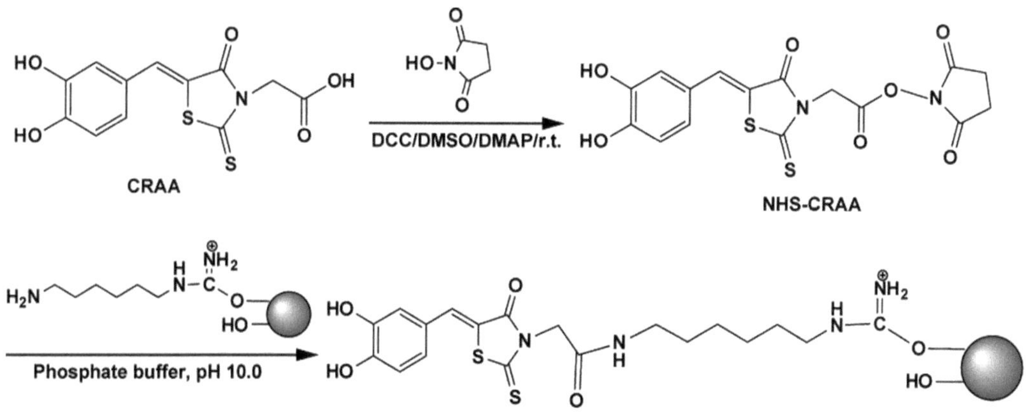

Fig. 2. Synthesis of the CRAA-based affinity resin.

1. A reaction mixture composed of 9.6 g 3-rhodanine acetic acid, 7.6 g 3,4-dihydroxybenzaldehyde (1.0:1.1), 8.2 g sodium acetate, and 150-mL acetic acid is heated to reflux and stirred for 6 h (see Note 1).
2. After cooling, yellow crystals form. The solution containing the crystals is poured into 150-mL cold water, filtered, and washed extensively with water.
3. CRAA is crystallized from acetic acid, yielding approximately 12.7 g product (82% yield) with m.p. 328–329°C. The product is dried in an oven at 100°C for 24 h, after which it is ready to use in the next step.

3.2. Preparation of the NHS Active Ester of CRAA

The second intermediate in the synthesis is the NHS active ester of CRAA: (5-[(3,4-dihydroxyphenyl)methylene]-4-oxo-2-thioxo-3-thiazolidineacetic *N*-hydroxysuccinimide ester) (19).

1. Under a N_2 atmosphere, the 50-mL CRAA/NHS reaction mixture is reacted at room temperature overnight under stirring using a magnetic stir plate.
2. The reaction is monitored by thin layer chromatography on silica gel plates developed with chloroform/methanol/acetic acid, and visualized under a 254 nm UV light (the R_f of CRAA is 0.39 and the product's R_f is 0.68) (see Note 2).
3. The DCU (dicyclohexylurea) is removed by vacuum filtration. The NHS–CRAA DMSO solution is used in the next step without further purification (see Note 3).

3.3. Synthesis of CRAA–Agarose Matrix

The final product of the synthesis is the CRAA–agarose matrix: (5-[(3,4-dihydroxyphenyl)methylene]-4-oxo-2-thioxo-3-thiazolidineacetic ω-aminohexylagarose amide) (20) (see Note 4).

1. The 50-mL NHS–CRAA ester DMSO solution is added dropwise into 100 mL of ω-aminohexylagarose suspended in coupling reaction buffer.

2. The pH of the reaction mixture is maintained at 10 (adjust with 2 M NaOH), and the reaction is allowed to progress at 7°C in a refrigerator overnight, with stirring using a magnetic stir plate.
3. The next day, 60 mL of quenching buffer is added to the reaction mixture to stop the reaction. The Tris amine reacts with unreacted NHS–CRAA active ester (see Note 5).
4. 47.7 g of sodium chloride is added to form a final 0.5 M saline solution. The liquid layer is decanted, and the labeled matrix is washed extensively with deionized water (see Note 6).
5. Approximately 10 mL of matrix is packed into a 1 cm × 20 cm column for column chromatography. The CRAA–agarose column is again washed extensively with deionized water before use.

3.4. Affinity Chromatography of the Dehydrogenase Subproteome (21)

1. The CRAA affinity column is equilibrated with Buffer A. Washing is done until the eluent is nearly colorless (CRAA is intensely colored).
2. The 0.5-mL protein sample is loaded onto the affinity column and washed with a large amount of Buffer A until no protein sample is detected in a Bradford assay (Bio-Rad). The wash buffer volume used is usually ten times the packing volume of the column (see Note 7).
3. The affinity column is eluted with Buffer B. The elution buffer volume is five times the bed volume of the column.
4. Fractions of 2 mL each are collected, then separated on an SDS-PAGE gel and stained using the SilverQuest® kit following the manufacturer's protocol (see Note 8).
5. The entire fraction, or bands extracted from SDS-PAGE gel, are subjected to proteolytic digestion with trypsin and MS/MS analysis (see next section).

3.5. Tandem MS Analysis to Identify Dehydrogenases in the Subproteome

1. Pooled fractions from the CRAA affinity column are concentrated using a Centricon filter with 10-kDa cutoff.
2. 100 μL of affinity purified protein mixtures are polymerized in the polyacrylamide gel mixture. With this mixture a 15% gel piece is formed. Polymerization is performed in the cap of an Eppendorf tube. The polymerized gel pieces are then transferred to the corresponding Eppendorf tube in 1 mL of 40% methanol and 7% acetic acid and incubated for 30 min.
3. Gel pieces are washed twice in water for 30 min each time while sonicating followed by two washes in 50% acetonitrile for 30 min each time while sonicating, and finally twice in 50% acetonitrile in 50 mM ammonium bicarbonate, pH 8.0.
4. Gel pieces are dried using a Speed Vac.

5. To each gel piece, 200 μL of the proteolysis mix is added and incubated overnight at 37°C.
6. Each gel piece with the digested proteins is extracted twice with 70% acetonitrile in 0.1% formic acid. From this step onward, use MS quality water.
7. Corresponding extracts of each gel are pooled together and dried. To each dried sample, the final extraction buffer is added. The samples are sonicated, and peptides are extracted using a C_{18} ZipTip.
8. Extracted peptides are collected in vial inserts and dried. To each dried sample, 5 μL of 0.1% formic acid in water containing 5% acetonitrile is added. Samples are now ready for injection into a LTQ LC-MS system.
9. The MS/MS data are collected and searched against the appropriate subset of the Uniprot database.

4. Notes

1. The small excess of 3,4-dihydroxybenzaldehyde can help to convert most rhodanine acetic acid to CRAA and itself can be readily removed by crystallization from acetic acid.
2. Synthesis of NHS–CRAA can also be monitored with ^{1}H-NMR of the reaction of NHS and CRAA in d6-DMSO at room temperature; make sure that there is at least 30% conversion of CRAA to product.
3. The NHS–CRAA active ester can be readily converted to an amide group by reaction with amines. Since the NHS–CRAA ester is moisture sensitive; it should be prepared right before the synthesis of the affinity matrix, and one should keep all glassware dry before it is covalently converted to the final affinity matrix.
4. Affi-Gel 10 (Bio-Rad®) is an NHS-activated agarose (or sepharose) matrix with a ten-atom carbon spacer and can be covalently linked to primary amines or hydroxyl groups to form the desired affinity matrix. It offers another way to build the affinity column with small molecules which have an $-NH_2$, –OH, or –SH group, rather than the $-CO_2^-$ that is present on the CRAA ligand. Cyanogen bromide activated resins are also useful for attaching ligands that contain amine groups, and resins with epoxide groups can be used for ligands with a range of nucleophilic functional groups.
5. Excess NHS–CRAA ester will react with the amine on the Tris buffer, and the product as well as free CRAA is washed away with a large amount of basic Buffer A.

6. The CRAA affinity matrix can be stored at 2–8°C for at least 6 months. The affinity matrix was stored in 0.5 M saline with 0.02% thimerosal. Avoid using any oxidizing reagents or strong base to rinse column, because CRAA is not stable to these reagents.
7. The optimal pH to incubate and elute dehydrogenases using CRAA is pH 7.8. One needs to balance considerations of stability of proteins at this pH, and the best performance (higher binding affinity) of CRAA at this pH where it is at least partly ionized (the catechol –OH is relatively acidic).
8. Depending on the concentration of protein samples, the eluted protein fractions may need to be concentrated before the SDS gel electrophoresis step.

Acknowledgments

We thank Dr. Bassam Wakim for assistance with mass spectrometry studies. This work was supported in part by GM085739 (NIH) and shared instrumentation grants S10 RR019012 (NIH) and CHE-0521323 (NSF).

References

1. Speers, A. E., and Cravatt, B. F. (2004) Chemical Strategies for Activity-Based Proteomics. *Chembiochem.* *5*, 41–47.
2. Tate, E. W. (2008) Recent Advances in Chemical Proteomics: Exploring the Post-Translational Proteome. *J. Chem. Biol.* *1*, 17–26.
3. Kruse, U., Bantscheff, M., Drewes, G., and Hopf, C. (2008) Chemical and Pathway Proteomics: Powerful Tools for Oncology Drug Discovery and Personalized Health Care. *Mol. Cell. Proteomics.* *7*, 1887–1901.
4. Rix, U., and Superti-Furga, G. (2009) Target Profiling of Small Molecules by Chemical Proteomics. *Nat. Chem. Biol.* *5*, 616–624.
5. Jeffery, D. A., and Bogyo, M. (2003) Chemical Proteomics and its Application to Drug Discovery. *Curr. Opin. Biotechnol.* *14*, 87–95.
6. Verhelst, S. H., and Bogyo, M. (2005) Chemical Proteomics Applied to Target Identification and Drug Discovery. *BioTechniques.* *38*, 175–177.
7. Fonovic, M., and Bogyo, M. (2008) Activity-Based Probes as a Tool for Functional Proteomic Analysis of Proteases. *Expert Rev. Proteomics.* *5*, 721–730.
8. Sieber, S. A., and Cravatt, B. F. (2006) Analytical Platforms for Activity-Based Protein Profiling--Exploiting the Versatility of Chemistry for Functional Proteomics. *Chem. Commun. (Camb).* *22*, 2311–2319.
9. Sem, D. S., Bertolaet, B., Baker, B., Chang, E., Costache, A. D., Coutts, S., Dong, Q., Hansen, M., Hong, V., Huang, X., Jack, R. M., Kho, R., Lang, H., Ma, C. T., Meininger, D., Pellecchia, M., Pierre, F., Villar, H., and Yu, L. (2004) Systems-Based Design of Bi-Ligand Inhibitors of Oxidoreductases: Filling the Chemical Proteomic Toolbox. *Chem. Biol.* *11*, 185–194.
10. Kho, R., Baker, B. L., Newman, J. V., Jack, R. M., Sem, D. S., Villar, H. O., and Hansen, M. R. (2003) A Path from Primary Protein Sequence to Ligand Recognition. *Proteins.* *50*, 589–599.
11. Kho, R., Newman, J. V., Jack, R. M., Villar, H. O., and Hansen, M. R. (2003) Genome-Wide Profile of Oxidoreductases in Viruses, Prokaryotes, and Eukaryotes. *J. Proteome Res.* *2*, 626–632.
12. Argyrou, A., Jin, L., Siconilfi-Baez, L., Angeletti, R. H., and Blanchard, J. S. (2006) Proteome-Wide Profiling of Isoniazid Targets

in Mycobacterium Tuberculosis. *Biochemistry.* *45*, 13947–13953.
13. El-Kabbani, O., Ruiz, F., Darmanin, C., and Chung, R. P. (2004) Aldose Reductase Structures: Implications for Mechanism and Inhibition. *Cell Mol. Life Sci.* *61*, 750–762.
14. Ramirez, M. A., and Borja, N. L. (2008) Epalrestat: An Aldose Reductase Inhibitor for the Treatment of Diabetic Neuropathy. *Pharmacotherapy.* *28*, 646–655.
15. Trayer I.P., and Trayer H.R. (1974) Affinity Chromatography of Nicotinamide Nucleotide-Dependent Dehydrogenases on Immobilized Nucleotide Derivatives. *Biochem. J.* ***141***, 775–787.
16. Ge, X., Olson, A., Cai, S., and Sem, D. S. (2008) Binding Synergy and Cooperativity in Dihydrodipicolinate Reductase: Implications for Mechanism and the Design of Biligand Inhibitors. *Biochemistry.* *47*, 9966–9980.
17. Ge, X., Wakim, B., and Sem, D. S. (2008) Chemical Proteomics-Based Drug Design: Target and Antitarget Fishing with a Catechol-Rhodanine Privileged Scaffold for NAD(P)(H) Binding Proteins. *J. Med. Chem.* ***51***, 4571–4580.
18. Kim, C. Y., Webster, C., Roberts, J. K., Moon, J. H., Alipio Lyon, E. Z., Kim, H., Yu, M., Hung, L. W., and Terwilliger, T. C. (2009) Analysis of Nucleoside-Binding Proteins by Ligand-Specific Elution from Dye Resin: Application to Mycobacterium Tuberculosis Aldehyde Dehydrogenases. *J. Struct. Funct. Genomics.*
19. Singh, R., and Mozzarelli, A. (2009) Cofactor Chemogenomics. *Methods Mol. Biol.* *575*, 93–122.
20. Katayama, H., and Oda, Y. (2007) Chemical Proteomics for Drug Discovery Based on Compound-Immobilized Affinity Chromato graphy. *J. Chromatogr. B. Analyt Technol. Biomed. Life. Sci.* ***855***, 21–27.
21. Vanin, E. F., and Ji, T. H. (1981) Synthesis and Application of Cleavable Photoactivable Heterobifunctional Reagents. *Biochemistry.* ***20***, 6754–6760.
22. Witzemann V., Muchmore D., and Raftery M.A. (1979) Affinity-Directed Cross-Linking of Membrane-Bound Acetylcholine Receptor Polypeptides with Photolabile Alpha-Bungarotoxin Derivatives. *Biochemistry.* ***18***, 5511–5518.
23. Morelli A., and Benatti U. (1974) Simple Chemical Synthesis of a Specific Effector for the Affinity Chromatography of Nicotinamide Adenine Dinucleotide Phosphate-Dependent Dehydrogenases. *Ital. J. Biochem.* *23*, 279–291.

Chapter 12

Probing the Specificity of Protein–Protein Interactions by Quantitative Chemical Proteomics

Duangnapa Kovanich, Thin Thin Aye, Albert J.R. Heck, and Arjen Scholten

Abstract

Chemical proteomics is a versatile tool to investigate protein–small molecule interactions, but can be extended to probe also secondary binding investigating small molecule–protein 1–protein 2 interactions, providing insight into protein scaffolds. This application of chemical proteomics has in particular been applied extensively to cyclic nucleotide (cAMP, cGMP) signaling. cAMP regulates cellular functions primarily by activating cAMP-dependent protein kinase (PKA). Compartmentalization of PKA plays an important role in the specificity of cAMP signaling events and is mediated by interaction of the regulatory subunit (PKA-R) with A-kinase anchoring proteins (AKAPs), which often form the core of even larger protein machineries. The selective binding of AKAPs to one of the major isoforms PKA-R type I (PKA-RI) and PKA-R type II (PKA-RII) is an important feature of cAMP/PKA signaling. However, this specificity is not well established for most AKAPs. Here, we describe a chemical proteomics approach that combines cAMP-based affinity chromatography with quantitative mass spectrometry to investigate PKA-R isoform/AKAP specificity directly in lysates of cells and tissues of any origin. With this tool, several novel PKA-R/AKAP specificities can be easily resolved.

Key words: Chemical proteomics, cAMP pull-down, PKA, AKAP, Quantitative mass spectrometry

1. Introduction

A large number of hormones and neurotransmitters utilize cyclic nucleotides, adenosine- and guanosine 3′,5′ cyclic monophosphate (cAMP, cGMP), as an intracellular second messenger (1). cAMP signaling plays an important role in a plethora of cellular processes and its principle intracellular target is cAMP-dependent protein kinase (PKA). Compartmentalization of PKA plays a key role in the specificity of cAMP-mediated signaling events. This specificity is achieved by binding of PKA's regulatory subunit

Gerard Drewes and Marcus Bantscheff (eds.), *Chemical Proteomics: Methods and Protocols*, Methods in Molecular Biology, vol. 803, DOI 10.1007/978-1-61779-364-6_12, © Springer Science+Business Media, LLC 2012

(PKA-R) to the large and diverse family of protein scaffolds, called A-kinase anchoring proteins (AKAPs). The binding affinities of PKA for AKAPs differ significantly for the two main classes of PKA-R isoforms, PKA-RI and PKA-RII. The AKAP family can roughly be divided into three groups on the basis of which PKA-R subunit(s) bind(s) to them. A first category is formed by AKAPs that bind rather specifically to PKA-RII (for a review, see ref. 2). Also several dual-specificity AKAPs have been described, such as D-AKAP1 (3) and D-AKAP2 (4), which can anchor both PKA-RI and PKA-RII, although still with rather different affinities. Only a few RI-specific AKAPs have been identified. Two examples include AKAP-CE which was identified in *Caenorhabditis elegans* (5) and SPHKAP (6) which is the first PKA-RI-specific AKAP identified in mammalians. Related to cAMP/PKA-mediated signaling, also cGMP-mediated signaling occurs in cells, mainly proceeding via the cGMP-dependent protein kinase (PKG), which as PKA also is compartmentalized.

Most cAMP-mediated signaling proteins have a relative low abundance when compared to the abundant house-keeping proteins. Therefore, specific enrichment techniques are required when these are to be studied by proteomics techniques. In recent years, cAMP- and cGMP-based affinity enrichment techniques have been combined with sensitive mass spectrometric identification of the enriched proteins (7, 8) making a plethora of new information available on this important family of proteins. For instance, when applied to heart tissue, these analyses revealed the presence of at least 14 different AKAPs, but also other cyclic nucleotide-based signaling proteins could be enriched and identified, such as Epac, different phosphodiesterases, as well as different PKG isoforms.

Synthetic cAMP analogs have been developed as tools to unravel cAMP signal transduction pathways (9). A prime interest in PKA-RI- and PKA-RII-specific analogs exists; however, none have been discovered thus far. In addition, analogs that can differentiate between PKA and other cAMP-based signaling proteins have been described, of which 8pCPT-2′-O-Me-cAMP is a prime example as it only activates Epac, and not PKA (10). We recently screened the characteristics of the immobilized analog of the Epac-compound, 8-AHA-2′-O-Me-cAMP, and found by quantitative mass spectrometry that it binds to PKA-RI about three- to four- fold better than to PKA-RII. Here we describe in detail how the binding characteristics of different immobilized cAMP analogs can be utilized to investigate PKA–AKAP interaction specificity directly in the lysate of cells and tissues. In a proof of principle study, this method could enrich for more than ten AKAPs and provide specificity information on all of them. Interestingly, for three of the enriched AKAPs their PKA-R specificity was established for the first time (11). Soon thereafter, the technique revealed the unique PKA-RI binding characteristics of the novel AKAP SPHKAP(6).

2. Materials

2.1. Cell and Tissue Lysis

1. Protease inhibitor cocktail (Complete mini, Roche Diagnostics).
2. Phosphatase inhibitor cocktail II (Sigma).
3. Phosphate buffer saline (PBS): Prepare 10× stock with 1.37 M NaCl, 27 mM KCl, 100 mM Na_2HPO_4, 18 mM KH_2PO_4, and adjust to pH 7.4 with HCl if necessary. Prepare working solution by dilution of one part with nine parts Milli-Q water.
4. Lysis buffer: PBS buffer, 0.1% Tween 20, protease inhibitor cocktail (one tablet per 15 ml buffer), and 150 μl of phosphatase inhibitor cocktail II (Sigma) (see Note 1).

2.2. cAMP-Affinity Pull-Down

1. 8AHA-cAMP-agarose beads (C8 beads) and 8AHA-2′-O-Me-cAMP-agarose beads (C8_OCH_3 beads) (Biolog, CAT. NO. A028 and A057, respectively): Prepare 25% bead slurry in Milli-Q water containing 0.1% NaN_3 and stored at 4°C (see Note 2).
2. 100 mM Adenosine 5′-diphosphate (ADP) sodium salt and guanosine 5′-diphosphate (GDP) sodium salt (Sigma-Aldrich) are dissolved in Milli-Q water and stored at 4°C.
3. Washing buffer: 10 mM ADP and GDP in PBS. Prepare for single use by dilution of 100 mM ADP, GDP with Lysis buffer.

2.3. Dual-Protease In-Solution Digestion

1. 50 mM Ammonium bicarbonate (ABC) in Milli-Q water.
2. 8 M Urea in 50 mM ABC (see Note 3).
3. LysC and Trypsin (Roche Diagnostics) are dissolved at 0.1 μg/μl in 50 mM acetic acid and stored in single use aliquots at −80°C.
4. Dithiothreitol (DTT) and iodoacetamide (IAA) are dissolved at 200 mM in 50 mM ABC and stored in single use aliquots at −20°C.

2.4. On-Column Isotopic Dimethyl Labeling

1. Visiprep DL Vacuum manifold system (Sigma-Aldrich).
2. HPLC solvent A: 0.1 M acetic acid in Milli-Q water.
3. HPLC solvent B: 0.1 M acetic acid in 80% acetonitrile.
4. 50 mM Sodium phosphate buffer pH 7.5: Prepare stock solutions of 50 mM NaH_2PO_4 and 50 mM Na_2HPO_4. Prepare working solution by mixing 2 ml of 50 mM NaH_2PO_4 with 7 ml of 50 mM Na_2HPO_4.
5. 250 μl of 4% (V/V) formaldehyde (CH_2O, 37%, Sigma-Aldrich) in Milli-Q water (see Note 4).
6. 250 μl of 4% (V/V) D-formaldehyde (CD_2O, 98%, Isotec) in Milli-Q water (see Note 4).

7. 600 mM Sodium cyanoborohydride ($NaBH_3CN$, Fluka) in Milli-Q water (see Note 5).
8. Light labeling reagent: mix 4.5 ml of 50 mM Sodium phosphate buffer (pH 7.5) with 250 μl of 4% formaldehyde and 250 μl of 600 mM $NaBH_3CN$.
9. Heavy labeling reagent: mix 4.5 ml of 50 mM Sodium phosphate buffer (pH 7.5) with 250 μl of 4% D-formaldehyde and 250 μl of 600 mM $NaBH_3CN$.
10. Sep-Pak C18 cartridges (Waters).

2.5. Strong Cation Exchange Fractionation

1. Two Zorbax BioSCX-Series II columns (0.8 mm (i.d.) × 50 mm (l), 3.5 μm, Agilent Technologies).
2. A suitable HPLC, e.g., consisting of FAMOS autosampler (Dionex/LC Packings), Shimadzu LC-9A binary pump, and SPD-6A UV-detector (Shimadzu).
3. Strong cation exchange (SCX) Solvent A: 20% acetonitrile, 0.05% formic acid, pH 3.0 in Milli-Q water.
4. SCX Solvent B: 500 mM KCl in 20% acetonitrile and 0.05% formic acid, pH 3.0 in Milli-Q water.

2.6. LC-MS/MS Analysis

1. Suitable high-resolution mass spectrometer, e.g., LTQ-Orbitrap equipped with an electrospray ion source (Thermo Electron) coupled online to a nanoflow HPLC system, e.g., Agilent 1200 series LC system.
2. Trapping column, 20 mm, 100 μm i.d. packed with Aqua C18 reversed-phase material (5 μm, Phenomenex).
3. Separation column, 400 mm, 50 μm i.d. packed with ReproSil-Pur C18-AQ reversed-phase material (3 μm, Dr. Maisch GmBH).
4. Distally coated fused-silica emitter (360 μm (o.d.); 20 μm (i.d.); tip inner diameter, 10 μm, New Objective).
5. HPLC Solvent A: 0.1 M acetic acid in Milli-Q water.
6. HPLC Solvent B: 0.1 M acetic acid in 80% acetonitrile in Milli-Q water.

2.7. Data Analysis

1. BioWorks (Thermo Electron) is used to extract mass spectrometric raw data.
2. Mascot (http://www.matrixscience.com) is used as search engine for protein identification.
3. MSQuant (12) is used for quantification. MSQuant is an open-source software for quantitative proteomics and can be downloaded from http://www.msquant.alwaysdata.net/.

3. Methods

For the affinity purification of cAMP-interacting proteins, various synthetic cAMP analogs are available. These can be immobilized to beads using different coupling positions on the nucleotide and different linkers. For instance, here we used agarose beads on which cAMP is immobilized via an 1,6-diaminohexyl (AHA) spacer attached to the 8-position on the imidazole ring of the adenine moiety. Two types of beads are used 8AHA-cAMP-agarose beads, further referred to as C8 and 8AHA-2′-O-Me-cAMP-agarose beads, further referred to as C8_OCH$_3$, in which the hydroxyl group at the 2′ position on the ribose is substituted by a methoxy group. The chemical structures of the two immobilized cAMP analog beads are depicted in Fig. 1. Initial chemical proteomics experiments showed that C8 has equal affinity for both PKA-RI

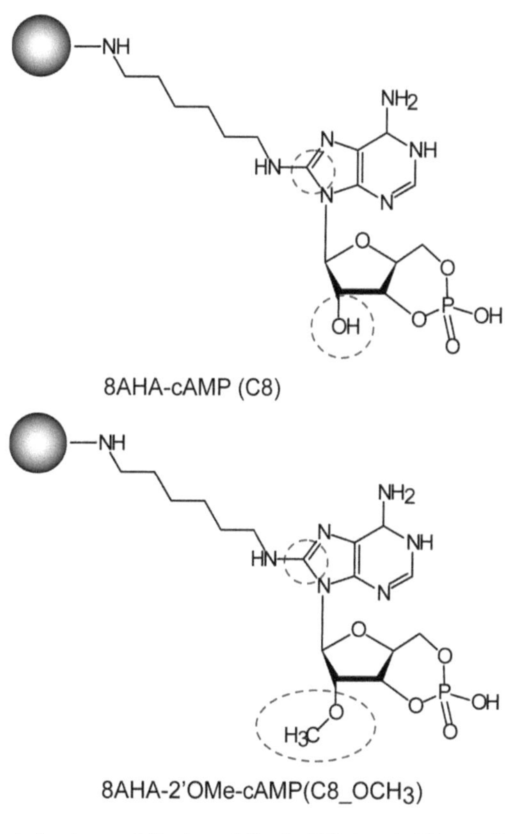

Fig. 1. Chemical structures of the immobilized cAMP beads C8 and C8_OCH$_3$. cAMP derivatives are immobilized via an 1,6-diaminohexyl (AHA) spacer that is attached to the 8-position of the imidazole ring of the adenine moiety. The hydroxyl group at the 2′ position on the ribose of the C8 beads (*left*) is substituted with a methoxy (OCH$_3$) group to generate C8_OCH$_3$ beads (*right*).

and PKA-RII, whereas C8_OCH_3 has a three- to four- fold preference for PKA-RI. This difference in affinity is used as a discriminating factor to evaluate the specificity of enriched AKAPs. Following the affinity pull-downs with both beads, all enriched proteins are digested and chemically labeled with stable isotopes using formaldehyde-based chemical dimethylation of primary amines (13). Then both labeled samples are mixed in a 1:1 ratio and analyzed by mass spectrometry.

Stable isotope dimethyl labeling utilizes isotopomers of formaldehyde and sodium cyanoborohydride to incorporate dimethyl labels at the ε-amino groups of all lysines and at the N-terminus of all proteins and/or peptides. Using two different formaldehyde isotopomers, peptide doublets are generated that exhibit a mass difference of at least 4 Da, easily distinguishable in the mass spectrometer. Peptides which are enriched by C8 are "heavy" isotope labeled and peptides which are enriched by C8_OCH_3 are "light" labeled. The intensity ratio between the "heavy" and "light" peaks represents their relative abundance in each of the pull-downs. The ratios averaged over all peptides observed for each enriched AKAP are calculated and used as an indicator of its preference for PKA-RI and/or PKA-RII isoforms.

3.1. Preparation of Cell and Tissue Lysate for cAMP-Affinity Pull-Down

3.1.1. Cell Lysis

1. Harvest the cells at a density of 1.5×10^6 cells/ml by centrifugation (30 min, 1,000 × *g*). Discard supernatant and wash the pellet once with ice-cold PBS (see Note 6). The yield varies with cell type. $1–10 \times 10^6$ cells can yield ~1 mg total protein.
2. Resuspend the pellet in 1 ml ice-cold Lysis buffer. Transfer the cell suspension to a pre-chilled glass homogenizer.
3. Lyse the cells using dounce homogenization on ice for 1–2 min (~20–30 strokes). Transfer the lysate to a pre-chilled Eppendorf tube.
4. Centrifuge the lysate at 20,000 × *g* at 4°C. Carefully transfer a soluble fraction to a new pre-chilled falcon tube. Resuspend the pellet in 1-ml ice-cold Lysis buffer and repeat the dounce homogenization in step 3.
5. Repeat centrifugation in step 4. Combine both supernatants from steps 3 and 4.
6. Wash the pellet from step 5 with 1 ml ice-cold Lysis buffer. Repeat centrifugation and combine the soluble fraction from this step to the one from step 5.
7. Measure the protein concentration of the soluble fraction, e.g., with a Bradford assay.

3.1.2. Tissue Lysis

1. Pre-chill a steel mortar, pestle, and spoon in liquid nitrogen.
2. Ground about 1 g of frozen tissue in the cold steel mortar with the cold pestle.

3. Use the cold spoon to transfer the pulverized tissue to a pre-chilled Eppendorf tube.
4. Add 1 ml of ice-cold Lysis buffer to the pulverized tissue, leave at room temperature for 5 min, and for another 10 min on ice.
5. Centrifuge the lysate at 20,000 × *g* and 4°C. Carefully transfer the supernatant to a new pre-chilled falcon tube. The insoluble pellet can be washed several times as described in steps 3–5 of Subheading 3.1.1.
6. Measure the protein concentration of the soluble fraction. The lysate is now ready for the pull-down assay (see Note 6).

3.2. cAMP-Affinity Pull-Down

1. C8 and C8_OCH_3 pull-downs are performed separately. Use a lysate with 10 mg total protein for each pull-down. The used amount depends somewhat on the abundance of PKA in the lysates and the number of washing steps incorporated.
2. Prior to the pull-down assay, supplement the lysate with 500 μl of 100 mM ADP and GDP to achieve a final concentration of 10 mM ADP and GDP to the lysate. Then the lysate is topped up with ice-cold PBS to 5 ml total volume to achieve a final protein concentration of 2 mg/ml. Then incubate at 4°C under agitation for at least 30 min (see Note 7).
3. Transfer 400 μl bead slurry (100 μl dry beads) into an Eppendorf tube and wash the beads by adding 500 μl Lysis buffer to the slurry (see Note 8).
4. Centrifuge for 1 min at 1,000 × *g* and 4°C. Discard the supernatant and repeat the washing step (see Note 9).
5. Transfer the beads to the ADP, GDP-treated lysate in a protein:beads ratio of 10 mg protein:100 μl of dry beads and incubate at 4°C under agitation for at least 2 h.
6. Centrifuge the beads, remove the unbound fraction, and wash the bead bound fraction two times with 1 ml of Washing buffer.
7. Repeat the washing and centrifugation steps with 2 ml Lysis buffer for four times (see Note 7).
8. Transfer the beads to a new Eppendorf tube. Remove the last wash as much as possible by pipetting with a thin-bore gel-loader tip. The bead bound fraction is now ready for dual-protease digestion, or if required for SDS-PAGE gel loading (see Note 9).

3.3. Dual-Protease In-Solution Digestion

1. Elute the bound proteins from the beads by adding 100 μl 8 M urea.
2. Reduce disulfide bonds of proteins with 2 mM DTT by adding 4 μl 200 mM DTT and incubate at 56°C for 15 min.

3. Alkylate the cysteine residues with 4 mM IAA by adding 8 μl 200 mM IAA to the digest and incubate at room temperature for 30 min in the dark.
4. Add 2 μl LysC solution. Incubate the protein mixture at 37°C for 4 h (see Note 10).
5. Dilute the LysC digested solution by adding 300 μl of 50 mM ABC to get a final concentration of 2 M urea.
6. Repeat the second-step digestion with trypsin by adding 2 μl trypsin and incubate at 37°C overnight.
7. Collect the supernatant (digested peptides) by centrifugation at 1,000 × *g* and transfer to a new Eppendorf tube. The digest is now ready for labeling.

3.4. On-Column Isotopic Dimethyl Labeling

1. These instructions assume the use of a Visiprep DL Vacuum manifold system. The Visiprep system contains a valve system that allows for precise flow control. It is essential to achieve a stable flow at 0.5 ml/min during sample loading, labeling, and elution steps (steps 5, 7, and 9).
2. Wash two Sep-Pak columns with 2 ml acetonitrile.
3. Equilibrate the Sep-Pak columns twice with 2 ml of HPLC Solvent A.
4. Acidify the digests from the C8 and C8_OCH_3 pull-down experiments by adding 1 ml of HPLC Solvent A.
5. Load each sample onto a separate Sep-Pak column.
6. Desalt the samples by washing the Sep-Pak columns with 2 ml of HPLC Solvent A.
7. Flush the Sep-Pak columns five times with 1 ml of respective labeling reagent. Light labeling reagent is flushed though the sample that originated from C8 pull-down, whereas heavy labeling reagent is flushed through the samples originated from C8_OCH_3 pull-down.
8. Wash the Sep-Pak columns with 2 ml of HPLC Solvent A to remove Sodium phosphate buffer.
9. Elute and collect the labeled samples from the Sep-Pak columns with 500 μl of HPLC Solvent B.
10. Mix the differentially labeled samples (i.e., "Heavy labeled" C8 and "Light labeled" C8_OCH3) in a 1:1 ratio.
11. Dry the mixed sample using a SpeedVac (1,500 rpm, room temperature) (see Note 11).
12. Reconstitute the labeled sample in 10 μl of SCX solvent A. The sample is now ready for SCX fractionation.

3.5. Strong Cation Exchange Fractionation

1. These instructions assume the use of Zorbax BioSCX-Series II columns (0.8 mm (i.d.) × 50 mm (l), 3.5 μm). SCX is performed using two in-line coupled Zorbax BioSCX-Series II

columns, FAMOS autosampler, Shimadzu LC-9A binary pump, and SPD-6A UV-detector. SCX is performed at pH 3 at which acidic residues (Asp and Glu) are uncharged to allow all peptides to bind to the column.

2. Inject 10 μl labeled sample by FAMOS autosampler.
3. Load the injected sample to the column with SCX Solvent A at a flow rate of 100 μl/min for 5 min.
4. Elute with a 1% per minute linear gradient of SCX Solvent B at a flow rate of 50 μl/min for 45 min. Then equilibrate the column with SCX Solvent A for 10 min.
5. A total of 45 SCX fractions (1 min each, 50 μl elution volume) are collected either manually or by using a suitable fraction collector.
6. Dry all fractions in a SpeedVac (1,500 rpm, room temperature) and resuspend in 20 μl 10% formic acid. The samples are now ready to subject to LC-MS/MS analysis.

3.6. LC-MS/MS Analysis

1. These instructions assume the use of a nanoflow liquid chromatography setup, directly coupled to a LTQ-Orbitrap mass spectrometer equipped with an electrospray ion source for MS analysis.
2. A volume of 10 μl of resuspended SCX fractions is used for subsequent nanoLC-LTQ-Orbitrap-MS (Thermo).
3. An Agilent 1200 series LC system is equipped with a 20 mm Aqua C18 trapping column (100 μm (i.d.), packed in-house) and a 400 mm ReproSil-Pur C18-AQ analytical column (50 μm (i.d.), packed in-house).
4. Trapping is performed at 5 μl/min for 10 min in Solvent A, and elution is achieved with a gradient of 10–35% Solvent B in 45 min in a total analysis time of 60 min. The flow rate is passively split to 100 nl/min for peptide separation. Nanospray is achieved using a distally coated fused-silica emitter biased to 1.7 kV.
5. The LTQ-Orbitrap mass spectrometer is operated in data-dependent mode, automatically switching between MS and MS/MS. Full-scan MS spectra (350–1,500 *m/z*) are acquired in the FT-Orbitrap with a resolution of $R = 60{,}000$ at 350 *m/z* after accumulation to a target value of 500,000 ions in the linear ion trap. The two most intense peaks above a threshold of 500 are fragmented in the linear ion trap using normalized collision energy of 35 after accumulation to a target value of 30,000 ions.

3.7. Data Analysis

1. Process MS/MS spectra collected from each LC-MS/MS run with Bioworks. Combine the data of the individual LC-MS/MS run into a single mascot generic file using MGF combiner (MSQuant).

2. Perform Mascot searching on the generated MGF file against an appropriate database using MASCOT (see Note 12).
3. Searching criteria include carbamidomethylation on cysteine residues as a fixed modification. Oxidation (M), light-dimethyl (K- and N-term), and heavy-dimethyl (K- and N-term) as variable modifications. Allow two missed cleavages and set a peptide mass tolerance to 10 ppm and the MS/MS mass tolerance to 0.9 Da.
4. The search result from Mascot is converted into .HTML file format using a standard Web browser (e.g., Internet Explorer). Use the .HTML file, in conjunction with the .RAW file generated by the mass spectrometer, as an input for the protein quantitation software MSQuant (12). MSQuant is needed to be customized for dimethyl labeling. An example of a screenshot of the quantitation window in MSQuant is shown in Fig. 2. Examples of MS spectra and extracted ion chromatograms of peptides enriched from pull-downs with two beads in HEK293 cell lysate are shown in Fig. 3.

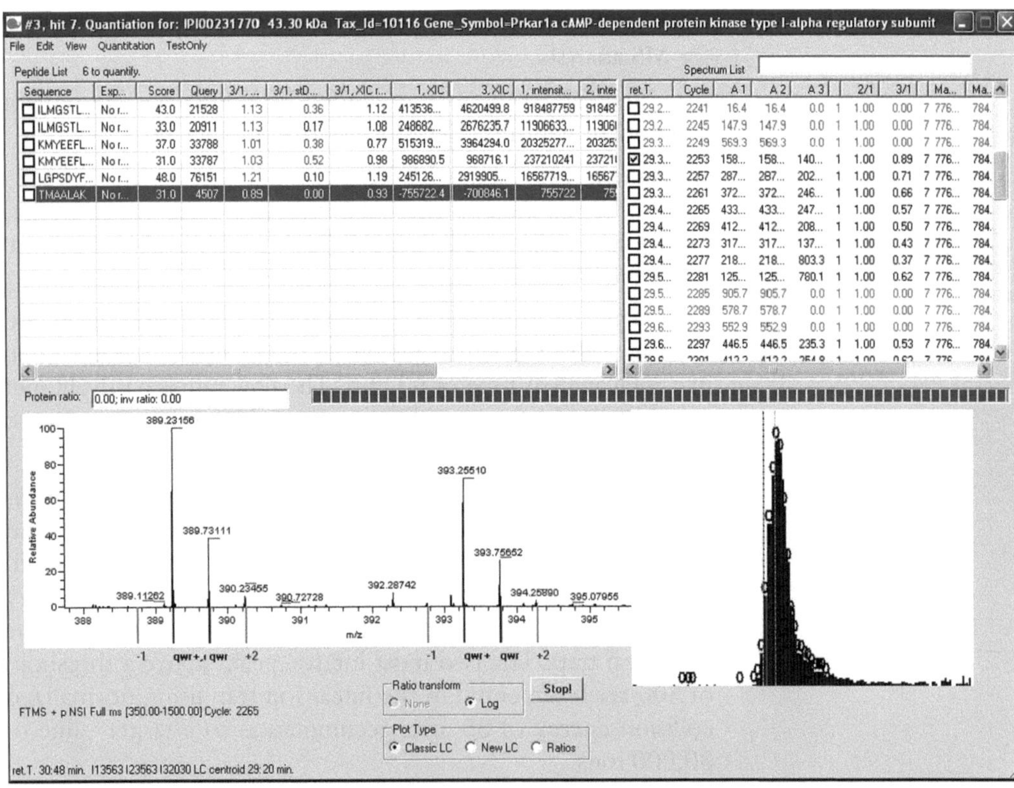

Fig. 2. Screenshot of the dimethyl-labeling quantitation window in MSQuant. This window can be divided into four panels. A list of identified peptides of a protein that can be quantified is shown in the *upper left panel* and a list of MS scans with the retention time, scan number, intensities, and ratio indicated is shown in the *upper right panel*. A raw mass spectrum of each scan can be visualized in the *lower left panel* by activating (e.g., double clicking) a scan. The *lower right panel* shows a graph where the peak area of the peptides is plotted against the retention time.

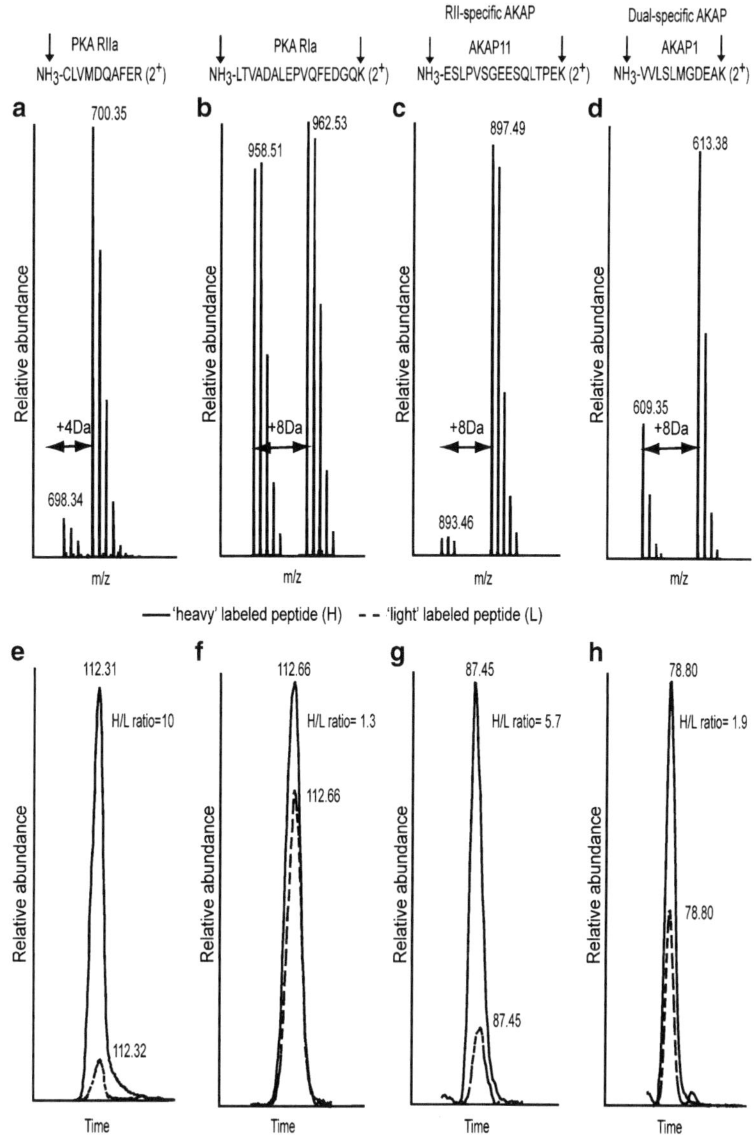

Fig. 3. An illustrative example of the subsequent quantitative MS analysis. MS spectra is shown in *Upper panels* (**a–d**) and extracted ion chromatograms (*lower panels*, **e–h**) of peptides originating from proteins in the affinity-enriched HEK293 cell lysate are depicted. MS spectra are presented in which "light" ($C8\_OCH_3$, *left*) and "heavy" (C8, *right*) peptides form peptide pairs that can be distinguished. Each dimethyl labeling event induces a typical 4 Da mass difference, so 4 Da (*m/z* difference of 2) for doubly charged peptides that have no lysine residues, and an 8 Da mass difference (4 *m/z* difference) is observed for doubly charged peptides with one lysine residue. Labeling sites are indicated by the *arrows*. In the *lower panel*, the extracted ion chromatograms of the heavy-labeled peptides (peptides which are enriched by C8) are in *solid line* and light-labeled peptides (peptides which are enriched by $C8\_OCH_3$) are in *dashed line*. Each individual peptide pair is used for the assessment of differential binding affinities. (This research was originally published in ref. 11 with permission from the American Society for Biochemistry and Molecular Biology).

5. Use MSQuant to extract and integrate ion chromatograms of all peptides. Area under the curve (AUC) between the monoisotopic peaks of "light" and "heavy" is used to calculate a ratio of the peptide pair. MSQuant does that for every charge state of a peptide pair. The constituted peptide ratio is calculated by averaging peptide ratios of all charge states and reported with their respective standard deviations. A protein ratio is calculated by averaging of all constituted peptide ratios originating from the protein. The protein ratio is used as a specificity ratio. An example of a summary of all specificity ratios obtained for each PKA isoform and all detected AKAPs is shown in Fig. 4.

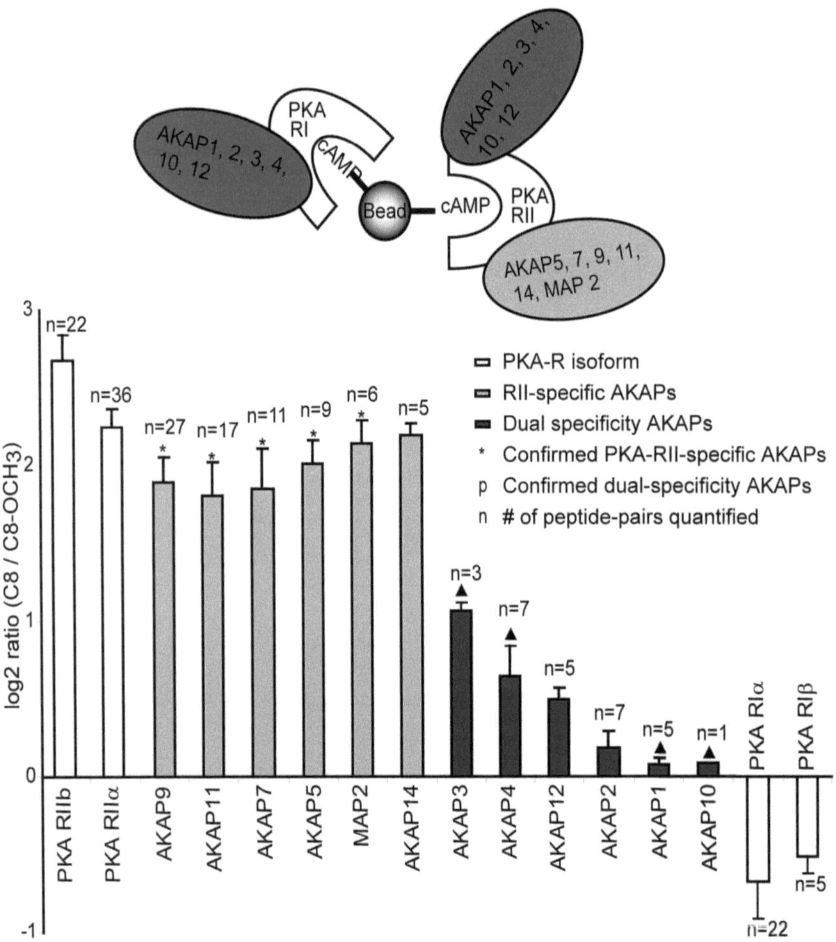

Fig. 4. Summary of all specificity ratios obtained for each PKA isoform and all detected AKAPs. Overview of averaged binding specificity ratios of all AKAPs affinity purified from HEK293, and RCC10 cells, rat lung, and rat testis tissue. The protein ratios are calculated by pooling all peptide pairs detected in each tissue for a certain protein. The overall standard deviation is calculated over all spectra in the pool. For comparison, all averaged ratios are depicted on a log2 scale. The number of spectra, which were used for the average ratios, are shown. The PKA-RII isoforms show a clear positive C8/C8_OCH_3 ratio, indicative of their preference for the C8-resin, while PKA-RI isoforms show clear negative ratios, meaning they are more abundant in the C8_OCH_3 pull-down. As expected, the ratio of known PKA-RII-specific AKAPs exactly follows the ratio of PKA-RII. In contrast, AKAP1, referred to as D-AKAP1 and AKAP10, referred to as D-AKAP2 for their dual specificity, have specificity ratio of approximately 1 (0 on the log2 scale) which can be rationalized by their ability to bind to both PKA-RI and PKA-RII isoforms. (This research was originally published in ref. 11 with permission from the American Society for Biochemistry and Molecular Biology).

4. Conclusion

Here, we describe a chemical proteomics approach targeting proteins involved in cAMP-mediated signaling, whereby both primary nucleotide interactors and secondary binders can be enriched for. This approach is highly relevant as the specificity of cAMP signaling is mediated by interaction of the regulatory subunit (PKA-R) with AKAPs, which often form the core of even larger protein machineries. As a proof of the potential, we utilized this technique to monitor the specificity of SPHKAP in mammalian heart and spleen tissues. Consistent with biochemical *in vitro* data, SPHKAP displayed similar ratios as PKA-RIα in both tissues, leading SPHKAP to be considered as the first mammalian AKAP that preferentially binds to PKA-RIα (6).

5. Notes

1. PBS buffer with 0.1% Tween 20 can be prepared for multiple uses and stored at 4°C as the buffer should be cold when used. However, the Lysis buffer should be prepared fresh for single use by adding protease and phosphatase inhibitor cocktails only very prior to use in order to keep 100% efficiency of the inhibitors.
2. 0.1% NaN_3 is necessary for preventing the bacterial and fungus growth in the bead slurry.
3. It is recommended to prepare 8 M urea in 50 mM ABC fresh for single use to avoid urea precipitation. It is possible to store the rest of the solution at RT under agitation.
4. Try to avoid putting the pipette tip or the poly pipette directly into a formaldehyde stock bottle. It is recommended to use a glass pipette to transfer the desired amount from the stock bottle into an Eppendorf for single use.
5. Labeling reagent mixtures should be kept at 4°C and not stored longer than 24 h to ensure labeling efficiency. It is recommended to prepare fresh sodium cyanoborohydride for single use as it is very water reactive and easy to decompose on exposure to moist air or water. Excess sodium cyanoborohydride can be discarded by neutralizing with NaOH.
6. Cell pellet can be stored at −80°C for future use. To avoid dissociation of protein complexes, tissue lysate should be stored at high protein concentration in 50% glycerol at −20°C until use. Before pull-down, dilute the lysate to get a protein concentration at 2 mg/ml and a glycerol concentration at 10% or below.

7. It is necessary to supplement the lysate prior to the pull-down as well as the Washing buffer of the first two washes after the pull-down with 10 mM ADP and GDP to reduce interactions of noncyclic nucleotide binding proteins to the resin (7, 8). It is also essential to further wash the immobilized fraction at least three to four times with a large volume of Lysis buffer. We recommend to use at least 1–2 ml buffer per 100 μl of beads for each washing step.
8. Cut off the end of a pipette tip when handling cAMP-agarose beads to avoid disruption of the beads and to assist in measuring out the correct volume of beads. It is also important to centrifuge cAMP beads at maximally 1,000 × *g* to avoid loss of binding.
9. An alternative to in-solution digestion is SDS-PAGE gel loading. The enriched proteins can be eluted from the beads by boiling the beads at 95°C in an equal volume of SDS-PAGE loading buffer for 5 min. After that, the eluate can be subjected to an SDS-PAGE gel, followed by in-gel digestion and LC-MS/MS analysis or Western blot analysis.
10. For a single cAMP pull-down with 10 mg protein, expected bound fraction is 20 μg. A protease:protein ratio of 1:50 to 1:100 (w/w) is recommended for digestion.
11. The labeled sample should not dry completely as it is difficult to reconstitute. The best thing is to dry the sample until 1–2 μl are left. After that the sample can be stored at -20°C until MS analysis.
12. For samples of human origin, it is recommended to use the Swiss-Prot/Uniprot database as its current annotation of ~23,000 entries is the most nonredundant. For rat and mouse samples, no clear preference for a database to be used can be stated.

Acknowledgments

The authors would like to thank Toon A.B. van Veen for lysate preparation. This work was supported by the Netherlands Proteomics Centre (D.K., T.A.A., A.J.R.H., and A.S.), the Utrecht Institute of Pharmaceutical Sciences (D.K. and A.S.), and a Focus and Massa grant from Utrecht University (A.S.).

References

1. Bar, H. P., and Hechter, O. (1969) Adenyl cyclase and hormone action. I. Effects of adrenocorticotropic hormone, glucagon, and epinephrine on the plasma membrane of rat fat cells, *Proc Natl Acad Sci USA* ***63***, 350–356.
2. Wong, W., and Scott, J. D. (2004) AKAP signalling complexes: focal points in space and time, *Nat Rev Mol Cell Biol* ***5***, 959–970.
3. Huang, L. J., Durick, K., Weiner, J. A., Chun, J., and Taylor, S. S. (1997) Identification of a novel protein kinase A anchoring protein that binds both type I and type II regulatory subunits, *J Biol Chem* ***272***, 8057–8064.
4. Wang, L., Sunahara, R. K., Krumins, A., Perkins, G., Crochiere, M. L., Mackey, M., Bell, S., Ellisman, M. H., and Taylor, S. S. (2001) Cloning and mitochondrial localization of full-length D-AKAP2, a protein kinase A anchoring protein, *Proc Natl Acad Sci USA* ***98***, 3220–3225.
5. Angelo, R., and Rubin, C. S. (1998) Molecular characterization of an anchor protein (AKAPCE) that binds the RI subunit (RCE) of type I protein kinase A from Caenorhabditis elegans, *J Biol Chem* ***273***, 14633–14643.
6. Kovanich, D., van der Heyden, M. A., Aye, T. T., van Veen, T. A., Heck, A. J., and Scholten, A. Sphingosine kinase interacting protein is an A-kinase anchoring protein specific for type I cAMP-dependent protein kinase, *Chembiochem* ***11***, 963–971.
7. Scholten, A., Poh, M. K., van Veen, T. A., van Breukelen, B., Vos, M. A., and Heck, A. J. (2006) Analysis of the cGMP/cAMP interactome using a chemical proteomics approach in mammalian heart tissue validates sphingosine kinase type 1-interacting protein as a genuine and highly abundant AKAP, *J Proteome Res* ***5***, 1435–1447.
8. Scholten, A., van Veen, T. A., Vos, M. A., and Heck, A. J. (2007) Diversity of cAMP-dependent protein kinase isoforms and their anchoring proteins in mouse ventricular tissue, *J Proteome Res* ***6***, 1705–1717.
9. Poppe, H., Rybalkin, S. D., Rehmann, H., Hinds, T. R., Tang, X. B., Christensen, A. E., Schwede, F., Genieser, H. G., Bos, J. L., Doskeland, S. O., Beavo, J. A., and Butt, E. (2008) Cyclic nucleotide analogs as probes of signaling pathways, *Nat Methods* ***5***, 277–278.
10. Vliem, M. J., Ponsioen, B., Schwede, F., Pannekoek, W. J., Riedl, J., Kooistra, M. R., Jalink, K., Genieser, H. G., Bos, J. L., and Rehmann, H. (2008) 8-pCPT-2′-O-Me-cAMP-AM: an improved Epac-selective cAMP analogue, *Chembiochem* ***9***, 2052–2054.
11. Aye, T. T., Mohammed, S., van den Toorn, H. W., van Veen, T. A., van der Heyden, M. A., Scholten, A., and Heck, A. J. (2009) Selectivity in enrichment of cAMP-dependent protein kinase regulatory subunits type I and type II and their interactors using modified cAMP affinity resins, *Mol Cell Proteomics* ***8***, 1016–1028.
12. Mortensen, P., Gouw, J. W., Olsen, J. V., Ong, S. E., Rigbolt, K. T., Bunkenborg, J., Cox, J., Foster, L. J., Heck, A. J., Blagoev, B., Andersen, J. S., and Mann, M. MSQuant, an open source platform for mass spectrometry-based quantitative proteomics, *J Proteome Res* ***9***, 393–403.
13. Boersema, P. J., Raijmakers, R., Lemeer, S., Mohammed, S., and Heck, A. J. (2009) Multiplex peptide stable isotope dimethyl labeling for quantitative proteomics, *Nat Protoc* ***4***, 484–494.

Chapter 13

Fluorescence-Based Proteasome Activity Profiling

Annemieke de Jong, Karianne G. Schuurman, Boris Rodenko, Huib Ovaa, and Celia R. Berkers

Abstract

With the proteasome emerging as a therapeutic target for cancer treatment, accurate tools for monitoring proteasome (inhibitor) activity are in demand. In this chapter, we describe the synthesis and use of a fluorescent proteasome activity probe that allows for accurate profiling of proteasomal activity in cell lysates, intact cells, and murine and human patient-derived material, with high sensitivity using SDS-PAGE. The probe allows for direct scanning of the gel for fluorescent emission of the distinct proteasomal subunits and circumvents the use of Western blot analysis. Due to its suitable biochemical and biophysical properties, the fluorescent probe can also be used for confocal laser scanning microscopy and flow cytometry-based experiments.

Key words: Proteasome, Inhibition, Activity profiling, Fluorescent probe, Proteasome activity assay

1. Introduction

The use of proteasome inhibitors in the clinic for cancer treatment (1) has validated the proteasome as therapeutic target. The proteasome is responsible for the degradation of misfolded and redundant proteins and of key regulatory proteins, involved in many cellular processes such as proliferation and survival (2, 3). Inhibition of the proteasome causes disruption of many of these processes, eventually leading to cell death (1, 4). The proteasome inhibitor bortezomib (1, 5) (Fig. 1) is currently used for the treatment of multiple myeloma (6) and mantle cell lymphoma (7). A number of second-generation proteasome inhibitors, which differ in their mode of inhibition and subunit specificity, are currently in clinical trials. Eukaryotic 26S proteasomes consist of a 20S core and one or two 19S regulatory caps. The 19S regulatory caps are involved in

Gerard Drewes and Marcus Bantscheff (eds.), *Chemical Proteomics: Methods and Protocols*, Methods in Molecular Biology, vol. 803, DOI 10.1007/978-1-61779-364-6_13, © Springer Science+Business Media, LLC 2012

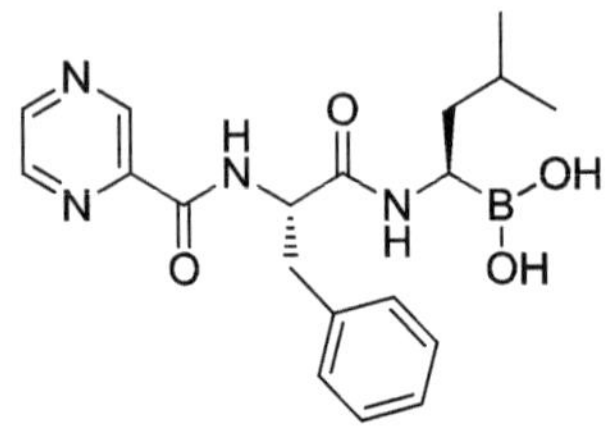

Bortezomib

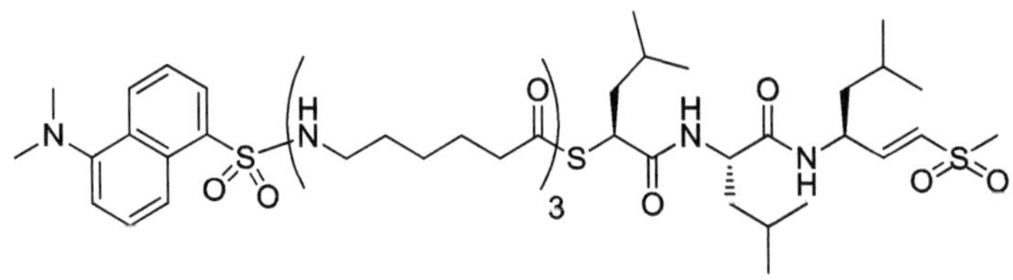

Dansyl-Ahx_3Leu_3VS

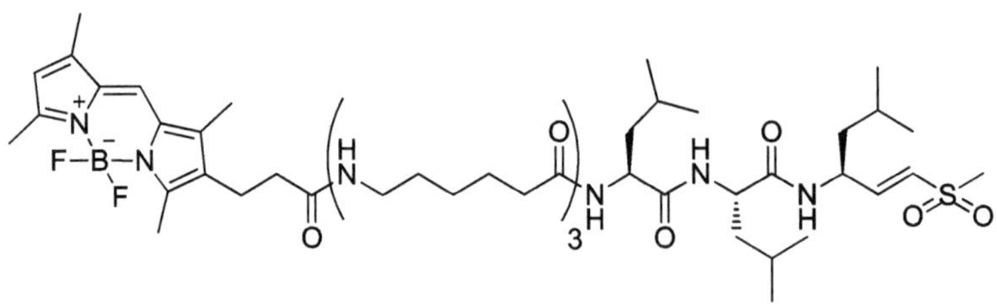

Me_4BodipyFL-Ahx_3Leu_3VS

Fig. 1. Structures of bortezomib, dansyl-Ahx_3Leu_3VS, and Me_4BodipyFL-Ahx_3Leu_3VS.

the recognition and unfolding of polyubiquitinated substrates, while the catalytic activity takes place in the 20S core (3). The 20S proteasome is a barrel-shaped protein complex composed of four stacked rings that consist of seven subunits each. The two outer rings are made up of α-subunits and interact with the 19S regulatory caps. The two inner rings consist of β-subunits, from which three constitutive subunits, termed β1, β2, and β5, are responsible for proteolysis. The β1, β2, and β5 subunits each provide a distinct protease activity, the caspase-like (cleavage after acidic residues), tryptic-like (cleavage after basic residues), and chymotryptic-like (cleavage after hydrophobic residues) activity, respectively. Upon interferon-γ stimulation, three additional immunoproteasome subunits (β1i, β2i, and β5i) are expressed, which replace the constitutive β1, β2, and β5 subunits to form the immunoproteasome. Immunoproteasomes are thought to have altered catalytic activity favoring production of antigenic peptides and are mainly expressed in lymphoid tissues, e.g., spleen, thymus, and lymph nodes (8). Different cells can express different ratios of constitutive and immunoproteasome subunits (9, 10). Variations in proteasomal composition affect substrate specificity and sensitivity to proteasome inhibition. To predict the sensitivity of patients to proteasome inhibitors, accurate tools are required that can correlate proteasome composition and the extent of proteasome inhibition to treatment response. Reagents that can be used to profile proteasome activity are valuable research tools and hold promise as diagnostic reagents.

Techniques that are commonly used to monitor proteasome activity include the application of fluorogenic substrates (11, 12), small molecule-based activity assays (13–18), and models based on recombinant reporter proteins (19–21). Traditionally, fluorogenic substrates are used to measure the activity of different proteasome active sites, but most fluorogenic substrates cannot be used in cells, and prior cell lysis is required before activity measurements can be performed. Reporter proteins can be used in living cells, but their use is limited to genetically altered cells or organisms. In addition, their activity readout depends on the balance between synthesis and degradation of fusion proteins, which involves many cellular factors other than the proteasome, including the rate of fusion-protein synthesis.

The first small molecule-based activity assay for profiling of the specificity of proteasome inhibitors in living cells was reported in 2005 (13). This dansylated vinylsulfone based probe (Fig. 1) contains a proteasome-targeting motif, α,β-unsaturated vinyl sulfone (VS) group that covalently reacts through a Michael addition with the γ-hydroxyl of the N-terminal threonine residue of all catalytic β-subunits of the proteasome (22), resulting in the formation of a β-sulfonyl ether linkage. Antibodies against the dansyl moiety were used for the detection of labeled active subunits by Western blot analysis (13). Subsequent replacement of the dansyl group by high quantum yield fluorophores allowed for direct scanning of the SDS-PAGE gel for fluorescence emission of fluorescently labeled subunits (14, 18). Due to their favorable biochemical and biophysical properties, Bodipy-based proteasome activity probes can be used to monitor proteasome activity in cell extracts, living cells, and murine tissues using a range of techniques including SDS-PAGE (14), confocal laser scanning microscopy (14, 18), and flow cytometry (14). The advantage of flow cytometry-based assays is that they can be used to measure both proteasome activation and proteasome inhibition in large numbers of samples. Information on the activity of distinct subunits on the other hand can only be obtained in SDS-PAGE-based profiling experiments.

A typical SDS-PAGE-based profiling experiment is shown in Fig. 2. Lysates of MelJuso (human melanoma; Fig. 2a) and THP-1 (Human monocytic leukemia, Fig. 2b) cells were incubated with the proteasome inhibitors MG132 (25 μM), bortezomib (0.1 μM and 1 μM), epoxomicin (1 μM), or a DMSO control. Subsequently, samples were incubated with fluorescent proteasome activity probe to label proteasome subunits, proteins were separated by

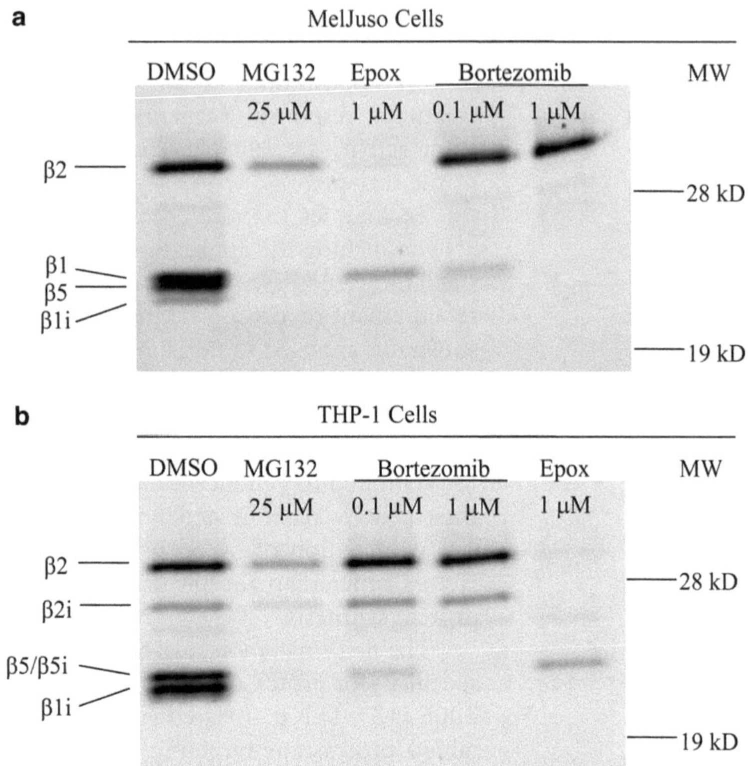

Fig. 2. Gel images showing the active proteasome subunit labeling in MelJuso (**a**) and THP-1 (**b**) cell lysates that were preincubated with the indicated concentrations of MG132, epoxomicin or bortezomib, followed by incubation with 1 μM $Me_4BodipyFL\text{-}Ahx_3Leu_3VS$.

SDS-PAGE and the resulting gel was scanned for fluorescence emission. Upon incubation of proteasome with probe, all active proteasome subunits become fluorescently labeled. Prior inhibition of a subunit with a proteasome inhibitor prevents probe binding, resulting in the disappearance of a fluorescent band on the gel. Therefore, the measured fluorescence intensity directly correlates to the activity of the labeled β-subunit. In the DMSO controls, all active subunits are labeled, and the composition of proteasome in these cells was visualized (Fig. 2). Incubation with different proteasome inhibitors resulted in the disappearance of particular bands on the gels, indicative of the subunit specificity of these inhibitors.

Figure 3 shows the results of a typical fluorescence assisted cell sorting (FACS) assay, in which the fluorescence intensity is plotted versus the cell count. MelJuso cells were incubated with 1 μM MG132, followed by incubation with probe (Fig. 3, blue curve). As controls, cells were incubated with probe only (Fig. 3, red curve) or not incubated (Fig. 3, black curve). Addition of probe results in a shift of the cell population toward higher fluorescence intensity. Upon preincubation with MG132, the peak shifts back to lower fluorescence intensity, indicating inhibition. As some

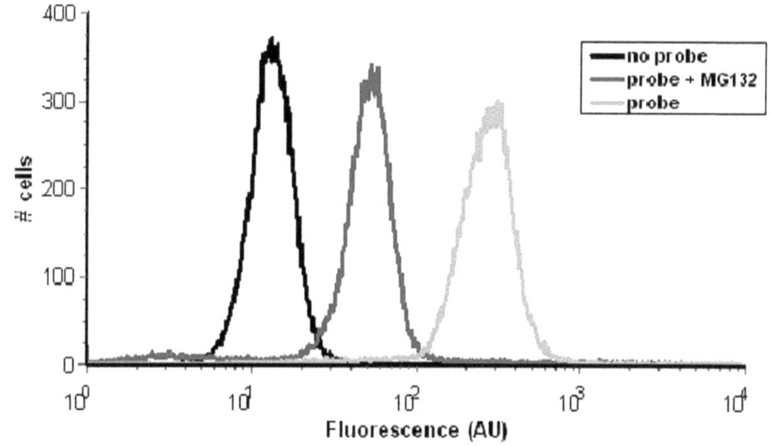

Fig. 3. Flow cytometry histograms showing the fluorescence intensity in MelJuso cells that were nonincubated (*black*), incubated with 200 nM Me_4BodipyFL-Ahx_3Leu_3VS (*red*), or preincubated with 1 μM MG132, followed by incubation of 200 nM Me_4BodipyFL-Ahx_3Leu_3VS (*blue*).

probe always aspecifically sticks to cell membranes, the signal intensity at 1 μM MG132 is taken as 100% inhibition, whereas the signal intensity in probe only treated cells is taken as 0% inhibition.

In this chapter, we describe the synthesis of the Bodipy-based fluorophore Me_4BodipyFL-*N*-hydroxy-succinimidyl (NHS) ester in six steps. The Bodipy fluorophore was chosen for its ease of synthesis from commercially available reagents on a large scale and for its fluorescein-like spectral properties (excitation maximum of 515 nm and emission maxima of 519 nm), which make it widely applicable to common fluorescence-based methodologies. In addition, we describe the synthesis of the proteasome-targeting moiety Ahx_3Leu_3VS and subsequent coupling with Me_4BodipyFL-NHS ester yielding the fluorescent probe Me_4BodipyFL-Ahx_3Leu_3VS (Fig. 1). Finally, optimized procedures to profile the effects of proteasome inhibitors in cell lysates, living cells, and murine tissues using this fluorescent proteasome activity probe in both SDS-PAGE and FACS-based assays are described.

2. Materials

2.1. Synthesis of the Fluorescent Probe Me_4BodipyFL-Ahx_3Leu_3VS

2.1.1. Synthesis of Me_4BodipyFL-NHS Ester

1. *Tert*-butyl acetoacetate, acetic acid, sodium nitrite.
2. Methyl 4-acetyl-5-oxohexanoate, zinc dust, sodium acetate, ethyl acetate (EtOAc), magnesium sulfate ($MgSO_4$).
3. Trifluoroacetic acid (TFA), triethylorthoformate, toluene, EtOAc.
4. Ethanol (absolute), 2,4-dimethylpyrrole, HCl in dioxane (4 N), diethyl ether.

5. 1,2-Dichlorobenzene, triethylamine, boron trifluoride etherate, hexanes, EtOAc, toluene.
6. 1 M Lithium hydroxide solution, EtOAc, 0.1 M HCl, $MgSO_4$, acetic acid.
7. Dichloromethane (CH_2Cl_2), *N*-hydroxysuccinimide, 1-ethyl-3-(3-dimethylaminopropyl) carbodiimide hydrochloride (EDCI), 4-dimethylaminopyridine (DMAP), hexanes, EtOAc.

2.1.2. Synthesis of Me_4BodipyFL-Ahx_3Leu_3VS

1. Fmoc-L-leucine-PEG-polystyrene resin (Applied Biosystems), Fmoc-leucine-OH (Fmoc-Leu-OH, Novabiochem), Fmoc-aminohexanoic acid (Fmoc-Ahx-OH, Novabiochem), piperidine, benzotriazol-1-yl-oxytripyrrolidinophosphonium hexafluorophosphate (PyBop, coupling reagent), *N,N*-diisopropylethylamine (DIPEA), *N*-methyl-2-pyrrolidone (NMP).
2. Di-*tert*-butyl-dicarbonate, DIPEA, dimethylformamide (DMF), CH_2Cl_2, methanol (MeOH), acetic acid.
3. (S;*E*)-5-methyl-1-(methylsulfonyl)hex-1-en-3-amine (leucinyl vinyl sulfone, LeuVS); EDCI; DIPEA; DMF; CH_2Cl_2; MeOH.
4. TFA; toluene; diethyl ether.
5. Me_4-BodipyFL-NHS ester (7); DIPEA; DMF.

2.1.3. Analytical Procedures

1. Acros silica gel (0.030–0.075 mm) for flash chromatography.
2. Silica-coated plastic sheets (Merck silica gel F_{254}) for thin-layer chromatography (TLC).
3. LC/MS analysis: Waters LCT mass spectrometer in line with a Waters 2795 HPLC system and a Waters 2996 photodiode array detector. Reversed-phase runs were performed on a 3 μm Atlantis T3, C18 RP, 2.1 × 100 mm column (Waters) using gradient elution with H_2O/0.1% formic acid as solvent A and acetonitrile/0.1% formic acid as solvent B at a flow rate of 0.4 mL/min.
4. Preparative HPLC: Waters 1525 EF HPLC system in line with a Waters 2487 dual λ absorbance detector using gradient elution with H_2O/0.05% TFA as solvent A and acetonitrile/0.05% TFA as solvent B. Preparative runs were performed on a 10 μm 19 × 250 mm Atlantis dC18 column (Waters) at a flow rate of 18 mL/min.
5. Nuclear magnetic resonance (NMR) spectra were recorded using a Bruker Avance 300 (^{1}H: 300 MHz; ^{13}C: 75 MHz) spectrometer.
6. Fluorescence spectra were measured on a customized fluorimeter set-up using a mercury vapor lamp at 72 W, appropriate gratings, and a photomultiplier at 1,000 V (Photon Technology International). Excitation spectra were recorded at 100 nM in water at 550 nm emission, while emission spectra were recorded in water at 480 nm excitation.

2.2. Profiling of Proteasome Activity Using SDS-PAGE-Based Assays

2.2.1. In Vitro Profiling of Proteasome Activity in Cell Lysates

1. Cell line of choice cultured in appropriate medium, e.g., DMEM (Dulbecco's modified Eagle's medium) for adherent cell lines, and RPMI 1640 (Roswell Park Memorial Institute) medium for suspension cell lines supplied with fetal calf serum (FCS) and antibiotics (100 U/mL penicillin and 100 μg/mL streptomycin).
2. Phosphate-buffered saline (PBS), Trypsin solution (0.25%, Invitrogen) for adherent cells.
3. HR lysis buffer (see Note 1): 50 mM Tris–HCl (pH 7.4), 5 mM $MgCl_2$, 250 mM sucrose, 1 mM DTT (added fresh from a 1 M stock solution before use), 2 mM ATP (added fresh from a 0.5 M stock solution before use). Prepare HR buffer lacking DTT and ATP, filter over a 0.22 μm filter (e.g., MILLEX®GS, Millipore), and store at 4°C. Supplement the amount of HR buffer needed for a single experiment (typically 1 mL) with ATP and DTT before use (see Note 2).
4. Bradford reagent (Biorad).
5. Dimethylsulfoxide (DMSO) (optional).
6. 50× Stock solutions of proteasome inhibitors in DMSO. Store at −20°C.
7. 50 μM Stock solution of Me_4BodipyFL-Ahx_3Leu_3VS in DMSO. Store at −20°C.

2.2.2. Labeling of Active Proteasome Subunits in Living Cells

1. Cell line of choice cultured in appropriate medium (see Subheading 2.2.1).
2. 500× Stock solutions of proteasome inhibitors in DMSO. Store at −20°C.
3. 50 μM Stock solution of Me_4BodipyFL-Ahx_3Leu_3VS in DMSO. Store at −20°C.
4. 5 mM Stock solution of MG132 (Sigma) in DMSO. Store at −20°C.
5. PBS, Trypsin solution (0.25%, Invitrogen) for adherent cells.
6. NP40 lysis buffer (see Note 1): 50 mM Tris–HCl (pH 7.4), 150 mM NaCl, 1% NP40. Prepare NP40 lysis buffer, filter over a 0.22 μm filter (e.g., MILLEX®GS 0.22 μm Filter unit, Millipore), and store at 4°C.
7. Bradford reagent (Biorad).

2.2.3. Ex Vivo Profiling of Proteasome Subunit Activity in Murine Tissues

1. Freshly dissected mouse tissue of choice.
2. HR lysis buffer (see item 3 in Subheading 2.2.1).
3. Glass beads (≤106 μm, acid washed, Sigma).
4. Bradford reagent (Biorad).
5. 50 μM Stock solution of Me_4BodipyFL-Ahx_3Leu_3VS in DMSO. Store at −20°C.

2.2.4. Gel Electrophoresis and In-Gel Fluorescence Readout

1. Precast gel system (NuPAGE, Invitrogen).
2. NuPAGE® Novex 12% Bis-Tris Gel, 1.0 mm (Invitrogen), NuPAGE® MOPS SDS Running buffer (Invitrogen), NuPAGE® Antioxidant (Invitrogen), SeeBlue® Plus2 Pre-Stained Standard (Invitrogen).
3. 3× Reducing sample buffer (see Note 1). For 1.2 mL: 900 μL 4× NuPAGE® LDS Sample buffer (Invitrogen), 90 μL β-mercaptoethanol (Sigma), 210 μL water. Store at room temperature.
4. ProXPRESS 2D Proteomic imaging system (Perkin Elmer) for the determination of In-gel fluorescence intensity. Fluorescence intensities are quantified using the analysis software TotalLab.

2.3. Flow Cytometry

1. MelJuso cells (human melanoma) cultured in DMEM supplied with FCS and antibiotics (100 U/mL penicillin and 100 μg/mL streptomycin).
2. 96-Well flat bottom tissue culture plates (BD Falcon).
3. DMSO, PBS, Trypsin solution (0.25%; Invitrogen).
4. 2.5 mM Stock solutions of the desired compounds in DMSO. Store at −20°C.
5. 0.4 mM Stock solution of MG132 (Sigma) in DMSO. Store at −20°C.
6. 50 μM Stock solution of Me_4BodipyFL-Ahx_3Leu_3VS in DMSO. Store at −20°C.
7. FACS buffer: PBS containing 2% FCS. Store at 4°C.
8. Fixation buffer: PBS containing 2% formaldehyde. Store at room temperature.
9. FACSCalibur (BD Biosciences, 488 nm laser).

3. Methods

3.1. Synthesis of the Fluorescent Proteasome Probe Me_4BodipyFL-Ahx_3Leu_3VS

3.1.1. Synthesis of Me_4-BodipyFL-NHS Ester (Scheme 1)

1. To a mixture of *tert*-butyl acetoacetate (2.80 g, 17.4 mmol) in acetic acid (4 mL), stirred on an ice bath, add dropwise a solution of sodium nitrite (1.40 g, 20.3 mmol) in deionized water over a 5 min period. Stir for an additional 16 h at 4°C (e.g., in a cold room) (23).
2. Slowly, add the resulting solution containing oxime 1 in five portions to a solution of methyl 4-acetyl-5-oxohexanoate (5 mL, 28.6 mmol) in acetic acid (15 mL) kept at 65°C. Simultaneously, add a mixture of zinc dust (3 g, 45.9 mmol)

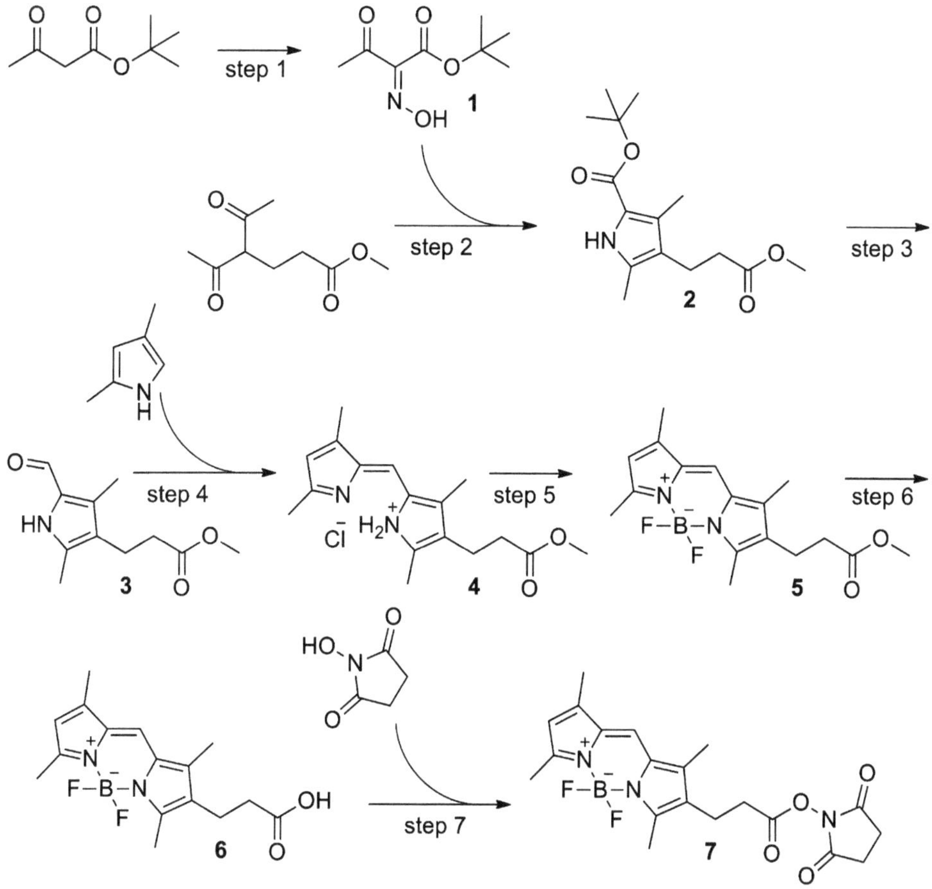

Scheme 1. Synthesis of Me_4BodipyFL-NHS ester 7. Step numbers correspond to step numbers outlined in Subheading 3.1.1

and sodium acetate (3 g, 36.6 mmol) in five portions. Stir the mixture for 2 more hours at 65°C, then pour it into ice water (250 mL) and leave slowly stirring for 16 h, allowing the mixture to warm up to room temperature. Filter off the precipitate, including product and zinc residues. Take up the residue in EtOAc (100 mL) and dry the organic layer with $MgSO_4$. Filter off the $MgSO_4$ and concentrate the filtrate in vacuo to dryness to yield tetrasubstituted pyrrole **2** as a light brown solid (2.87 g, 10.2 mmol, 59% yield). The crude product can be used without further purification in the next step.

3. Dissolve pyrrole **2** (1 g, 3.55 mmol) in TFA (20 mL) and stir for 20 min on ice. Slowly, add triethylorthoformate (2 mL, 12 mmol) and stir the solution gently at 0°C for 15 min. Add water (2 mL) and stir for 10 min at 0°C. Coevaporate the solution with toluene (60 mL) in vacuo to dryness at room temperature.

Purify the crude product by flash column chromatography (24) using EtOAc as the eluent to yield aldehyde **3** as a brown solid (469 mg, 2.25 mmol, 63% yield).

4. Dissolve aldehyde **3** (469 mg, 2.25 mmol) in absolute ethanol (8 mL) and cool to 0°C. Add 2,4-dimethylpyrrole (256 mg, 2.69 mmol) under an argon atmosphere and add 4 mL of cold 4 N HCl in dioxane. Stir for 15 min at 0°C and then transfer the solution to a 50-mL Falcon tube and centrifuge at 1,500 × *g* for 10 min at 4°C. Discard the supernatant, resuspend the resulting orange solid in cold diethylether (40 mL), and centrifuge again. Decant the supernatant and dry the residue under a stream of nitrogen to yield crude dipyrrole **4** as an orange solid, which is used in the next step without further purification.
5. Suspend dipyrrole **4** in 1,2-dichlorobenzene (20 mL) under an argon atmosphere and add triethylamine (1 mL, 7.16 mmol) and boron trifluoride diethyl etherate (1 mL, 8.12 mmol) in concert. Stir the resulting metallic solution for 30 min at 100°C and then allow the mixture to cool to room temperature. Dilute the solution with dry hexanes (25 mL) and apply directly onto a silica gel column for purification by flash column chromatography using 25% EtOAc in toluene as the eluent to afford Me_4BodipyFL-methyl ester **5** as an orange solid (481 mg, 1.44 mmol, 64% yield starting from aldehyde **3**).
6. Dissolve Me_4BodipyFL-methyl ester **5** (400 mg, 1.24 mmol) in ethanol (6 mL) and add a solution of 1 M lithium hydroxide in water (1.3 mL). Stir the solution until monitoring by TLC (eluent: EtOAc containing 0.1% acetic acid) reveals complete conversion. Add EtOAc (40 mL) and 0.1 M HCl (30 mL) and separate the two layers. Dry the organic layer with $MgSO_4$ and concentrate in vacuo to dryness to obtain an orange solid.
7. Purify by flash column chromatograpy using EtOAc containing 0.1% acetic acid as the eluent to yield Me_4BodipyFL-OH **6** as an orange solid (306 mg, 0.95 mmol, 77% yield). ^{1}H NMR ($CDCl_3$): δ 7.09 (s, 1H), 6.10 (s, 1H), 2.66 (dd, J=8.2 Hz, J=7.1 Hz, 2H), 2.59 (s, 6H), 2.60–2.53 (m, 2H), 2.31 (s, 3H), 2.27 (s, 3H). ^{13}C NMR ($CDCl_3$): δ 178.2 (CO), 156.5 (C_q), 155.3 (C_q), 141.0 (C_q), 138.1 (C_q), 133.3 (C_q), 132.6 (C_q), 127.8 (C_q), 119.7 (CH), 118.9 (CH), 33.9 (CH_2), 19.2 (CH_2), 14.6 (CH_3), 12.7 (CH_3), 11.2 (CH_3), 9.6 (CH_3). Fluorescence data: λ_{ex} = 515 nm, λ_{em} = 519 nm, see Fig. 4.
8. Dissolve Me_4BodipyFL-OH **6** (306 mg, 0.95 mmol) in CH_2Cl_2 (60 mL) and add *N*-hydroxy succinimide (120 mg, 1.05 mmol), coupling reagent EDCI (201 mg, 1.05 mmol), and coupling catalyst DMAP (11 mg, 0.1 mmol). Stir the mixture for 16 h at room temperature and add 0.05 M HCl (40 mL) and separate the two layers. Dry the organic layer with $MgSO_4$ and concentrate in vacuo to dryness to obtain an orange solid.

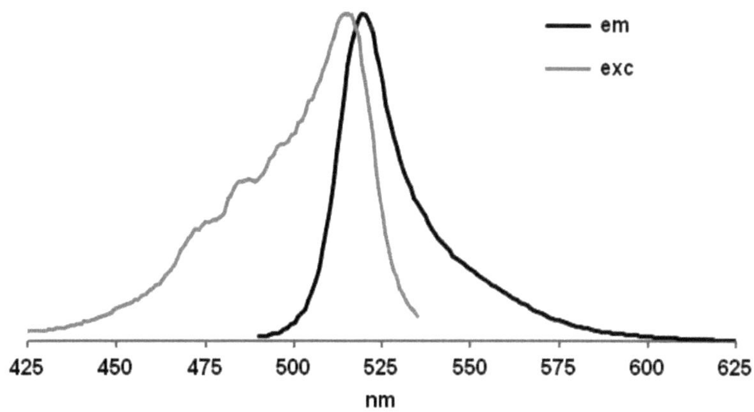

Fig. 4. Fluorescence spectrum of Me_4BodipyFL-OH measured in water, displaying $\lambda_{exc} = 515$ nm and $\lambda_{em} = 519$ nm. Derivatives of this dye display similar fluorescence spectra.

9. Purify by flash column chromatography using EtOAc:hexanes (2:3 v/v) as the eluent to obtain Me_4BodipyFL-NHS ester 7 as a dark orange solid (242 mg, 0.85 mmol) in 61% yield. ^{1}H NMR ($CDCl_3$): δ 7.02 (s, 1H), 6.02 (s, 1H), 2.85–2.78 (m, 6H), 2.75–2.70 (m, 2H), 2.50 (s, 6H), 2.22 (s, 3H), 2.19 (s, 3H). ^{13}C NMR ($CDCl_3$): δ 169.0 (2× CO), 167.7 (CO), 156.9 (C_q), 154.8 (C_q), 141.3 (C_q), 138.0 (C_q), 133.4 (C_q), 132.5 (C_q), 126.7 (C_q), 119.9 (CH), 119.0 (CH), 31.1 (CH_2), 25.6 (2× CH_2), 19.2 (CH_2), 14.6 (CH_3), 12.6 (CH_3), 11.2 (CH_3), 9.6 (CH_3). Fluorescence data: $\lambda_{ex} = 513$ nm, $\lambda_{em} = 517$ nm.

3.1.2. Synthesis of Me_4BodipyFL-Ahx_3Leu_3VS

1. The synthesis of Me4BodipyFL-Ahx3Leu3VS is depicted in Scheme 2 (see Note 3). Fmoc-L-Leu-PEG-polystyrene Wang resin (1 g, 0.17 mmol equivalents) (see Note 4) is subjected to four coupling cycles, in which deprotection of the Fmoc group with piperidine/NMP (1:4 v/v; 10 mL/g of dry resin) is followed by a coupling cycle with three equivalents of Fmoc-protected amino acid, three equivalents of DIPEA, and three equivalents of PyBop coupling reagent in NMP (10 mL/g of dry resin). After each step, the resin is thoroughly washed with NMP (5 × 10 mL). Coupling steps are performed with Fmoc-Leu-OH (1×) and Fmoc-Ahx-OH (3×), sequentially (see Note 5) (14).
2. After the final coupling step, remove the Fmoc group with piperidine/NMP (1:4 v/v), to afford solid supported **8**, and subsequently cleave the peptide from the resin by treating the resin with 100% TFA (10 mL) for 30 min. Coevaporate the cleavage mixture with toluene (40 mL) to obtain H_2N-Ahx_3Leu_2-OH 9 (100 mg, 0.17 mmol) in >90% purity as judged by LC/MS analysis. The product is used in the next step without further purification.

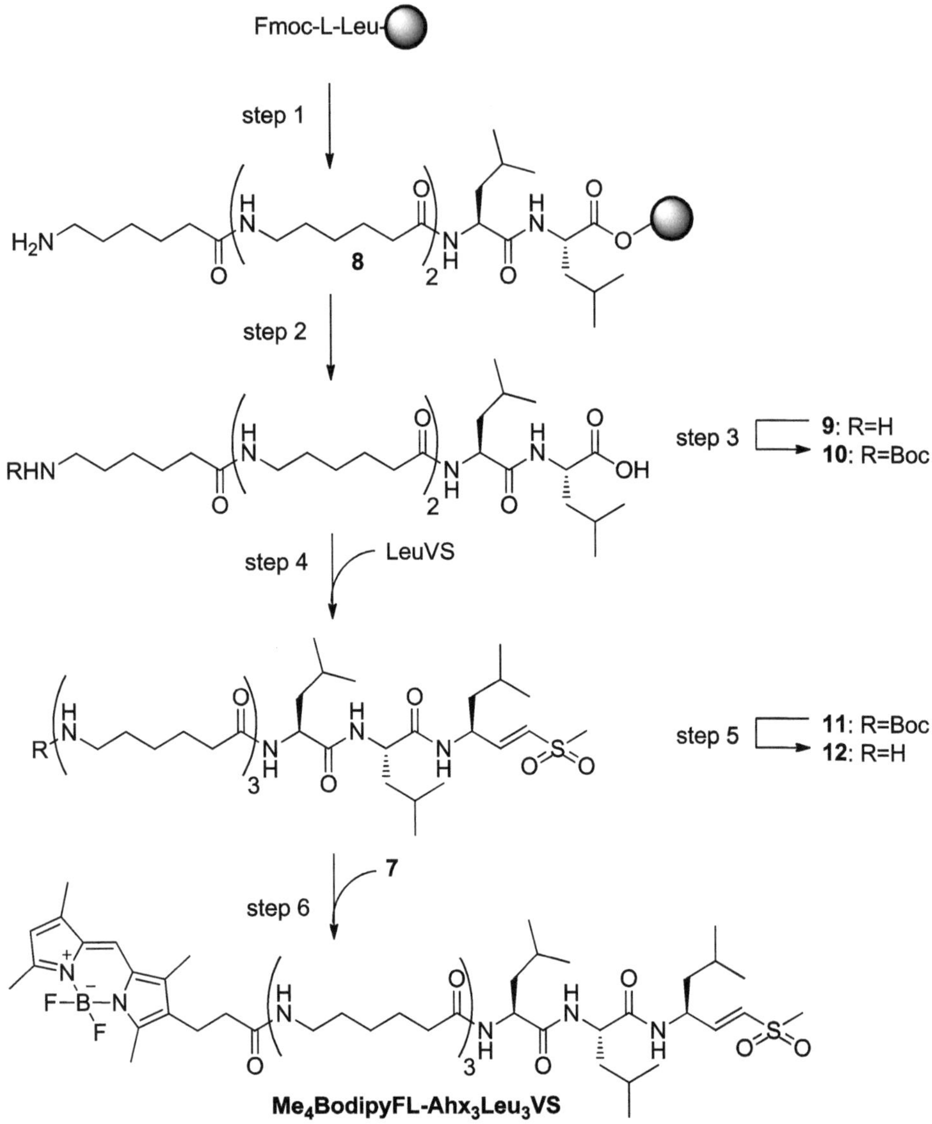

Scheme 2. Synthesis of $Me_4BodipyFL\text{-}Ahx_3Leu_3VS$. Step numbers correspond to step numbers outlined in Subheading 3.1.2.

3. Dissolve $H_2N\text{-}Ahx_3Leu_2\text{-}OH$ **9** (100 mg, 0.17 mmol) in DMF (2 mL) and add DIPEA (148 μL, 0.85 mmol) and di-*tert*-butyl-dicarbonate (44 mg, 0.20 mmol). Stir the resulting suspension for 16 h at room temperature and then concentrate in vacuo to dryness to yield an off-white solid. Purify by flash column chromatography using a gradient of 10–20% MeOH in CH_2Cl_2 containing 0.1% acetic acid as the eluent to obtain $Boc\text{-}Ahx_3Leu_2OH$ 10 as a white solid (116 mg, 0.17 mmol) in >98% yield.
4. L-Leucinyl vinyl sulfone (LeuVS) was prepared as reported (22, 25–27). Dissolve LeuVS (50 mg, 0.26 mmol) in DMF

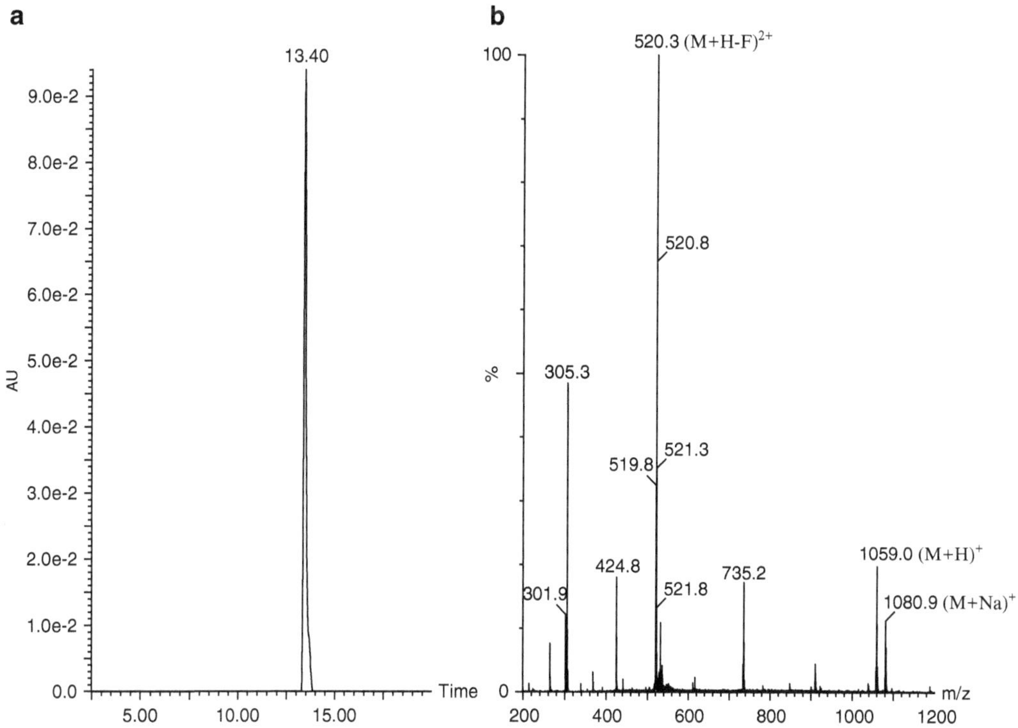

Fig. 5. Reversed-phase chromatogram (**a**) and MS spectrum (**b**) of Me_4BodipyFL-Ahx_3Leu_3VS. MS (ESI): 1,058.66 $(M+H)^+$; 520.27 $(M+H\text{-}F)^{2+}$; 1,080.91 $(M+Na)^+$.

and add Boc-Ahx_3Leu_2OH **10** (116 mg, 0.17 mmol), DIPEA (89 μL, 0.51 mmol), and coupling reagent EDCI (50 mg, 0.26 mmol). Stir the solution for 3 h at room temperature and then concentrate in vacuo to dryness to yield an off-white solid. Purify by flash column chromatography using 10% MeOH in CH_2Cl_2 as the eluent to obtain Boc-Ahx_3Leu_3VS **11** as a white solid (103 mg, 0.12 mmol) in 71% yield.

5. In a 50-mL Falcon tube, dissolve BocAhx_3Leu_3VS **11** (12 mg, 14 μmol) in TFA (1 mL) and leave at room temperature for 30 min. Precipitate the product by adding cold diethylether (40 mL). Isolate the product by centrifugation at $1{,}000 \times g$ (low brake speed) for 5 min at 4°C to afford H_2N-Ahx_3Leu_3VS **12** as a white solid (10.6 mg, 14 μmol) in >98% yield.

6. Dissolve H_2N-Ahx_3Leu_3VS **12** (10.6 mg, 14 μmol) in DMF (1 mL) and add Me_4BodipyFL-NHS ester 7 (6.4 mg, 15.5 μmol) and DIPEA (24 μL, 0.14 mmol) (see Note 6). Stir the resulting solution at room temperature under an argon atmosphere for 16 h. Concentrate the solution in vacuo to dryness and purify by reversed-phase HPLC to obtain the fluorescent proteasome probe Me_4BodipyFL-Ahx_3Leu_3VS (6 mg, 5.7 μmol) in 40% yield (see Fig. 5 for LC/MS data). Fluorescence data: $\lambda_{ex} = 515$ nm, $\lambda_{em} = 519$ nm.

3.2. Profiling of Proteasome Activity Using SDS-PAGE Based Assays

3.2.1. In Vitro Profiling of Proteasome Activity in Cell Lysates

1. Grow the cell line of choice at 37°C and 5% CO_2 in a humidified incubator in the appropriate medium supplemented with 10% FCS until log-phase (suspension cells) or until 80% confluency is reached (adherent cells). Medium may be supplemented with antibiotics.
2. Harvest suspension cells by centrifugation. To this end, transfer cells to a Falcon tube and centrifuge at 1,200 × *g* for 2 min at 4°C (see Note 7). Discard the supernatant and resuspend the cell pellet in 10–20 pellet volumes of PBS. Transfer cells to an Eppendorf tube and pellet cells by centrifugation at 1,200 × *g* for 2 min at 4°C. Discard the supernatant. Harvest adherent cells by trypsinization. To this end, aspirate the medium, wash cells with PBS and aspirate. Add just enough trypsin to cover the cells. As soon as cells detach, add medium containing 10% FCS to the cells to inactivate the trypsin and transfer the cells to a Falcon tube. Pellet cells by centrifugation at 1,200 × *g* for 2 min at 4°C. Discard the supernatant and resuspend the cell pellet in 10–20 pellet volumes of PBS. Transfer cells to an Eppendorf tube and pellet cells by centrifugation at 1,200 × *g* for 2 min at 4°C. Discard the supernatant.
3. Resuspend cells in 2 pellet volumes of cold HR buffer.
4. Lyse cells mechanically (e.g., by sonication using the Bioruptor according to the manufacturer's instructions). **Critical**: For proper labeling, do not add detergent to the lysis buffer. If detergent is absolutely required for proper cell lysis, keep the detergent concentration as low as possible.
5. Centrifuge for at 14,000 × *g* for 3 min at 4°C to remove membrane fractions and cell debris and transfer the supernatant to a fresh Eppendorf tube.
6. Determine protein concentrations using a Bradford assay. For incubation with proteasome inhibitors prior to proteasome labeling, proceed to step 7. To label the proteasome directly, proceed to step 9. **Critical**: Proceed with labeling directly and do not store nonincubated lysates. The quality of labeling will decrease if lysates are freeze–thawed.

   Option 1: Incubation with proteasome inhibitors followed by labeling with probe.

7. To study the effects of proteasome inhibitors, transfer 25 μg of lysate to a fresh Eppendorf tube and adjust the volume to 24.5 μL with HR buffer. Add 0.5 μL of a 50× stock solution of the desired proteasome inhibitor in DMSO to obtain the desired 1× concentration of proteasome inhibitor and a final protein concentration of 1 μg/μL. Include a reference sample to which 0.5 μL DMSO but no proteasome inhibitor is added. Vortex and incubate the samples for the desired time period at 37°C.

8. Add 0.5 μL of a 50 μM Me_4BodipyFL-Ahx_3Leu_3VS stock solution in DMSO to the samples to obtain a final concentration of 1 μM Me_4BodipyFL-Ahx_3Leu_3VS. Vortex and incubate for 1 h at 37°C. Proceed to Subheading 3.2.4.

   Pause point: At this point, incubated lysates can be snap-frozen in liquid N_2 and stored at −20°C until further use.

   Option 2: Incubation with probe only.

9. To label proteasome subunits directly, transfer 25 μg lysate to a fresh Eppendorf tube and adjust the volume to 24.5 μL with HR buffer. Add 0.5 μL of a 50 μM Me_4BodipyFL-Ahx_3Leu_3VS stock solution in DMSO to obtain a final probe concentration of 1 μM and a final protein concentration of 1 μg/μL. Vortex and incubate the sample for 1 h at 37°C. Proceed to Subheading 3.2.4.

   Pause point: At this point, incubated lysates can be snap-frozen in liquid N_2 and stored at −20°C until further use.

3.2.2. Labeling of Active Proteasome Subunits in Living Cells

Labeling of active proteasome subunits can be performed in suspension cells (proceed to step 1) or adherent cells (proceed to step 5).

Suspension cells

1. Grow the cell line of choice in appropriate medium containing 10% FCS at 37°C and 5% CO_2 in a humidified incubator until log-phase is reached. Medium may be supplemented with antibiotics.
2. Count cells, transfer cells to a Falcon tube, and pellet cells by spinning at 1,200 × *g* for 2 min at room temperature. Discard the supernatant.
3. Resuspend cells in fresh medium at 0.5×10^6–1.0×10^6 cells/mL.
4. Add 0.5 mL of cell suspension (0.25×10^6–0.5×10^6 cells) to each well of a 24-well plate (or as many wells as needed). For incubation with proteasome inhibitors prior to proteasome labeling, proceed to step 7. To label proteasome only, directly proceed to step 8.

Adherent cells

5. Grow the cell line of choice in a 24-well plate in appropriate medium supplemented with 10% FCS at 37°C and 5% CO_2 in a humidified incubator until 80% confluency is reached. Medium may be supplemented with antibiotics.
6. Aspirate the medium and add 0.5 mL of fresh medium to the cells. For incubation with proteasome inhibitors prior to proteasome labeling, proceed to step 7. To label proteasome only, directly proceed to step 8.

7. To study the effects of proteasome inhibitors, add 1 μL of a 500× stock solution of the desired proteasome inhibitor in DMSO to each well to obtain the desired 1× concentration of proteasome inhibitor. Include a reference sample to which 1 μL DMSO but no proteasome inhibitor is added. Incubate the cells for the desired time period in an incubator.
8. Add 5 μL of a 50 μM Me_4BodipyFL-Ahx_3Leu_3VS stock solution in DMSO to each well to obtain a final probe concentration of 500 nM. Incubate the cells for 1 h in an incubator. If cells do not tolerate 1% DMSO, 1 μL of a 250 μM Me_4BodipyFL-Ahx_3Leu_3VS stock solution in DMSO can be added alternatively.
9. To block all remaining proteasome activity and prevent postlysis labeling events, add 1 μL of a 5 mM MG132 stock solution in DMSO to each well to obtain a final MG132 concentration of 10 μM. Incubate the cells for 1 h in an incubator. Step 9 can be omitted if no differences in labeling are observed between samples that are incubated with MG132 before cell harvest and samples that are harvested directly after step 8.
10. To harvest suspension cells: transfer the cells in each well to an Eppendorf tube and pellet cells by centrifugation at 1,200 × *g* for 2 min at 4°C. Discard the supernatant and resuspend the cell pellet in 1 mL PBS. Pellet cells by centrifugation at 1,200 × *g* for 2 min at 4°C. Discard the supernatant. To harvest adherent cells: aspirate the medium, wash the cells by adding 1 mL of PBS to each well, and aspirating the PBS. Add 50 μL trypsin to each well. As soon as cells detach, add 1 mL of fresh medium containing 10% FCS to the cells to inactivate the trypsin. Transfer the cells in each well to an Eppendorf tube and pellet cells by centrifugation at 1,200 × *g* for 2 min at 4°C. Discard the supernatant and resuspend the cell pellet in 1 mL PBS (see Note 7). Pellet the cells by centrifugation at 1,200 × *g* for 2 min at 4°C. Discard the supernatant.

    Pause point: At this point, incubated cell pellets can be snap-frozen in liquid N_2 and stored at −20°C until further use.
11. Resuspend cells in 1–2 pellet volumes of cold NP40 lysis buffer and lyse for 30 min at 4°C (see Note 7).
12. Centrifuge at 14,000 × *g* for 3 min at 4°C to remove membrane fractions and cell debris. Transfer the supernatant to fresh Eppendorf tube.
13. Determine protein concentrations using a Bradford assay.
14. Transfer 25 μg lysate to a fresh Eppendorf tube and adjust the final volume to 25 μL with NP40 lysis buffer to obtain a final protein concentration of 1 μg/μL. Proceed to Subheading 3.2.4.

    Pause point: At this point, cell lysates can be snap-frozen in liquid N_2 and stored at −80°C until further use.

3.2.3. Ex Vivo Profiling of Proteasome Subunit Activity in Murine Tissues

1. Remove the tissue types to be analyzed and rinse with PBS.
2. Grind the tissue, transfer to an Eppendorf tube, and add 1–2 volumes of cold HR buffer.
3. Add 1 volume of glass beads (≤106 μm, acid washed, Sigma) to the tissue. Lyse cells mechanically by vortexing at high speed for 45 min at 4°C (see Note 7). **Critical**: For proper labeling, do not add detergent to the lysis buffer. If detergent is absolutely required for proper cell lysis, keep the detergent concentration as low as possible.
4. Remove beads, membrane fractions, and cell debris by centrifugation at 14,000 × *g* for 5 min at 4°C and transfer the supernatant to a fresh Eppendorf tube.
5. Determine protein concentrations using a Bradford assay. **Critical**: Proceed with labeling directly and do not store non-incubated lysates. The quality of labeling will decrease if lysates are freeze–thawed.
6. Transfer 25 μg lysate to a fresh Eppendorf tube and adjust the volume to 24.5 μL with HR buffer. Add 0.5 μL of a 50 μM Me_4BodipyFL-Ahx_3Leu_3VS stock solution in DMSO to obtain a final probe concentration of 1 μM and a final protein concentration of 1 μg/μL. Vortex and incubate the samples for 1 h at 37°C. Proceed to Subheading 3.2.4.

   Pause point: At this point, incubated lysates can be snap-frozen in liquid N_2 and stored at −20°C until further use.

3.2.4. Gel Electrophoresis and In-Gel Fluorescence Readout

1. These instructions assume the use of the NuPAGE precast gel system (Invitrogen) and precast mini gels to separate proteins on SDS-PAGE. If a mini gel does not separate the individual proteasome subunits sufficiently, a larger gel system can be used, e.g., the PROTEAN II xi Cell system from Biorad (16 × 20 cm glass plates, 12.5% separating gel, 4% stacking gel.) When using this system, increase the sample volume to 50 μL per sample. Do not change the final concentrations of proteasome inhibitors, protein, and probe in the samples. Load 20–30 μL sample per well and run at 10 mA for 16 h. Subsequently, increase the current to 35 mA, wait until the blue front has run off the gel and run for another hour before removing the gel from the system. The protocol described below can also be adapted to other gel systems. Use common protocols for denatured samples. Separate proteins using a 12 or 12.5% separating gel and a 4% stacking gel.
2. Add 12.5 μL 3× reducing sample buffer (see Subheading 2.2.4 for buffer recipe) to each sample obtained in Subheadings 3.2.1–3.2.3. The volume of these samples should be 25 μL. Vortex and denature by boiling the sample for 10 min at 70°C. Centrifuge at 14,000 × *g* for 1 min at room temperature.

Denatured samples can be stored at −20°C for later use. **Critical**: When using the NuPAGE gel system also use the NuPAGE LDS Sample buffer to prepare the 3× reducing sample buffer. The use of a different reducing sample buffer leads to improper running of the gel, resulting in fuzzy and unfocussed bands.

3. Assemble the NuPAGE gel unit using a precast NuPAGE 12% Bis-Tris gel according to the manufacturer's instructions. Add 1× MOPS buffer to both the inner and outer chamber of the gel unit. Add 125 μL antioxidant to the inner gel chamber only. Load 10 μL of denatured sample per well. Keep one well free and load this well with 6.5 μL prestained molecular weight marker (e.g., SeeBlue® Plus2 Pre-Stained Standard from Invitrogen). Load 3 μL of 3× reducing sample buffer to any remaining wells.
4. Run the gel at 170–180 V. **Critical**: For proper separation of the β1i and β5 subunits, run the gel until the15 kDa protein in the molecular weight marker (Lysozyme in the SeeBlue® Plus2 Pre-Stained Standard) is at the bottom of the gel.
5. Remove the gel from the cassette and image the wet gel slab for 10–120 s using a fluorescence imager containing an appropriate filter set (excitation at 480 nm, emission at 530 nm).
6. Analyze images using appropriate software to quantify fluorescence intensities.

3.3. High-Throughput FACS-Based Proteasome Activity Assay

1. Seed MelJuso cells in a 96-well flat bottom tissue culture plate in DMEM supplemented with 10% FCS and antibiotics. Add 10,000 cells in a total volume of 100 μL per well (see Note 8).
2. Place cells at 37°C and 5% CO_2 in a humidified incubator and let cells attach for 16–24 h.
3. To screen compounds for their effects on proteasome activity, add 1 μL of 2.5 mM stock solutions of the desired compounds in DMSO to 99 μL medium to obtain 5× compound solutions. To obtain 5× control solutions, add 1 μL of a 0.4 mM stock solution of MG132 in DMSO (positive control, 100% inhibition) or 1 μL DMSO only (negative control, 0% inhibition) to 99 μL medium.
4. Remove medium from the cells, add 80 μL of fresh medium to each well and 20 μL of the 5× solutions of compounds and controls obtained in step 3 (then the final concentration of compound in each well is 5 μM, final concentration of MG132 is 800 nM). Incubate cells for 16 h in an incubator.
5. Make a 42× dilution in DMEM of a 50 μM Me_4BodipyFL-Ahx_3Leu_3VS DMSO stock solution to obtain a 1.2 μM probe solution (e.g., for one 96-well plate, add 48 μL of a 50 μM

Me_4BodipyFL-Ahx_3Leu_3VS DMSO stock solution to 1,952 μL DMEM). Add 20 μL of the 1.2 μM probe solution to each well to obtain a final probe concentration of 200 nM. Incubate cells for 2 h in an incubator.

6. Discard the supernatant, add 200 μL PBS to each well to wash the cells, and discard the PBS.
7. Harvest cells. Add 20 μL trypsin to each well. As soon as cells detach, add 55 μL FACS buffer and 25 μL fixation buffer to each well to obtain a final formaldehyde concentration of 0.5% (see Subheading 2.3 for buffer recipes). Fix cells by shaking the plate on a shaker for at least 20 min.

   Pause point: At this point, fixed cells can be stored at 4°C until further use.
8. Measure intracellular fluorescence in the cells by flow cytometry. Fluorescence is measured in the Fl1 channel (530/30 filter).

4. Notes

1. All buffers are prepared in water that has a resistivity of 18.2 MΩ cm (MilliQ water).
2. Store stock solutions of ATP and DTT in water in aliquots at -20°C.
3. A shorter synthetic route to the synthesis of Me_4BodipyFL-Ahx_3Leu_3VS involves the use of a hyper acid-labile resin and introduction of the Me_4BodipyFL fluorophore on solid phase. This route, however, requires larger amounts of fluorophore NHS ester. Synthesize the peptide as described in step 1, starting from preloaded hyper acid-labile resin. After the final coupling step, remove the Fmoc group and wash the resin thoroughly. Subsequently, resuspend the resin in NMP (10 mL/g of resin) and add five equivalents of Me_4BodipyFL-NHS ester and five equivalents of DIPEA and allow the reaction to proceed for 16 h. Wash the resin thoroughly and cleave the peptide from the resin using 2% TFA in CH_2Cl_2 or 25% HFIP (or according to protocol of the manufacturer of the hyper acid-labile resin) in CH_2Cl_2 and allow the cleavage to take place for 20 min. Precipitate the product by adding 10 volume equivalents of cold diethylether/pentane (3:1 v/v) and isolate by centrifugation for 5 min at 1,000 × *g* (low brake speed). Wash the pellet three times by adding cold diethylether/pentane (3:1 v/v) and centrifuging at 800 × *g* (high brake speed). Continue with step 4 to obtain the final probe. Steps 5 and 6 are omitted in this procedure.

4. The use of low capacity and preloaded PEG polystyrene resin is recommended.

5. To circumvent the introduction of a Boc group on the N-terminus in solution as described in step 3, it is also possible to use Boc-6-aminohexanoic acid (Boc-Ahx-OH) instead of Fmoc-Ahx-OH, as a building block in the last coupling cycle. Synthesize the peptide as described in step 1, starting from preloaded hyper acid-labile resin. Couple Boc-Ahx-OH to the resin in the final coupling step. Wash the resin thoroughly and cleave the peptide from the resin using 2% TFA in CH_2Cl_2. Concentrate the compound under reduced pressure. Purify the compound as described in step 3 and continue with step 4.

6. Other commercially available NHS-activated dyes may also be used here, circumventing the synthesis of the Me_4BodipyFL-NHS ester. We recommend the use of BodipyFL-NHS ester or BodipyTMR-NHS ester (Invitrogen) for optimal labeling results in both cell lysates and intact cells (14).

7. Cell lysates or samples containing protease activity were kept on ice unless indicated otherwise.

8. This protocol describes a high-throughput FACS-based proteasome activity assay in a 96-well format using MelJuso cells. The protocol can easily be adapted to 384-well or 24-well format by changing the amounts of cells and volumes described in Section 3.3. Do not change the final concentrations of compounds and probe. The protocol can be performed with other adherent cells and can be adapted to suspensions cells. When using suspension cells, perform wash steps by centrifugation (1,200 × *g* for 5 min, then discard the supernatant and resuspend cells).

Acknowledgements

Henk Hilkmann is acknowledged for peptide synthesis and the authors thank Dr Kees Jalink for assistance with fluorescence spectroscopy. C.R.B. is supported by KWF grant 2005-3368.

References

1. Richardson, P. G., Mitsiades, C., Hideshima, T., and Anderson, K. C. (2006) Bortezomib: proteasome inhibition as an effective anti-cancer therapy. *Annu. Rev. Med* **57**, 33–47.
2. Etlinger, J. D. and Goldberg, A. L. (1977) A soluble ATP-dependent proteolytic system responsible for the degradation of abnormal proteins in reticulocytes. *Proc. Natl Acad. Sci. USA* **74**, 54–58.
3. Glickman, M. H. and Ciechanover, A. (2002) The ubiquitin-proteasome proteolytic pathway: destruction for the sake of construction. *Physiol Rev.* **82**, 373–428.

4. Concannon, C. G., Koehler, B. F., Reimertz, C., Murphy, B. M., Bonner, C., Thurow, N. et al. (2007) Apoptosis induced by proteasome inhibition in cancer cells: predominant role of the p53/PUMA pathway. *Oncogene* **26**, 1681–1692.
5. Hideshima, T., Richardson, P., Chauhan, D., Palombella, V. J., Elliott, P. J., Adams, J. et al. (2001) The proteasome inhibitor PS-341 inhibits growth, induces apoptosis, and overcomes drug resistance in human multiple myeloma cells. *Cancer Res* **61**, 3071–3076.
6. Kane, R. C., Bross, P. F., Farrell, A. T., and Pazdur, R. (2003) Velcade: U.S. FDA approval for the treatment of multiple myeloma progressing on prior therapy. *Oncologist* **8**, 508–513.
7. Kane, R. C., Dagher, R., Farrell, A., Ko, C. W., Sridhara, R., Justice, R. et al. (2007) Bortezomib for the treatment of mantle cell lymphoma. *Clin Cancer Res* **13**, 5291–5294.
8. Stohwasser, R., Standera, S., Peters, I., Kloetzel, P. M., and Groettrup, M. (1997) Molecular cloning of the mouse proteasome subunits MC14 and MECL-1: reciprocally regulated tissue expression of interferon-gamma-modulated proteasome subunits. *Eur J Immunol.* **27**, 1182–1187.
9. Dahlmann, B., Ruppert, T., Kuehn, L., Merforth, S., and Kloetzel, P. M. (2000) Different proteasome subtypes in a single tissue exhibit different enzymatic properties. *J Mol Biol* **303**, 643–653.
10. Dahlmann, B., Ruppert, T., Kloetzel, P. M., and Kuehn, L. (2001) Subtypes of 20S proteasomes from skeletal muscle. *Biochimie* **83**, 295–299.
11. Elliott, P. J., Soucy, T. A., Pien, C. S., Adams, J., and Lightcap, E. S. (2003) Assays for proteasome inhibition. *Methods Mol Med* **85**, 163–172.
12. Lightcap, E. S., McCormack, T. A., Pien, C. S., Chau, V., Adams, J., and Elliott, P. J. (2000) Proteasome inhibition measurements: clinical application. *Clin Chem* **46**, 673–683.
13. Berkers, C. R., Verdoes, M., Lichtman, E., Fiebiger, E., Kessler, B. M., Anderson, K. C. et al. (2005) Activity probe for in vivo profiling of the specificity of proteasome inhibitor bortezomib. *Nat. Methods* **2**, 357–362.
14. Berkers, C. R., van Leeuwen, F. W., Groothuis, T. A., Peperzak, V., van Tilburg, E. W., Borst, J. et al. (2007) Profiling proteasome activity in tissue with fluorescent probes. *Mol Pharm.* **4**, 739–748.
15. Bogyo, M., Shin, S., McMaster, J. S., and Ploegh, H. L. (1998) Substrate binding and sequence preference of the proteasome revealed by active-site-directed affinity probes. *Chem Biol* **5**, 307–320.
16. Kessler, B. M., Tortorella, D., Altun, M., Kisselev, A. F., Fiebiger, E., Hekking, B. G. et al. (2001) Extended peptide-based inhibitors efficiently target the proteasome and reveal overlapping specificities of the catalytic beta-subunits. *Chem Biol* **8**, 913–929.
17. Ovaa, H., van Swieten, P. F., Kessler, B. M., Leeuwenburgh, M. A., Fiebiger, E., van den Nieuwendijk, A. M. et al. (2003) Chemistry in living cells: detection of active proteasomes by a two-step labeling strategy. *Angew Chem Int Ed Engl* **42**, 3626–3629.
18. Verdoes, M., Florea, B. I., Menendez-Benito, V., Maynard, C. J., Witte, M. D., van der Linden, W. A. et al. (2006) A fluorescent broad-spectrum proteasome inhibitor for labeling proteasomes in vitro and in vivo. *Chem Biol* **13**, 1217–1226.
19. Dantuma, N. P., Lindsten, K., Glas, R., Jellne, M., and Masucci, M. G. (2000) Short-lived green fluorescent proteins for quantifying ubiquitin/proteasome-dependent proteolysis in living cells. *Nat Biotechnol* **18**, 538–543.
20. Lindsten, K., Menendez-Benito, V., Masucci, M. G., and Dantuma, N. P. (2003) A transgenic mouse model of the ubiquitin/proteasome system. *Nat Biotechnol* **21**, 897–902.
21. Luker, G. D., Pica, C. M., Song, J., Luker, K. E., and Piwnica-Worms, D. (2003) Imaging 26S proteasome activity and inhibition in living mice. *Nat Med* **9**, 969–973.
22. Bogyo, M., McMaster, J. S., Gaczynska, M., Tortorella, D., Goldberg, A. L., and Ploegh, H. (1997) Covalent modification of the active site threonine of proteasomal beta subunits and the Escherichia coli homolog HslV by a new class of inhibitors. *Proc Natl Acad Sci USA* **94**, 6629–6634.
23. Smith, K. M. and Pandey, R. K. (1983) New Efficient Total Syntheses of Derivatives of Protoporphyrin-IX Bearing Deuturated Methyl Groups. **20**, 1383–1388.
24. Still, W. C., Kahn, M., and Mitra, A. (1978) Rapid Chromatographic Technique for Preparative Separations with Moderate Resolution. *Journal of Organic Chemistry* **43**, 2923–2925.
25. Palmer, J. T., Rasnick, D., Klaus, J. L., and Bromme, D. (1995) Vinyl sulfones as mechanism-based cysteine protease inhibitors. *J Med Chem* **38**, 3193–3196.

26. Rydzewski, R., Burrill, L., Mendonca, R., Palmer, J., Rice, M., Tahilramani, R. et al. (2006) Optimization of Subsite Binding to the 5 Subunit of the Human 20S Proteasome Using Vinyl Sulfones and 2-Keto-1,3,4-oxadiazoles: Syntheses and Cellular Properties of Potent, Selective Proteasome Inhibitors. *Journal of Medicinal Chemistry* **49**, 2953–2968.
27. Wang, G., Mahesh, U., Chen, G. Y., and Yao, S. Q. (2003) Solid-phase synthesis of peptide vinyl sulfones as potential inhibitors and activity-based probes of cysteine proteases. *Org Lett* **5**, 737–740.

Chapter 14

Chemical Cross-Linking and High-Resolution Mass Spectrometry to Study Protein–Drug Interactions

Mathias Q. Müller and Andrea Sinz

Abstract

This method describes the combination of chemical cross-linking and high-resolution mass spectrometry for analyzing conformational changes in target proteins that are induced by drug binding. Our approach is exemplified for detecting conformational changes within the peroxisome proliferator-activated receptor alpha upon binding of low-molecular weight compounds, proving that our strategy provides a basis to efficiently characterize target protein–drug interactions.

Key words: Protein–drug interaction, Chemical cross-linking, Isotope labeling, Mass spectrometry

1. Introduction

The rate of drug discovery is greatly dependent on the development and improvement of rapid and reliable analytical methods for screening target protein–drug interactions. In those cases where high-resolution methods for structural analysis are applicable, such as X-ray crystallography and NMR spectroscopy, the solved three-dimensional structure of a protein with its bound ligand gives insights into stable interactions within the complex. Theoretical modeling might reveal further interactions using the known three-dimensional structure as a starting point.

We describe an alternative approach for analyzing protein–drug interactions based on building up a set of structurally defined interactions by covalently connecting pairs of functional groups within a protein using chemical cross-linking. Recently, chemical cross-linking of proteins has been combined with a proteolytic digestion of the protein and a subsequent mass spectrometric analysis of the created cross-linked peptides (1–5).

Gerard Drewes and Marcus Bantscheff (eds.), *Chemical Proteomics: Methods and Protocols*, Methods in Molecular Biology, vol. 803, DOI 10.1007/978-1-61779-364-6_14,

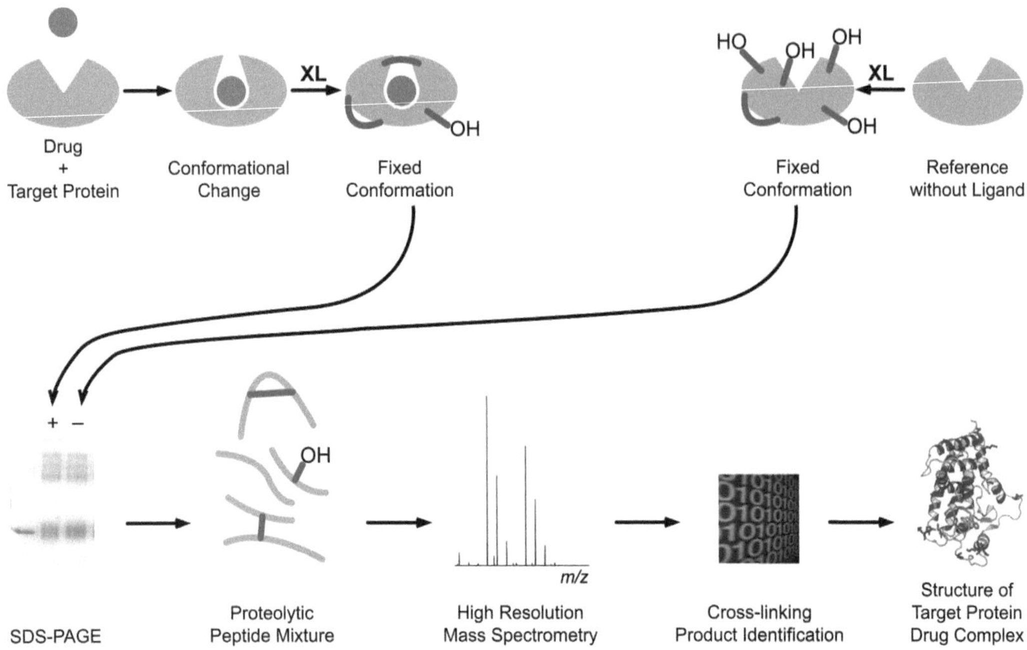

Fig. 1. Analytical strategy for analyzing conformational changes in proteins upon ligand binding by chemical cross-linking and high-resolution mass spectrometry.

This strategy shown in Fig. 1 is demonstrated using the example for the peroxisome proliferator-activated receptor alpha (PPARα) in the presence of an antagonist, an agonist and free PPARα, respectively, as negative control (6). The peroxisome proliferator-activated receptors (PPARs) are ligand-activated transcription factors that belong to the nuclear receptor protein family. Three subtypes of PPARs (α, δ (or β, Nuc-1), and γ) have been identified so far (7). PPARs are activated by fatty acids and eicosanoids and are also targets for antidiabetic and antidyslipidemic drugs (8–10).

The cross-linking reactions are conducted using isotope-labeled amine-reactive *N*-hydroxysuccinimide (NHS) esters *Bis*(sulfosuccinimidyl)glutarate (BS^2G) and *Bis*(sulfosuccinimidyl) suberate (BS^3) bridging distances of approximately 7.7 and 11.4 Å, which are employed as 1:1 mixture of nondeuterated (D_0) and four times (D_4) deuterated derivatives in order to facilitate a subsequent MS identification of cross-linking products (11, 12). The reaction mechanism is shown in Fig. 2.

Cross-linked target proteins are separated by one-dimensional gel electrophoresis (SDS-PAGE, Fig. 3). Gel bands of interest are excised from the gel and *in-gel* digested with trypsin and GluC. Extracted peptide mixtures are separated by Nano-HPLC online coupled to a nano-ESI-LTQ-Orbitrap mass spectrometer.

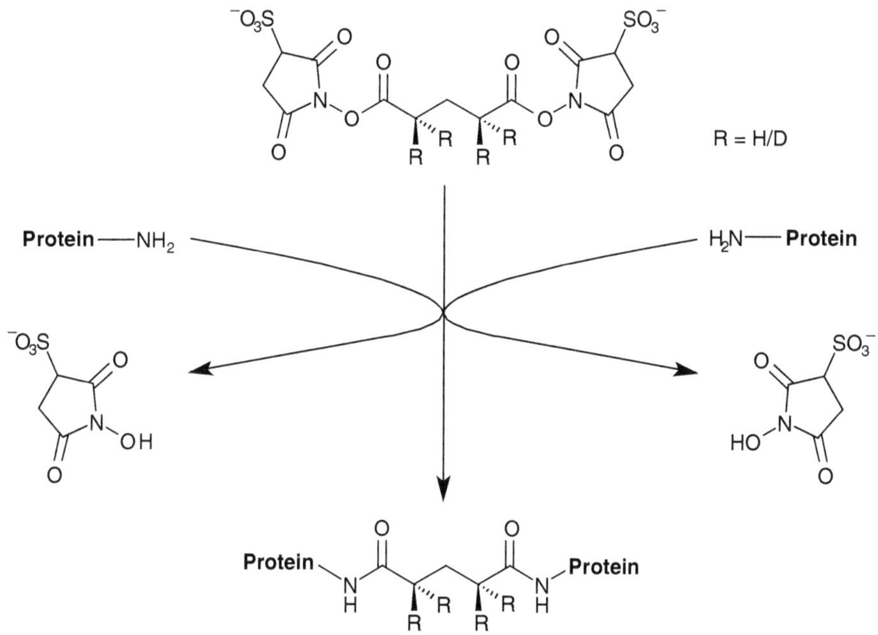

Fig. 2. Reaction mechanism of homobifunctional cross-linking with *N*-hydroxysuccinimide (NHS) esters.

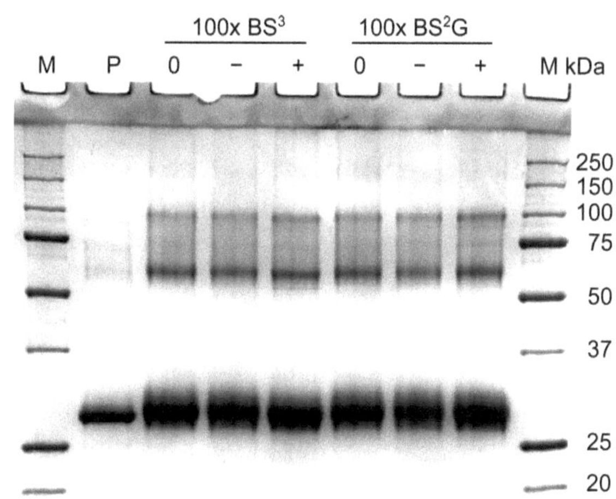

Fig. 3. Separation of cross-linking reaction mixtures (60-min reaction time, 200-pmol protein per lane) by SDS-PAGE, exemplified for peroxisome proliferator-activated receptor alpha (PPARα)/ligand complexes (6). M: molecular weight marker; P: PPARα before cross-linking experiments (13); 0: cross-linking in the absence of ligands; –: cross-linking in the presence of an antagonist GW6471 (14); +: cross-linking in the presence of the putative agonist YS81 (15).

The location of the created cross-links within the protein analyzed by high-resolution mass spectrometry allows drawing conclusions on the altered distance geometries and as a consequence, on conformational changes within a protein that are induced by drug binding.

2. Materials

2.1. Chemical Cross-Linking

1. Cross-linking reagents: BS^3-D_0, BS^3-D_4, BS^2G-D_0, and BS^2G-D_4 (store dry under argon at −20°C, Thermo Fisher Scientific, formerly Pierce, see Note 1).
2. Neat dimethylsulfoxide (DMSO, see Note 2).
3. Cross-linking buffer: 20 mM HEPES buffer, pH 7.4 (see Note 3).
4. Quenching solution: 1 M ammonium bicarbonate.
5. Protein of interest stored in cross-linking buffer, concentration 10 μM.
6. 5 mM ligand stock solution (dissolved in H_2O, DMSO, or MeOH).

2.2. SDS-Polyacrylamide Gel Electrophoresis

1. Milli-Q water (see Note 4).
2. Resolving gel: 1.5 M Tris–HCl, pH 8.8.
3. Stacking gel: 0.5 M Tris–HCl, pH 6.8.
4. 40% (v/v) acrylamide/bisacrylamide solution [37.5:1 (v/v)] and *N,N,N′,N′*-tetramethylethylenediamine (TEMED, both VWR, stored at 4°C).
5. Laemmli buffer and 10% (v/v) SDS solution (store at room temperature, both Bio-Rad).
6. Ammonium persulfate (APS, Sigma): prepare 10% (w/v) solution in Milli-Q water and immediately freeze in single use (100 μL) aliquots at −20°C.
7. Running buffer: 10× concentrated TGS (Tris–glycin–SDS) buffer (Roth, store at room temperature).
8. Unstained molecular weight markers (SM0661, Bio-Rad).
9. Colloidal Coomassie staining solution (16): Solution A: 5 % (w/v) Coomassie Brilliant Blue G250 suspended in Milli-Q H_2O. Solution B: 20 g/L *ortho*-phosphoric acid, 100 g/L ammonium sulfate, dissolved in Milli-Q H_2O (see Note 5).
10. Fixing solution: 40% (v/v) methanol, 10% (v/v) acetic acid.

2.3. Proteolytic In-Gel Digestion

1. Trypsin sequencing grade (Roche Diagnostics). Storage: Dissolve 25 μg lyophilized trypsin in 50 μL 1 mM HCl, prepare aliquots containing 1 μg/2 μL trypsin, quick-freeze in liquid nitrogen, and store at −80°C.
2. GluC sequencing grade (Roche Diagnostics, see Note 6).
3. Destaining solutions: Acetonitrile (ACN) LiChroSolv® (VWR), 50% (v/v) ACN, 100 mM ammonium bicarbonate (Sigma) (see Note 7).

4. Reducing buffer: Freshly prepared 10 mM dithiothreitol (DTT) in 100 mM ammonium bicarbonate.
5. Alkylating buffer: Freshly prepared 55 mM iodoacetamide (Sigma) in 100 mM ammonium bicarbonate.
6. Acetonitrile (ACN) LiChroSolv® (VWR).
7. Trifluoroacetic acid (TFA, Sigma).
8. Digestion buffer, freshly prepared: dissolve one trypsin (or GluC) aliquot in 80 μL 50 mM ammonium bicarbonate, store on ice.
9. Extraction solution: 50% (v/v) ACN, 5% (v/v) TFA.

2.4. Nano-HPLC/Nano-ESI-LTQ-Orbitrap Mass Spectrometry

1. Instrumentation: Ultimate nano-HPLC system (Dionex Corp., Idstein, Germany) equipped with reversed phase C18 columns (precolumn: Acclaim PepMap, 300 μm × 5 mm, 5 μm, 100 Å, separation column: Acclaim PepMap, 75 μm × 150 mm, 5 μm, 100 Å, Dionex Corp., Idstein, Germany). The nano-HPLC system is directly coupled to the nano-ESI source (Proxeon, Odense, Denmark) of an LTQ-Orbitrap XL hybrid mass spectrometer (Thermo Fisher Scientific, Bremen, Germany).
2. Solvent A: 5% (v/v) acetonitrile (ACN), 0.1% (v/v) formic acid (FA) prepared in Milli-Q H_2O (see Note 4).
3. Solvent B: 80% (v/v) ACN, 0.1% (v/v) FA.

2.5. Software for Data Analysis

1. Data acquisition: XCalibur 2.0.7 (Thermo Fisher Scientific) in combination with DCMS link 2.0 (Dionex).
2. Preparing mass lists: Proteome Discoverer 1.0 (Thermo Fisher Scientific).
3. Identification of cross-linking products: GPMAW (General Protein Mass Analysis for Windows, version 8.2, Lighthouse Data, Odense, Denmark, http://www.gpmaw.com) (17).

3. Methods

3.1. Chemical Cross-Linking

1. Prepare an equimolar mixture of target protein and drug (ranging from 2 to 10 μM each, final concentration) in the cross-linking buffer (e.g., 20 mM HEPES buffer) in case a 1:1 complex is studied. For equilibration of the protein/drug complex, the mixture is incubated for 30 min at room temperature using an incubation shaker. For each reaction condition at varying protein/ligand concentrations, cross-linking reagents or cross-linker excesses, a new Eppendorf tube should be used. A final volume of 200 μL per reaction condition is adequate.
2. Freshly prepare cross-linker stock solutions (200 mM) in neat DMSO.

3. Add 1:1 mixtures of nondeuterated (D_0) and four times deuterated (D_4) cross-linkers BS^3 and BS^2G (see Note 8), respectively, in 50 and 100 M excess of cross-linker over the target protein/ligand complex. Allow the reaction to proceed for 30 and 60 min (see Note 9).
4. For reaction quenching, remove a 200-μL aliquot from the cross-linking reaction mixture after 30 min, transfer it into a new Eppendorf tube, and add 4 μL of 1 M ammonium bicarbonate solution (20 mM final concentration). The remaining cross-linking reaction mixture is quenched in the same fashion after 60 min.
5. For storage, quick-freeze the samples in liquid nitrogen and store at −20°C before MS analyses are performed.

3.2. SDS-PAGE

1. The instructions for SDS-PAGE are based on the use of a Bio-Rad Mini-PROTEAN® 3 cell or Mini-PROTEAN® Tetra cell system.
2. Prepare two 0.75-mm resolving gels (12%) by mixing 3 mL acrylamide/bisacrylamide solution, 2.5 mL 1.5 M Tris–HCl, pH 8.8, 100 μL 10% (w/v) SDS, and adding Milli-Q H_2O to a total volume of 10 mL. Start the polymerization reaction by adding 10 μL TEMED and 50 μL 10% (w/v) APS. Pour the gel into the tray, leaving space for a stacking gel, and overlay with Milli-Q H_2O. The gel should be polymerized after 30 min.
3. Remove the water and rinse the top of the gel twice with Milli-Q H_2O.
4. Prepare the stacking gel (4%) by mixing 625 μL acrylamide/bisacrylamide solution, 1,250 μL 0.5 M Tris–HCl, pH 6.8, and 50 μL 10% (w/v) SDS, and by adding Milli-Q H_2O to a total volume of 5 mL. Polymerization is initiated by adding 10 μL TEMED and 25 μL 10% (w/v) APS.
5. Pour the stacking gel onto the polymerized resolving gel and immediately insert a 10-well comb. The gel should be polymerized after 45 min (see Note 10).
6. Prepare the Colloidal Coomassie Blue staining solution at least 2 h before use (7). Per gel, suspend 500 μL of solution A into 25 mL of solution B and shake gently.
7. Take 20 μL of each cross-linking sample (absolute protein amount should be at least 5 μg per sample), reduce the sample volume in a vacuum concentrator to 10 μL, add 10 μL Laemmli buffer containing, and incubate at 95°C for 5 min.
8. Prepare the running buffer by diluting 100 mL of the ten times concentrated running buffer with 900 mL Milli-Q H_2O and mix gently.

9. Add the running buffer to the upper and lower chambers of the gel unit and load 20 μL of each sample in a well. In each of the two outer wells, 5 μL of unstained molecular weight marker is prepared.
10. Connect the gel chamber to a power supply. Run the gel at 100 V for 15 min, then at 200 V for 35–40 min until the tracking dye has reached the rim of the gel.
11. Remove the gels from the glass plates, transfer each into a single staining box, and rinse twice with Milli-Q H_2O. Add 25 mL of fixing solution and incubate for 1 h at room temperature on a shaker.
12. Discard the fixing solution and replace it by 25 mL of Colloidal Coomassie Blue staining solution. Incubate overnight at room temperature on a shaker.
13. Discard the staining solution and destain with Milli-Q H_2O.

3.3. In-Gel Digestion

Working under a laminar flow box wearing gloves and sleeve protectors is recommended to avoid keratin and dust contamination of the samples. The laminar flow box should include an incubation shaker suitable for Eppendorf tubes.

A. *Excise protein bands from the SDS gel*
   1. Wash the Coomassie-stained gel twice for 10 min with Milli-Q H_2O on a shaker.
   2. Excise gel bands of interest using a sterile scalpel and transfer gel bands to a clean glass plate. Cut the gel band into cubes of approximately 1 mm^3 and transfer them into an Eppendorf tube.

B. *Wash and destain gel pieces*
   3. Wash the gel pieces once with 100 μL Milli-Q H_2O, shake for 10 min, and discard the supernatant using a 100-μL pipet tip.
   4. Wash twice with 100 μL 50% (v/v) acetonitrile (ACN). Shake for 10 min and discard the liquid.
   5. Add 80 μL ACN, shake for 5 min. The gel pieces become white and stick to each other.
   6. Discard ACN, add 80 μL 100 mM ammonium bicarbonate. Shake for 5 min.
   7. Add 80 μL ACN and shake for another 10 min.
   8. Discard all liquid.
   9. Dry gel pieces for 20 min in a vacuum concentrator.

C. *Reduce and alkylate proteins (required only for cysteine-containing proteins)*

10. Soak gel pieces in 80 µL reducing buffer for 45 min at 56°C (*reduction step*).
11. Bring the tubes to room temperature.
12. Discard the liquid and replace it by 80 µL alkylating buffer. Incubate the solution for 30 min at room temperature in the darkness (*alkylation step*).
13. Discard the supernatant and wash the gel pieces according to steps 6–9.

D. *In-gel digestion*

14. Bring the gel pieces to complete dryness before adding the enzyme solution (trypsin or GluC in ammonium bicarbonate).
15. Per gel band, add trypsin in a ratio of 1:25 (enzyme:substrate, w/w) and incubate gel pieces for 15 min on ice. In case a double digestion is performed, add the second protease (e.g., GluC) in the same ratio and incubate gel pieces for another 15 min on ice.
16. Add 50 mM ammonium bicarbonate. The gel pieces have to be covered completely.
17. Incubate overnight at 37°C.

E. *Extract peptides*

18. Stop the digestion by adding 50 µL ACN/H_2O/TFA [47.5% (v/v)/47.5% (v/v)/5% (v/v), see Note 11].
19. Shake for 10 min, remove the supernatant, and transfer it to a new Eppendorf tube.
20. Repeat step 19 twice with 50 µL ACN/H_2O/TFA [47.5% (v/v)/47.5% (v/v)/5% (v/v)]. Pool supernatants of one gel band.
21. Concentrate supernatants in a vacuum concentrator to a volume of 5–10 µL. Do not concentrate to dryness in order to avoid sample loss.
22. Peptide mixtures are ready to be analyzed by nano-HPLC/nano-ESI-LTQ-Orbitrap mass spectrometry. Before MS analysis, samples might be stored at −20°C.

3.4. Nano-HPLC/Nano-ESI-LTQ-Orbitrap-MS

In our studies, fractionation of tryptic peptide mixtures of PPARα was carried out using an Ultimate nano-HPLC system (Dionex Corporation, Idstein, Germany) directly coupled to the nano-ESI source (Proxeon, Odense, Denmark) of an LTQ-Orbitrap XL hybrid mass spectrometer (Thermo Fisher Scientific, Bremen, Germany).

After washing the peptides on the precolumn for 15 min with water containing 0.1% (v/v) TFA, they are eluted and separated using gradients from 0 to 50% B (90 min), 50 to 100% B (1 min), and 100% B (5 min), where solvent A is 5% (v/v) ACN containing 0.1% (v/v) FA and solvent B is 80% (v/v) ACN containing 0.1% (v/v) FA. MS data are acquired over 122 min in data-dependent MS^2 mode: Each high-resolution full scan (*m/z* 300–2,000, resolution 60,000) in the orbitrap is followed by three product ion scans in the linear ion trap (LTQ) for the three most intense signals in the full-scan mass spectrum (isolation window 2 u). Dynamic exclusion (exclusion duration 180 s, exclusion mass width ±5 ppm) is enabled to allow detection of less abundant ions.

3.5. Data Analysis

Data analysis of cross-linked peptides is performed using the GPMAW software (8) allowing lysines and protein *N* termini as reaction sites of NHS-ester-based cross-linkers. Please note that trypsin will not cleave at modified lysines, therefore the number of missed cleavage sites has to be set to a higher value.

Application of isotope-labeled D_0/D_4 cross-linking reagents allows filtering of putative cross-linked candidates based on their characteristic isotope patterns (Fig. 4a). All monoisotopic deconvoluted *m/z* values (obtained by Proteome Discoverer software) showing a mass difference of 4 amu remain in the mass list. The remaining *m/z* values, which do not fulfill this criterion, are removed. The resulting theoretical mass lists are subsequently compared with experimental mass lists obtained by high-resolution mass spectrometry, allowing a maximum mass deviation of 3 ppm. As a result, a list of potential cross-linked peptides is automatically created, for which tandem MS data have to be evaluated. GPMAW supports the calculation of theoretical fragment ions of cross-linking products (Fig. 4b). Nevertheless, full mass spectra as well as tandem MS data have to be checked manually.

Depending on the different conformational state of the target protein induced by ligands, i.e., agonists or antagonists, different cross-linking products are identified within the protein as is shown for PPARα/ligand complexes in Table 1. For example, Lys-222 is connected with Lys-349 (*m/z* 1379.7401) – this cross-link exclusively found in presence of the antagonist GW6471 demonstrates a large conformational change upon drug binding. Lys-349 positioned close to the C-terminal activation function helix AF2 (amino acids 458–467) flips over in direction toward Lys-222 and closes the ligand binding pocket of PPARα (6).

Additionally, the different patterns of lysines that are modified by partially hydrolyzed cross-linkers yield valuable information on the surface accessibilities of specific residues in the target protein (Table 2).

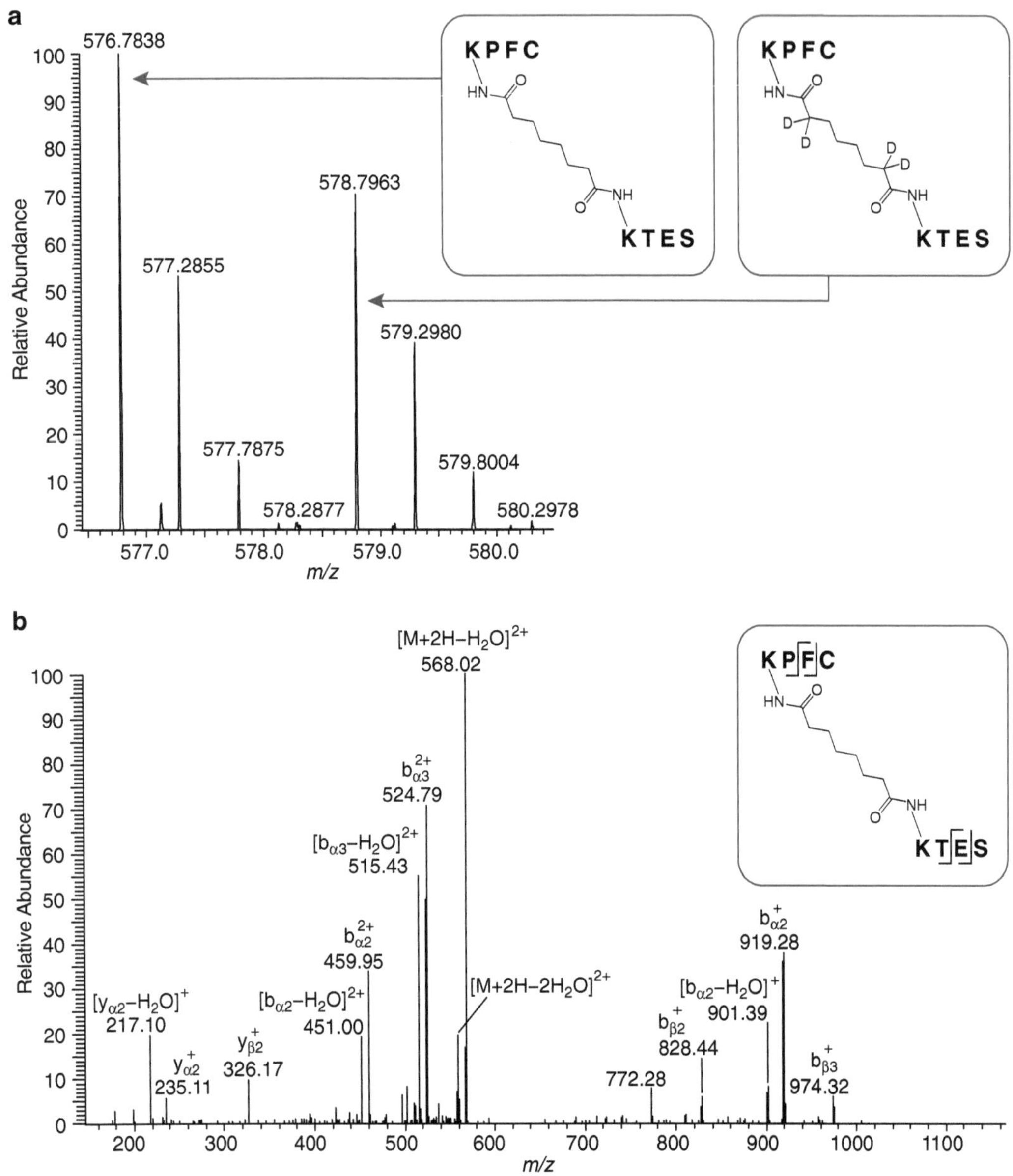

Fig. 4. (**a**) Full mass spectrum (analyzed in the orbitrap at high resolution $R=60{,}000$) of a putative cross-linking product showing a characteristic D_0/D_4 isotopic pattern. (**b**) Tandem mass spectrum [fragment ions analyzed in the LTQ] of the doubly charged intramolecular cross-linked product (*m/z* 576.78), in which lysine-349 is connected with lysine-449 of PPARα in the presence of the antagonist GW6471.

Table 1
Cross-linking products in PPARα/antagonist GW6471 and PPARα/agonist YS81 complexes with the amine-reactive cross-linkers BS^3 and BS^2G. Masses are given for D_0 cross-linkers

$[M+H]^+_{exp}$	$[M+H]^+_{theo}$	Ligand	XL[a]	Sequence[b]
1152.5601	1152.5605	–	BS^3	$\mathbf{K}_{349}$PFC + $\mathbf{K}_{449}$TES
1152.5600	1152.5605	GW6471	BS^3	$\mathbf{K}_{349}$PFC + $\mathbf{K}_{449}$TES
1152.5600	1152.5605	YS81	BS^3	$\mathbf{K}_{349}$PFC + $\mathbf{K}_{449}$TES
1379.7401	1379.7352	GW6471	BS^3	NFNMN$\mathbf{K}_{222}$VK + $\mathbf{K}_{449}$T
1389.7843	1389.7835	GW6471	BS^3	E$\mathbf{K}_{252}$TLVAK + $\mathbf{K}_{449}$TES
1110.5127	1110.5136	–	BS^2G	$\mathbf{K}_{349}$PFC + $\mathbf{K}_{449}$TES
1110.5123	1110.5136	GW6471	BS^2G	$\mathbf{K}_{349}$PFC + $\mathbf{K}_{449}$TES
1110.5129	1110.5136	YS81	BS^2G	$\mathbf{K}_{349}$PFC + $\mathbf{K}_{449}$TES
1347.7360	1347.7366	GW6471	BS^2G	E$\mathbf{K}_{252}$TLVAK + $\mathbf{K}_{449}$TES

All peptide sequences were confirmed by MS/MS data, cross-linked lysines are shown in bold.
[a]XL: cross-linker
[b]All cysteines (C) are carbamidomethylated

Table 2
Different hydrolyzed cross-linking products in ligand-free PPARα, PPARα/antagonist GW6471, and PPARα/agonist YS81 complexes with the cross-linkers BS^3 and BS^2G

$[M+H]^+_{exp}$	$[M+H]^+_{theo}$	Ligand	XL[a]	Sequence[b]
1454.7778	1454.7778	–	BS^3	DQVTLL$\mathbf{K}_{310}$YGVY
1454.7774	1454.7778	GW6471	BS^3	DQVTLL$\mathbf{K}_{310}$YGVY
1904.1223	1904.1225	GW6471	BS^3	QLVTEHAQLVQII$\mathbf{K}_{448}$K
1777.8477	1777.8465	–	BS^2G	IYEAYL$\mathbf{K}_{216}$NFNM_{ox}NK
2303.3093	2303.3092	–	BS^2G	V$\mathbf{K}_{224}$ARVILSGKASNNPPFVIH
1449.7191	1449.7155	YS81	BS^2G	NFNMN$\mathbf{K}_{222}$V$\mathbf{K}_{224}$AR
902.5190	902.5193	GW6471	BS^2G	E$\mathbf{K}_{252}$TLVAK
902.5191	902.5193	YS81	BS^2G	E$\mathbf{K}_{252}$TLVAK
634.3082	634.3083	–	BS^2G	EP$\mathbf{K}_{358}$F
634.3082	634.3083	YS81	BS^2G	EP$\mathbf{K}_{358}$F
1820.9068	1820.9069	GW6471	BS^2G	DRPGLLNVGHIE$\mathbf{K}_{399}$MQ
1590.8781	1590.8783	GW6471	BS^2G	DDIFLFP$\mathbf{K}_{425}$LLQK

Masses are given for D_0 cross-linkers. All sequences were confirmed by MS/MS data, modified lysines are shown in bold.
[a]XL: cross-linker
[b]Mox: oxidized methionine

4. Notes

1. Cross-linking reagents are sensitive to hydrolysis and therefore should be stored dry under inert gas.
2. Neat DMSO is required to prepare the NHS cross-linker stock solutions immediately before use in order to prevent hydrolysis of the reagents.
3. In most cases, the buffer conditions for storing the protein are suitable. The pH value should range between 7.4 and 8.0, which guarantees a higher yield of cross-linking products. Tris buffers interfere with amine-reactive cross-linking reagents and should be avoided. Please note that amine-reactive NHS esters also react with hydroxyl groups of serines, threonines, and tyrosines, albeit with a lower frequency compared to amine groups (18, 19).
4. Milli-Q H_2O and HPLC grade solvents are recommended to reduce contamination of cross-linking samples.
5. Complete destaining of the gels is important to obtain high-quality LC/MS data. Colloidal Coomassie Blue stain is more sensitive compared to conventional staining protocols with Coomassie Brilliant Blue R250.
6. Trypsin does not cleave C terminally to lysines that are modified by cross-linkers. Therefore, a second protease is required to achieve good proteolytic digestion yields, e.g., endoproteinase AspN (cleaves N terminally to aspartic and also glutamic acid) or GluC (cleaves C terminally to glutamic and also aspartic acid). AspN and GluC solutions in Milli-Q H_2O are stable for 1–2 days at 2–8°C. Freezing is not recommended.
7. Organic solvents must not be kept in plastics due to a potential contamination by polymers.
8. Using 1:1 mixtures of deuterated (D_0) and nondeuterated (D_4) cross-linking reagents facilitates identification of cross-linking products in high-resolution mass spectra by characteristic mass shifts of 4 amu. Nano-HPLC retention time shifts of peptides modified with the deuterated cross-linker should not be any issue. Separation of deuterated and nondeuterated peptides has never been observed, but intensity variation ±20% of the nondeuterated peptide signal compared with the deuterated variant is possible.
9. The following parameters are recommended to be optimized: molar excess of cross-linking reagents ranging from 20-fold till 200-fold for maximal yields of cross-linking products. Highly cross-linked proteins could hamper subsequent proteolysis and increase the possibility of yielding homo-oligomers of the protein.

Allow the cross-linking reaction to proceed between 5 and 120 min. A reaction time of 60 min gives optimal results in most cases.

10. Remaining SDS gels wrapped in wet precision wipes and packed in plastic bags keep fresh in the refrigerator at 4°C for 2 weeks. SDS gels that have been allowed to polymerize overnight at 4°C give optimal separation results of cross-linking products.
11. In case *offline*-ESI-MS experiments are planned, the use of formic acid (FA) should be preferred over TFA.

Acknowledgments

MQM is supported by the DFG-funded Graduiertenkolleg 1026 at the Martin-Luther-Universität Halle-Wittenberg. Dr. C. Ihling is acknowledged for valuable advice and assistance on the nano-HPLC/nano-ESI-MS system.

References

1. Young, M. M., Tang, N., Hempel, J. C., Oshiro, C. M., Taylor, E. W., Kuntz, I. D., et al. (2000) High throughput protein fold identification by using experimental constraints derived from intramolecular cross-links and mass spectrometry. *Proc Natl Acad Sci USA.* **97**, 5802–5806.
2. Sinz, A. (2003) Chemical cross-linking and mass spectrometry for mapping three-dimensional structures of proteins and protein complexes. *J Mass Spectrom.* **38**, 1225–1237.
3. Sinz, A. (2006) Chemical cross-linking and mass spectrometry to map three-dimensional protein structures and protein-protein interactions. *Mass Spectrom Rev.* **25**, 663–682.
4. Lamos, S., Krusemark, C., Mcgee, C., Scalf, M., Smith, L., Belshaw, P. (2006) Mixed isotope photoaffinity reagents for identification of small-molecule targets by mass spectrometry. *Angew Chem Int Ed.* **45**, 4329–4333.
5. Sinz, A. (2006) Isotope-Labeled Photoaffinity Reagents and Mass Spectrometry To Identify Protein-Ligand Interactions. *Angew Chem Int Ed.* **46**, 660–662.
6. Müller, M. Q., de Koning, L. J., Schmidt, A., Ihling, C., Syha, Y., Rau, O., et. al. (2009) An Innovative Method to Study Target Protein-Drug Interactions by Mass Spectrometry. *J. Med. Chem.* **52**, 2875–2879.
7. Michalik, L., Wahli, W. (1999) Peroxisome proliferator-activated receptors: three isotypes for a multitude of functions. *Curr Opin Biotechnol.* **10**, 564–570.
8. Plutzky, J. (2003) The potential role of peroxisome proliferator-activated receptors on inflammation in type 2 diabetes mellitus and atherosclerosis. *Am J Cardiol.* **92**, 34J–41J.
9. Evans, R. M., Barish, G. D., Wang, Y. (2004) PPARs and the complex journey to obesity. *Nat Med.* **10**, 355–361.
10. Issemann, I., Prince, R. A., Tugwood, J. D., Green, S. (1993) The peroxisome proliferator-activated receptor:retinoid X receptor heterodimer is activated by fatty acids and fibrate hypolipidaemic drugs. *J Mol Endocrinol.* **11**, 37–47.
11. Ihling, C., Schmidt, A., Kalkhof, S., Schulz, D. M., Stingl, C., Mechtler, K., et al. (2006) Isotope-labeled cross-linkers and Fourier transform ion cyclotron resonance mass spectrometry for structural analysis of a protein/peptide complex. *J Am Soc Mass Spectrom.* **17**, 1100–1113.
12. Schmidt, A., Kalkhof, S., Ihling, C., Cooper, D. M. F., Sinz, A. (2005) Mapping protein interfaces by chemical cross-linking and Fourier transform ion cyclotron resonance mass

spectrometry: application to a calmodulin/adenylyl cyclase 8 peptide complex. *Eur J Mass Spectrom.* **11**, 525–534.

13. Müller, M. Q., Roth, C., Sträter, N., Sinz, A. (2008) Expression and purification of the ligand-binding domain of peroxisome proliferator-activated receptor alpha (PPARalpha) *Protein Expr Purif.* **62**, 185–189.
14. Xu, H.E., Stanley, T. B., Montana, V. G., Lambert, M. H., Shearer, B. G., Cobb, J. E. et al. (2002) Structural basis for antagonist-mediated recruitment of nuclear co-repressors by PPARalpha. *Nature.* **415**, 813–817.
15. Rau, O., Syha, Y., Zettl, H., Kock, M., Bock, A., Schubert-Zsilavecz, M. (2008) alpha-Alkyl Substituted Pirinixic Acid Derivatives as Potent Dual Agonists of the Peroxisome Proliferator Activated Receptor Alpha and Gamma. *Arch Pharm.* **341**, 191–195.
16. Neuhoff, V., Arold, N., Taube, D., Ehrhardt, W. (1988) Improved staining of proteins in polyacrylamide gels including isoelectric focusing gels with clear background at nanogram sensitivity using Coomassie Brilliant Blue G-250 and R-250. *Electrophoresis.* **9**, 255–262.
17. Peri, S., Steen, H., Pandey, A. (2001) GPMAW – a software tool for analyzing proteins and peptides. *Trends Biochem Sci.* **26**, 687–689.
18. Kalkhof, S., Sinz, A. (2008) Chances and pitfalls of chemical cross-linking with amine-reactive N-hydroxysuccinimide esters *Anal Bioanal Chem.* **392**, 305–312.
19. Mädler, S., Bich, C., Touboul, D., Zenobi, R. (2009) Chemical cross-linking with NHS esters: a systematic study on amino acid reactivities. *J Mass Spectrom.* **44**, 694–706.

Chapter 15

Monitoring Ligand Modulation of Protein–Protein Interactions by Chemical Cross-Linking and High-Mass MALDI Mass Spectrometry

Natalia Gasilova and Alexis Nazabal

Abstract

Analyzing the effect of ligands on protein–protein interactions is important to better understand the cellular processes. In vitro characterization of these modulations remains challenging because of the drawbacks associated with the analysis of noncovalent interactions. To facilitate the analysis, stabilization of the protein complex by chemical cross-linking followed by High-Mass MALDI mass spectrometry is a recently developed method offering several advantages: No need for immobilization or special tags, the analysis is possible directly on wild-type protein complexes, no need for buffer exchange, large applicability range for any type of protein complex from 0 to 1,500 kDa. Using this method, we analyzed the effect of the inhibitors Nutlin-3a and Nutlin-3b on the protein complex MDM2-p53. Using this fast and sensitive method, the IC_{50} values of these inhibitors have been determined.

Key words: Protein interactions, Inhibitors, High-Mass MALDI-MS, Chemical cross-linking, IC_{50} values

1. Introduction

In cells, proteins are organized in a complicated and flexible network that can be regulated to ensure almost all cellular processes. When drugs are targeting proteins, they in fact affect the entire networks the targeted protein belongs to, having effects on protein–protein interactions. The direct analysis of protein interaction modulators is extremely important during the preclinical stage of drug discovery when hundreds of potential "lead compounds" must be investigated. These modulators may take the form of small molecules (typically with MW < 1,000 Da), recombinant proteins, recombinant peptides, antibodies, or antibody

Gerard Drewes and Marcus Bantscheff (eds.), *Chemical Proteomics: Methods and Protocols*, Methods in Molecular Biology, vol. 803, DOI 10.1007/978-1-61779-364-6_15, © Springer Science+Business Media, LLC 2012

fragments with the property to bind covalently or noncovalently to a target protein or protein complex. A number of modulators of protein interactions that potentially could become drug candidates have been described in the last decade (1–3). To characterize these modulators, biochemists are using conventional technologies including ELISA (enzyme linked immunosorbant assay) type assays, radioimmunoassays, surface plasmon resonance (SPR) technology, or fluorescence-based assays such as homogeneous time-resolved fluorescence (HTRF). Although mass spectrometry is a standard tool for the analysis of proteins, it is still challenging to use this technology for the analysis of protein–protein interactions and protein complexes. The main difficulty is the tendency of noncovalently interacting proteins to dissociate during the analysis. To characterize protein–protein interactions by mass spectrometry, electrospray ionization (ESI) is the preferred method as the sample can be analyzed in the presence of favorable buffers, maintaining the interactions stable (4). Nevertheless, finding favorable conditions to observe intact ions from protein complexes using ESI is time consuming and still a major bottleneck when trying to make these measurements in higher throughput. MALDI mass spectrometry-based analysis of intact protein complexes has recently gained interest when optimized cross-linking reagents (5) were combined with ultrahigh-mass detection systems (6, 7). Cross-linking of the protein complexes circumvents the problem of complex dissociation during sample preparation or laser desorption. The high-mass detection allows to detect the intact covalently linked protein complexes with nM sensitivity up to 1,500 kDa (8–10). Here, we introduce the use of this methodology for the characterization of MDM2-P53 protein complex inhibitors.

2. Materials

2.1. Recombinant Proteins and Inhibitors

1. Recombinant p53 protein (Active Motif, Rixensart, Belgium), lyophilized with 1× dilution buffer (20 mM Tris–HCl, pH 8.0, 20% glycerol, 100 mM KCl, 0.2 mM EDTA, 1 mM DTT), stored at -20°C.
2. Recombinant MDM2 protein (ab82080, Abcam, Cambridge, MA, USA), with buffer 20% glycerol, 20 mM Tris–HCl, 100 mM potassium chloride, 1 mM DTT, 0.2 mM EDTA, pH 8, stored at –20°C.
3. Nutlin-3a, 1 mg lyophilized (N 6287, Sigma-Aldrich, Buchs, Switzerland), stored at -20°C.
4. Nutlin-1, 1 mg lyophilized was obtained from Hybrigenics SA, Paris, France, stored at –20°C.

2.2. Cross-Linking Reagents and Buffers

1. R100 cross-linking mix (1,1′-(suberoyldioxy)bisazabenzotriazole (SBAT); di(3H-(1,2,3)triazolo(4,5-b)pyridin-3-yl)octanedioate; 1,1′-(suberoyldioxy)bisbenzotriazole (SBBT); di(1H-benzo(*d*)(1,2,3)triazol-1-yl)octanedioate; glutaroyldioxy bisazaben-zotriazole (GBAT); di(3H-(1,2,3)triazolo(4,5-*b*) pyridin-3-yl) glutarate), 16× 2 mg lyophilized, stored at 4°C (CovalX, Schlieren, Switzerland) (see Note 1).
2. Dimethylformamide (DMF), bio-analysis grade.

2.3. Sample Preparation

1. Cross-linking buffer: 5 mM sodium phosphate, 7.5 mM sodium chloride, pH 7.2, 0.05% sodium azide (purification buffer, K100 MALDI stabilization kit, CovalX, Schlieren, Switzerland), stored at 4°C.
2. Zeba Micro Spin Desalting Columns (87767, Pierce Biotechnology, Rockford, IL, USA).
3. Sinapinic acid MALDI matrix (N78887, Sigma-Aldrich, Buchs, Switzerland).
4. Matrix buffer: Acetonitrile (Mallinckrodt Baker, Teugseweg, The Netherland): distilled water 1:1, 0.1% trifluoroacetic acid (TFA; Mallinckrodt Baker, Teugseweg, The Netherlands) (see Note 2).

2.4. MALDI Mass Spectrometry

1. ABSciex 4800 MALDI-ToF/ToF mass spectrometer (ABSciex, Framingham, MA, USA).
2. HM2 TUVO High-Mass system (CovalX, Schlieren, Switzerland).
3. Complex Tracker 2.0 analysis software (CovalX, Schlieren, Switzerland).

3. Methods

In order to characterize the effect of the inhibitors Nutlin-1 and Nutlin-3a on the protein complex (p53·MDM2), the first step is to analyze the target complex using High-Mass MALDI mass spectrometry. To characterize this protein complex, different concentrations of the proteins p53 and MDM2 are analyzed using High-Mass MALDI mass spectrometry before and after cross-linking in order to determine the optimal protein concentrations for the study. The optimal concentrations for the target proteins are obtained when two conditions are verified simultaneously: (1) the control experiment (without cross-linking) is showing the different subunits of the protein complex (p53 and MDM2) with a minimal nonspecific clustering effect and (2) when the cross-link experiment

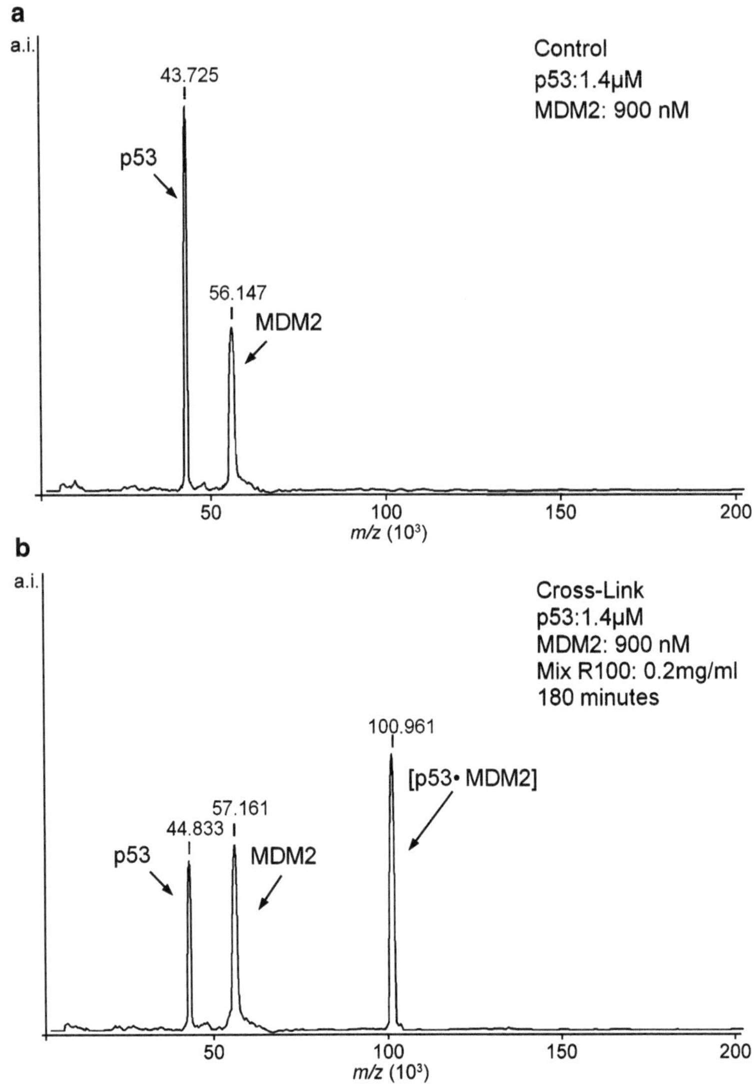

Fig. 1. High-Mass MALDI MS analysis of the protein complex MDM2-P53. (**a**) Mixture containing MDM2 (900 nM) and p53 (1.4 μM) was directly analyzed using High-Mass MALDI mass spectrometry (control experiment). (**b**) High-Mass MALDI MS analysis of the same sample after cross-linking (Mix R100, 0.2 mg/ml, 180-min incubation time). A specific protein complex is detected with $MH^+ = 100.961$ kDa.

(for the same protein concentration) is showing the highest ratio protein complex/protein subunits. When the optimal concentration of the protein is determined, a time course experiment is performed to determine the incubation time necessary to achieve the cross-linking reaction (Fig. 1).

When the optimal conditions of cross-linking are established for the complex (p53·MDM2), the protein mixture is incubated with the inhibitors Nutlin-1 or Nutlin-3. Different concentrations of the inhibitors (from 1 nM to 10 μM) are preincubated with

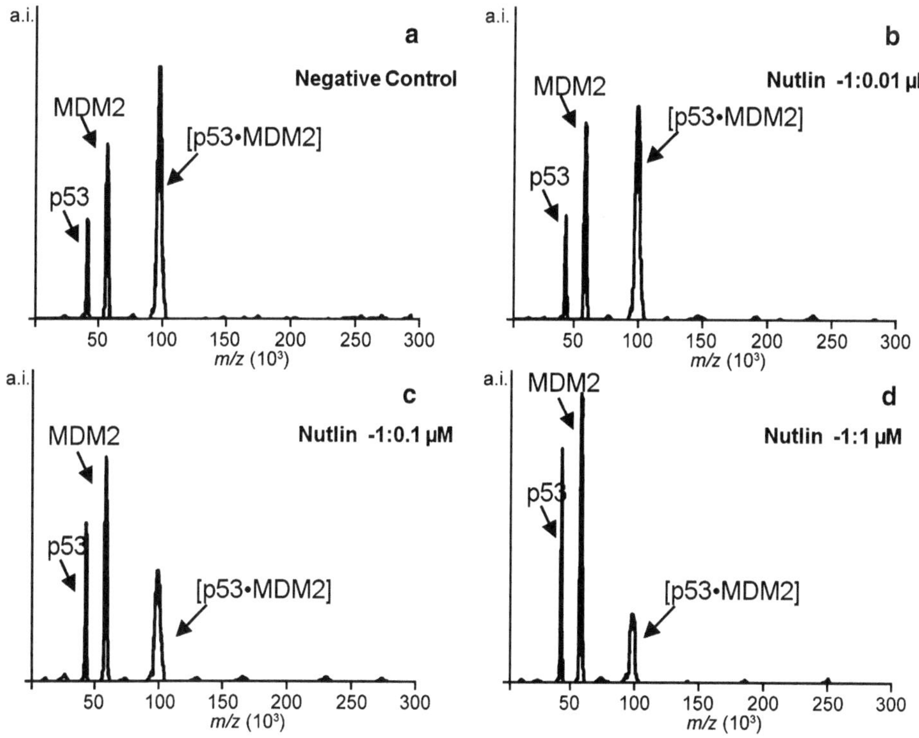

Fig. 2. High-Mass MALDI MS analysis of the protein complex (MDM2·P53) incubated with different concentrations of Nutlin-1 inhibitors. MDM2 (900 nM) was preincubated 60 min with different concentrations of Nutlin-1 (**a**) negative control; (**b**) 0.01 μM; (**c**) 0.1 μM, and (**d**) 1 μM. After incubation, MDM2 sample was mixed with p53 (1.4 μM) and submitted to cross-linking before High-Mass MALDI mass spectrometry analysis. With increasing concentration of Nultin-1, the relative intensity of the peak corresponding to the complex (MDM2·p53) is decreasing when compared with the relative intensity of p53 and MDM2.

MDM2 for different incubation times ranging from 30 to 360 min before mixing with p53 and submission to the cross-linking reagents. After the cross-linking incubation time, the samples are directly analyzed by High-Mass MALDI mass spectrometry. The effect of the inhibitors on the protein complex is monitored by integrating the peaks corresponding to the subunits p53 and MDM2 and the peak corresponding to the complex (p53·MDM2) (Fig. 2).

3.1. Protein Sample Preparation: P53/MDM2 Mix

1. The vial containing the protein p53 is thawn on the bench for 20 min. Lyophilized p53 (5 μg) is dissolved with 40 μl 1× dilution buffer in order to obtain 40 μl of a 2.9-μM solution.
2. The vial containing 10 μg of full-length human MDM2 is thawn on the bench for 20 min. The vial contains 50 μl of a 3.6-μM MDM2 solution.
3. Each vial is submitted to buffer exchange using Zeba Micro Spin desalting columns (0.5 ml columns). The column's bottom

closure is removed and the column is placed into a 1.5-ml microcentrifuge collection tube. Then, the column is centrifuged at 1,000 × *g* for 1 min to remove the storage solution. To equilibrate the column, 300 μl of the exchange buffer (5 mM sodium phosphate, 7.5 mM sodium chloride, pH 7.2, 0.05% sodium azide) is added on top of the resin bed and centrifuged at 1,000 × *g* for 1 min. After discarding the flow-through, the column is equilibrated two additional times. After centrifugation, the top of the column is blotted to remove excess liquid. The column is placed into a new collection tube and 40 μl of the protein sample p53 is applied to the top of the compact resin bed. The sample is centrifuged at 1,000 × *g* for 2 min and the flow-through that contains the sample is collected.

The same buffer exchange procedure is applied to 40 μl of the protein sample MDM2.

4. After buffer exchange, 5 μl aliquots of the p53 solution are pipetted into eight microcentrifuge vials.
5. After buffer exchange, 20 μl of the solution of MDM2 is pipetted into a separate 0.5-ml microcentrifuge tube. 10 μl of the cross-linking buffer (5 mM sodium phosphate, 7.5 mM sodium chloride, pH 7.2) is pipetted into seven additional 0.5-ml microcentrifuge tubes in order to prepare a dilution series of the MDM2 protein sample.
6. 10 μl of the MDM2 solution (from the microcentrifuge tube containing 20 μl) is pipetted and mixed with 10 μl of the cross-linking buffer prepared in a separate tube. This operation is repeated in series in order to obtain eight tubes of 10 μl containing dilutions 1, 1/2, 1/4, 1/8, 1/16, 1/32, 1/64, and 1/128 of the MDM2 protein samples. The diluted samples have expected concentrations of 3.6 μM (1), 1.8 μM (1/2), 900 nM (1/4), 450 nM (1/8), 225 nM (1/16), 112 nM (1/32), 66 nM (1/64), and 33 nM (1/128).
7. 5 μl of each diluted solution of the protein MDM2 is pipetted and mixed with the tubes containing 5 μl of the protein p53 prepared previously in order to obtain eight tubes containing mixtures of MDM2 and p53. The expected concentrations for the eight mixtures p53/MDM2 (10 μl final) are the following: 1.4 μM/1.8 μM, 1.4 μM/900 nM, 1.4 μM/450 nM, 1.4 μM/225 nM, 1.4 μM/112 nM, 1.4 μM /66 nM 1.4 μM/33 nM, and 1.4 μM/16 nM.

3.2. Sample Preparation for High-Mass MALDI Analysis: Control Experiments

1. 2 mg of sinapinic acid matrix is weighted and dissolved in 200 μl of matrix buffer (acetonitrile:distilled water; 1:1; 0.1% TFA) in order to obtain a solution of 10 mg/ml. The solution obtained is sonicated for 1 min at room temperature.

2. 1 µl of each p53/MDM2 mixture (eight different mixtures) is mixed with 1 µl of a sinapinic acid matrix solution.
3. After mixing, 1 µl of the mixture is directly spotted onto a standard MALDI plate.

3.3. Sample Preparation for High-Mass MALDI Analysis: Cross-Link Experiments

1. 2 mg of R100 mix is dissolved in DMF (1 ml) in order to obtain a solution of 2 mg/ml of cross-linker. The cross-linking solution obtained is vortexed for 1 min at room temperature.
2. Each p53/MDM2 mixture prepared for the control experiment (eight different mixtures, 9 µl left) is submitted to cross-linking by adding 1 µl of the cross-linking solution (2 mg/ml). The final volume for the cross-linking reaction is 10 µl.
3. After different incubation times at room temperature (20, 40, 60, 120, 180, and 360 min), 1 µl of each p53/MDM2 cross-linked mixtures (eight different mixtures) is mixed with 1 µl of a sinapinic acid matrix solution.
4. After mixing, 1 µl of the mixture is directly spotted onto a standard MALDI plate and immediately analyzed by High-Mass MALDI-ToF mass spectrometry (see Note 3).

3.4. High-Mass MALDI-ToF Mass Spectrometry Experiments

1. The High-Mass MALDI-ToF MS analysis has been performed in linear mode using the standard nitrogen laser.
2. The source parameters have been set as the following: ion source 1: 20 kV; ion source 2: 17 kV; lens voltage: 10.3 kV; delay extraction 600 ns; matrix suppression 5,000 Da.
3. The HM2 High-Mass detection settings were gain voltage: 3.14 kV; postacceleration voltage: 20 kV.
4. To calibrate the instrument, an external calibration with clusters of Insulin, BSA, and IgG has been applied.
5. For each spot, 500 laser shots are acquired using an automatic scan method. The spectrum acquired corresponds to the sum of the 500 laser shots.
6. Laser fluence is optimized to increase resolution of the peaks and reduce unspecific clustering effects (see Note 4). The laser fluence is set at the limit of the observed signal threshold.

3.5. Sample Preparation for Inhibition Experiments

1. 1 mg of the inhibitors Nultin-1 or Nutlin-3 are dissolved in 1 ml of DMSO in order to obtain two solutions with concentration 1,650 and 1,721 µM, respectively.

   5 µl of the solution of Nutlin-1 prepared is diluted with 820 µl of DMSO in order to obtain a solution of 10 µM of Nutlin-1. 5 µl of the solution of Nutlin-3 prepared is diluted with 855 µl of DMSO in order to obtain a solution of 10 µM

of Nutlin-3. Using the 10 μM solution of Nutlin-1 and Nutlin-3, two other solutions are prepared with concentration 1 and 0.1 μM by diluting the 10 μM solution 10× and 100×, respectively.

2. Six 0.5-ml microcentrifuge tubes containing 10 μl of MDM2 protein sample with a concentration of 2 μM are prepared following the protocol described in Subheading 3.1.
3. Six 0.5-ml microcentrifugation tubes containing 10 μl of p53 protein sample with a concentration of 2.6 μM are prepared following the protocol described in Subheading 3.1.
4. 1 μl of the solutions of MDM2 prepared is discarded (9 μl left) and 1 μl of Nutlin-1 or Nutlin-3 solutions with concentrations of 10, 1, or 0.1 μM is mixed with the MDM2 sample. The six samples prepared are containing MDM2 (1.8 μM) and Nutlin-1 or Nutlin-3 with final concentrations 1, 0.1, and 0.01 μM.
5. The samples containing MDM2 and the inhibitors Nutlin-1 or Nutlin-3 are incubated for 60 min at room temperature.
6. After the incubation time, 5 μl of each solution of MDM2 are mixed with 5 μl of the solution of p53 in order to obtain a mixture of MDM2/p53 with concentrations 900 nM/1.4 μM, respectively.
7. After mixing, the sample is immediately submitted to cross-linking following the protocol described in Subheading 3.3, steps 1 and 2.
8. The sample is incubated for 60 min to achieve the cross-linking reaction.
9. After the incubation time the sample is prepared for the High-Mass MALDI experiment following the protocol described in Subheading 3.3.
10. Immediately after crystallization, the samples are analyzed by High-Mass MALDI following the protocol described in Subheading 3.4.

3.6. Sample Preparation for IC_{50} Determination

1. Using the solutions of Nutlin-1 and Nutlin-3a prepared in Subheading 3.5, step 1 and having concentrations of 1,650 μM (Nutlin-1) and 1,721 μM (Nutlin-3a), 20 solutions of inhibitors with the concentrations: 0.1, 0.2, 0.4, 0.6, 0.8, 1, 2, 3, 4, 5, 6, 7, 8, 9, 10, 20, 40, 60, 80, and 100 μM are prepared. Final volume is 20 μl for each solution.
2. 120 microcentrifuge (0.5 ml) tubes containing 10 μl of MDM2 protein sample with a concentration of 2 μM are prepared following the protocol described in Subheading 3.1.

3. 120 microcentrifugation (0.5 ml) tubes containing 10 µl of p53 protein sample with a concentration of 2.6 µM are prepared following the protocol described in Subheading 3.1.
4. 1 µl of each solution of MDM2 prepared is discarded (9 µl left) and 1 µl of the inhibitors Nutlin-1 or Nutlin-3 with concentrations 0.1, 0.2, 0.4, 0.6, 0.8, 1, 2, 3, 4, 5, 6, 7, 8, 9, 10, 20, 40, 60, 80, and 100 µM are mixed with the MDM2 sample. The 40 samples are containing MDM2 (1.8 µM) and Nutlin-1 or Nutlin-3 with 20 different concentrations from 0.01 to 10 µM.
5. The analysis is repeated as triplicate (3× 40 samples: 120 samples containing MDM2 and the inhibitors).
6. The samples containing MDM2 and the inhibitors Nutlin-1 or Nutlin-3 are incubated for 60 min at room temperature.
7. After the incubation time, 5 µl of each solution of MDM2 are mixed with 5 µl of the solution of p53 in order to obtain a mixture of MDM2/p53 with concentrations 900 nM/1.4 µM, respectively.
8. After mixing, the sample is immediately submitted to cross-linking following the protocol described in Subheading 3.3.
9. The sample is incubated for 60 min to achieve the cross-linking reaction.
10. After the incubation time the sample is prepared for High-Mass MALDI experiment following the protocol described in Subheading 3.3.
11. Immediately after crystallization, the samples are analyzed by High-Mass MALDI following the protocol described in Subheading 3.4.

3.7. Data Analysis

1. After acquisition, the samples were loaded on the software Complex Tracker 2.0 for data analysis.
2. All cross-linking spectra were base line corrected and smoothed with the following parameter of the filter tool: top hat baseline correction, window size 10 kDa; Savitzky Golay smoothing, window size 10 kDa, polynomial degree 1.
3. The integration of the peaks corresponding to the protein p53, MDM2, and the protein complex (MDM2·p53) is calculated automatically using the integration tool.
4. The percentage of binding for the complex (MDM2·p53) is calculated with 100% reference as the control experiment without inhibitors (Fig. 3).

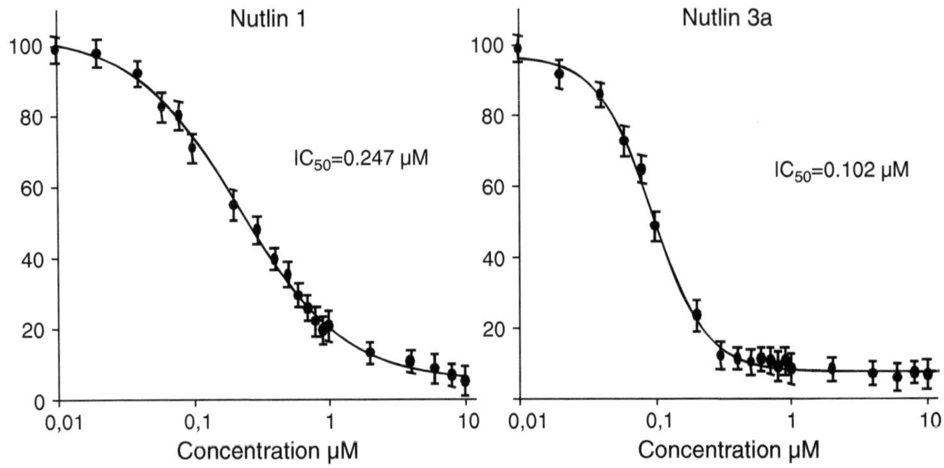

Fig. 3. IC_{50} value calculation for Nutlin-1 and Nutlin-3a inhibitors of the protein complex (MDM2·p53). MDM2 (900 nM) was preincubated 60 min with 20 different concentrations of Nutlin-1 or Nutlin-3, from 0.01 to 10 μM. After incubation, MDM2 sample was mixed with p53 (1.4 μM) and submitted to cross-link before High-Mass MALDI mass spectrometry analysis. The percentage of binding for each concentration of inhibitors is calculated based on the integration of the peaks detected (p53 monomer; MDM2 monomer and (MDM2·p53) complex) in comparison with the 100% binding reference (no inhibitor, negative control).

4. Notes

1. R100 cross-linking reagent is best stored at 4°C in a desiccator.
2. All solutions should be prepared in water that has a resistivity of 18.2 MΩ cm. This standard is referred to as "distillated water" in the text.
3. It is critical that the spots prepared for the cross-linking experiment are analyzed immediately after crystallization. The quality of the ionization starts to decrease (lower ratio signal/noise) when the samples are analyzed more than 1 h after crystallization of the cross-linked sample.
4. When analyzing protein samples with MALDI mass spectrometry, concentrated protein samples combined with high laser fluence induce an increase of unspecific protein interaction detection. These unspecific interactions are due to clustering effects that occur in the gas phase after laser desorption. In order to minimize these effects, the concentrations of the proteins are optimized and the laser fluence is adjusted close to the observed threshold of the signal.

Acknowledgments

The authors would like to thank Dr. Jean Christophe Rain from Hybrigenics SA (Paris) for discussion on the inhibition assay and encouragement from Professor Renato Zenobi (Department of Chemistry, Swiss Federal Institute of Technology-ETHZ, Zürich).

References

1. Sebolt-Leopold J.S. (2000) Development of anticancer drugs targeting the MAP kinase pathway. *Oncogene* 19, 6594–6599.
2. Stebbins C.E., Russo A.A., Schneider C., Rosen N., Hartl F.U., and Pavletich N.P. (1997) Crystal structure of an Hsp90-geldanamycin complex: targeting of a protein chaperone by an antitumor agent. *Cell* 89, 239–250.
3. Vassilev L.T., VU B.T., Graves B., Carvajal D., Podlaski F., Filipovic Z., Kong N., Kammlott U., Lukas C., Klein C., Fotouhi N., Liu E.A. (2004) In vivo activation of the p53 pathway by small-molecule antagonists of MDM2. *Science* 5659, 844–848.
4. Loo A.J. (2000) Electrospray ionization mass spectrometry: a technology for studying non covalent macromolecular complexes. *Int. Journal Mass Spectrom.* 200, 175–186.
5. Bich C., Maedler S., Chiesa K., DeGiacomo F., Bogliotti N., and Zenobi R. (2010) Detection of reactivity and applications of new amine reactive cross-linker for mass spectrometry detection of protein complexes. *Anal. Chem.* 82, 172–179.
6. Nazabal A., Wenzel R.J., Zenobi R. (2006) Immunoassays with direct mass spectrometric detection. *Anal. Chem.* 78, 3562–3570.
7. Bich C., Scott M., Panagiotidis A., Wenzel R.J., Nazabal A., Zenobi R. (2008) Characterization of antibody-antigen interactions: Comparison between surface plasmon resonance measurements and high-mass matrix assisted laser desorption/ionization mass spectrometry. *Anal. Biochem.* 375, 35–45.
8. Riek U., Scholz R., Konarev P., Rufer A., Suter M., Nazabal A., Ringler P., Chami M., Müller S.A., Neumann D., Forstner M., Hennig M., Engel A., Svergun D., Schlattner U., Wallimann T. (2008) Structural properties of AMP-activated protein kinase. Dimerization, molecular shape, and changes upon ligand binding. *J. Biol. Chem.* 283, 18331–18343.
9. Michel F., Crucifix C., Granger F., Eiler S., Mouscadet J-F., Korolev S., Agapkina J., Ziganshin R., Gottikh M., Nazabal A., Emiliani S., Benarous R., Moras D., Schultz P., and Ruff M. (2009) Structural basis for HIV-1 DNA integration in the human genome, role of the LEDGF/P75 cofactor. *The EMBO Journal* 28, 980–991.
10. Moullintraffort L., Bruneaux M., Nazabal A., Allegro D., Giudice E., Zal F., Peyrot V., Barbier P., Thomas D., Garnier C. (2010) Biochemical and Biophysical characterization of the Mg2+ induced 90 kDa heat shock protein oligomers. *J. Biol. Chem.* 283, 18331–18343.

Chapter 16

Time-Controlled Transcardiac Perfusion Crosslinking for In Vivo Interactome Studies

Amy Hye Won Jeon and Gerold Schmitt-Ulms

Abstract

The time-controlled transcardiac perfusion crosslinking (tcTPC) method differs from conventional perfusion fixation in that the crosslinking reagent is administered throughout the circulatory system for only a relatively short period of time, thereby allowing limited crosslinking to occur. Bait protein complexes are isolated by affinity capture (AC) under stringent conditions and are recovered from the AC matrix by acidic elution. Affinity-purified proteins are reduced, alkylated, and digested with a specific endoproteinase, such as trypsin. Subsequently, peptides are isotopically labeled, separated by reversed-phase chromatography and analyzed by quantitative tandem mass spectrometry (MS/MS). The proteins crosslinked to the bait protein during tcTPC are identified by database searches with conventional protein identification software. The tcTPC strategy offers unique advantages over alternative approaches for studying a subset of protein complexes which require a particular environment for their structural integrity, such as membrane protein complexes that are notorious for their tendency to dissociate upon detergent solubilization. The sensitivity and utility of this method are influenced by the spatial distribution of chemical groups within the bait protein complexes that can engage in productive crosslinks.

Key words: Time-controlled transcardiac perfusion crosslinking, Formaldehyde, Protein–protein interactions, Affinity capture, iTRAQ quantitation, Mass spectrometry

1. Introduction

Individual proteins do not act in isolation but engage in complex and dynamic interactions with other proteins to fulfill their diverse cellular roles (1–5). Numerous methods have been devised to provide insights into the specific interactions a protein engages in, commonly referred to as the interactome of a given protein (6–8). Increasingly, interactions of a protein of interest, hereafter referred to as the bait protein, are probed biochemically based on the

Gerard Drewes and Marcus Bantscheff (eds.), *Chemical Proteomics: Methods and Protocols*, Methods in Molecular Biology, vol. 803, DOI 10.1007/978-1-61779-364-6_16,

identification of proteins that co-purify with the bait protein during its isolation. In particular, the combination of protein tagging and affinity chromatography of protein complexes has proven to be powerful (9, 10). A subset of proteins, however, seem refractory to this kind of analysis. Notorious among these are membrane proteins, partly because no generic solubilization strategy can be devised for them and it cannot be predicted how well a given membrane protein complex will tolerate the exposure to a specific detergent (11, 12). Alternative approaches are also needed if the objective is to characterize the molecular environment of a protein of interest, including those proteins which reside in spatial proximity but do not engage in bona fide interactions.

The time-controlled transcardiac perfusion crosslinking (tcTPC) method stabilizes existing next neighbor relationships among proteins by formaldehyde crosslinking prior to the disruption of tissue integrity (Fig. 1) (13). Mild formaldehyde crosslinking has been employed extensively for the study of nucleosomal protein interactions (14–17). Some features that make formaldehyde crosslinking attractive are (1) the water solubility of the reagent; (2) the absence of reagent-induced rearrangements of the proteins; and (3) the crosslink bonds are short (2–3 Å), endure harsh treatments and are reversible (15). The tcTPC procedure is

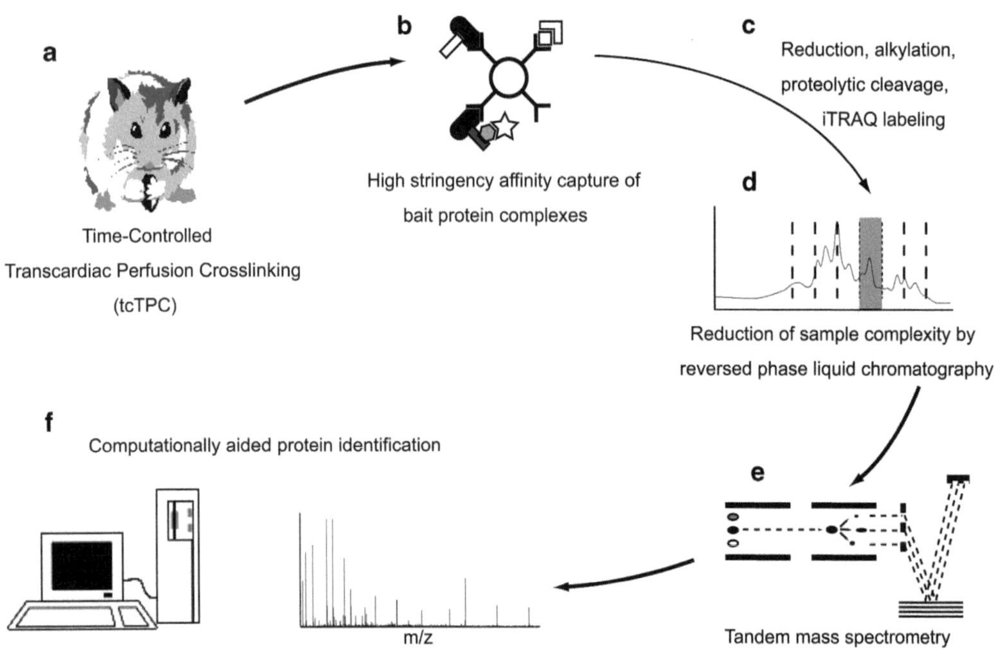

Fig. 1. Schematic outline of tcTPC method. (**a**) The formaldehyde crosslinking solution is pumped through the circulatory system of the mouse in a time-controlled manner. (**b**) Protein complexes are purified by stringent affinity capture, (**c**) then reduced, alkylated, digested, and conjugated to iTRAQ reagents. (**d**) Samples are combined and subjected to reversed phase liquid chromatography and (**e**) quantitative tandem mass spectrometry, which is followed by (**f**) computationally aided protein identification.

followed by rapid tissue dissection and an optimized sample work-up scheme that includes some modifications to conventional protocols for the co-affinity capture of proteins. The covalent stabilization of protein interactions translates into the ability to treat affinity-captured protein complexes with stringent salt and detergent washing steps. Compared with alternative protocols that do not make use of crosslinking or employ in vitro crosslinking, the move to in vivo crosslinking further reduces the risk of being misled in situations where nonphysiological interactors bind directly to the bait protein or to the affinity-purified protein complex only when present in an extract but do not physiologically interact with the bait protein when cellular integrity is maintained.

Nevertheless, unspecific interactors can be present in interactome samples as a result of (1) aggregated proteins in the sample that co-sediment with the affinity matrix, (2) proteins that bind directly to the affinity matrix, (3) proteins which under physiological conditions are found in a different cellular compartment than the bait protein, but have an intrinsic propensity to bind to the bait protein when present in an extract, (4) abundant cellular proteins that populate affinity purification eluate fractions when samples are subject to less than the most stringent washing conditions, (5) proteins that originate from the AC matrices themselves, e.g., if crude antibody preparations were coupled to chemically activated matrix beads, and finally (6) proteins such as trypsin, human skin, and hair proteins introduced into the sample during handling procedures. With this many possible sources of unspecific proteins, rather than aiming to eliminate all contaminants, a feasible objective is to minimize their occurrence and more importantly, to know their identities. To this end, we strategically embedded an isotopic labeling step at the peptide level, which affords the quantitative and comparative analysis of the sample of interest and a negative control sample. The following method represents the most recent version of a protocol we have gradually improved over the years.

2. Materials

The tcTPC method was developed in North America and, thus, reagents and equipment items listed in the following paragraphs were primarily sourced in this region. Due to the generic nature of most materials required, we expect that equivalent substitutes for all items can be purchased from alternative suppliers and manufacturers.

2.1. Biological Source Material

1. The tcTPC method was developed utilizing mice. However, the method can also be applied to rats, hamsters, and other small rodents and should be adaptable to any species that is

equipped with a circulatory system for the rapid delivery of formaldehyde to the tissue(s) of interest.

2. Negative controls: Excellent bait-specific and negative control samples which differ in the presence or absence of a specific bait protein can be obtained from the crossbreeding of animals which are heterozygote for the bait protein of interest. Alternative approaches to generating negative controls are available should no knockout animals be available (see Subheading 3.1).

2.2. Time-Controlled Transcardiac Perfusion Crosslinking

1. Deep Anesthesia Cocktail: 10:1 mix (w/w) of Ketamine (Bioniche Life Sciences, Belleville, ON) and Xylazine (Bayer, Leverkusen, Germany).
2. Purging Buffer: Phosphate-buffered saline (PBS) supplemented with 1 U/mL of Heparin (commercially available as a 1,000 U/mL stock solution, e.g., from American Pharmaceutical Parners, Schaumburg, IL).
3. tcTPC Solution: Freshly made solution of 2–4% formaldehyde (Bioshop Canada, Burlington, ON) in Purging Buffer (see Note 1).
4. Surgical instruments: Bonn Artery scissors (No. 14086-09, Fine Science Tools, Foster City, CA), Micro-Mosquito curved hemostat (No. 13011-12, Fine Science Tools), 20-gauge needle with barrel tip and Luer lock hub (No. 18060-20, Fine Science Tools), angled dissector scissors (No. 14082-09, Fine Science Tools).

2.3. Generation of Soluble Extract

1. Rod homogenizer (PowerGen 120; Thermo Fischer Scientific, Waltham, MA).
2. Homogenization Buffer: 150 mM NaCl, 50 mM NH_4Cl, 100 mM Tris–HCl (pH 8.0), 1× Complete Protease Inhibitor Cocktail (Roche, Basel, Switzerland) (see Note 2).
3. Extraction Buffer: 150 mM NaCl, 1% deoxycholate, 1% NP-40, 20 mM Tris–HCl, pH 8.0 (see Note 3).

2.4. Generation of Affinity Matrix

1. AC matrix: Affi-Gel 10 agarose (Bio-Rad, Hercules, CA), Protein A or G agarose (Sigma-Aldrich, St. Louis, MO) or alternative magnetic bead matrix.
2. Affinity-purified bait-specific antibody.
3. Pierce BCA Assay kit (Thermo Fisher Scientific).
4. High-Salt Wash Buffer: 1 M NaCl, 20 mM Tris–HCl, pH 8.0.
5. Pre-Elution Buffer: 10 mM HEPES, pH 8.0 (see Note 4).
6. Acidic Elution Buffer: 15% acetonitrile (ACN), 0.2% trifluoroacetic acid (TFA), pH 1.9.
7. Buffer A: 1% NP-40, 150 mM NaCl, 25 mM HEPES, pH 7.5.

2.5. Affinity-Purification of Bait Protein Complexes

1. High Salt Buffer A: 1% NP-40, 500 mM NaCl, 25 mM HEPES, pH 7.5.
2. Handee Spin Cup Column (HSCC) (Thermo Fisher Scientific).
3. 0.65 mL Safe Seal Microcentrifuge Tubes (PGC Scientific, Frederick, MD).
4. Storage Buffer: 0.05% NaN_3 in PBS.

2.6. Generation of iTRAQ-Labeled Tryptic Digest

All solutions listed in this section are made fresh immediately prior to use.

1. Denaturation Buffer: 9 M urea in water (see Note 5).
2. pH Adjustment Buffer: 1 M HEPES, pH 8.0.
3. Reducing Agent: 75 mM tris (2-carboxyethyl) phosphine (TCEP) (Sigma-Aldrich) stock solution in water.
4. Alkylation Agent: 150 mM 4-vinylpyridine (Sigma-Aldrich) stock solution in water.
5. Trypsin Solution: Dissolve side chain-modified, L-1-tosylamide-2-phenylethyl chloromethyl ketone (TPCK)-treated porcine trypsin in water to a concentration of 0.4 μg/μL (Promega, Fitchburg, WI).
6. iTRAQ Reagent: Dissolve individual isobaric tags for relative and absolute quantitation (iTRAQ) labeling reagents (Applied Biosystems, Foster City, CA) in 70 μL ethanol, vortex for 1 min, and subject to brief spin to collect liquid at bottom of tube.

2.7. Tandem Mass Spectrometry

1. $ZipTip_{C18}$ tips (Millipore, Billerica, MA).

3. Methods

3.1. Design of Negative Control

Whenever bait knockout animals are available we recommend the use of an approach we refer to as the bait exclusion strategy (Fig. 2a). In this strategy, side-by-side affinity purifications are carried out from starting materials which differ in the presence or absence of the bait protein. Because all steps during the AC procedure employ the same reagents, the bait exclusion method makes it easy to attribute differences in the interactome lists of sample and control to the expression versus "knockout" of the protein of interest. While the conceptual simplicity of this approach is compelling, the experimenter needs to be cognizant that a protein may nonetheless end up in the bait protein-specific eluate as a result of its ability to bind nonspecifically to the bait or its physiological interactors. In instances where no knockout animals are available,

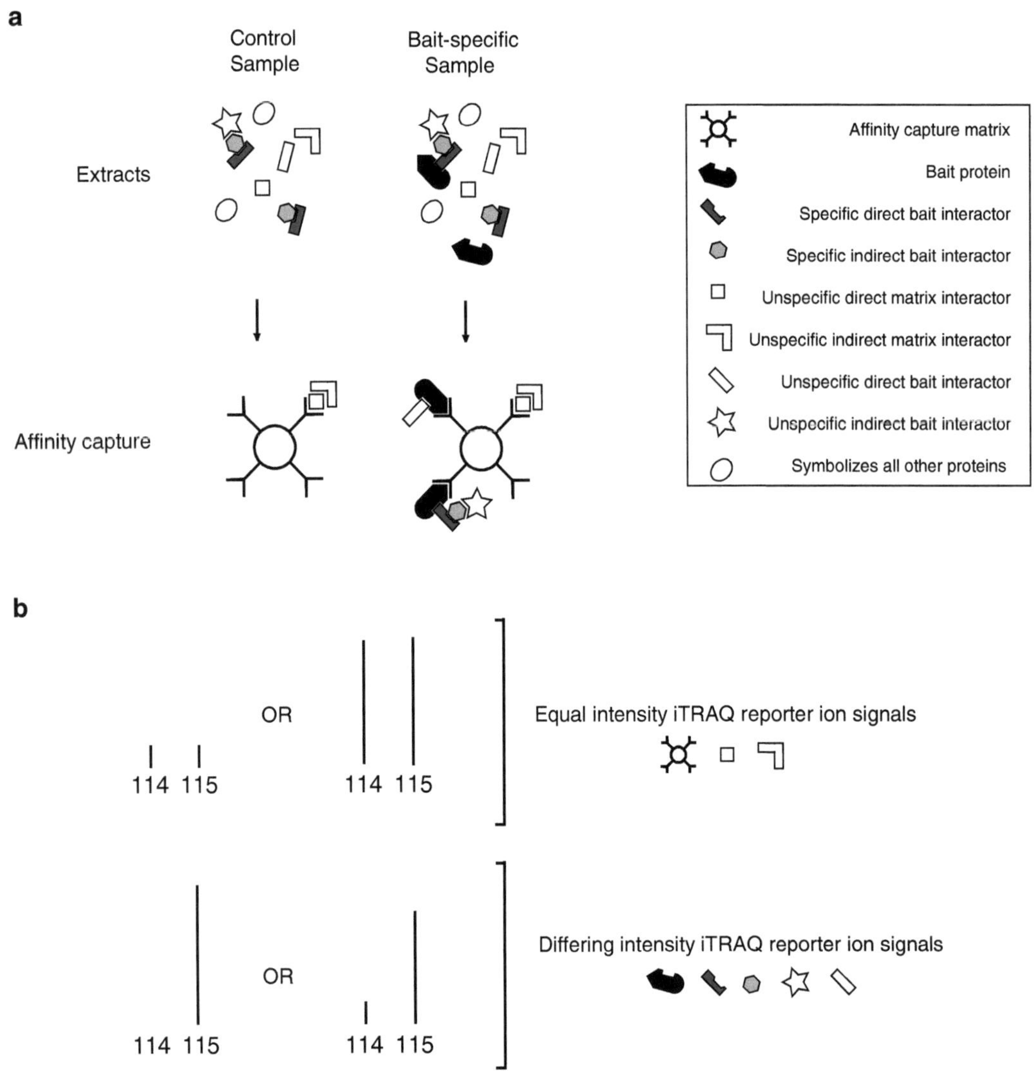

Fig. 2. Cartoon depicting bait exclusion strategy for the identification of unspecific interactors. (**a**) The Bait Exclusion strategy employs side-by-side affinity capture from starting materials which differ in the presence or absence of the bait protein. As a result, bait-specific interactors represent a subset of proteins exclusively found in eluate fractions derived from the bait-containing biological source material. (**b**) Expected iTRAQ reporter ion signature profiles assuming peptides in the control and bait-specific eluates were labeled with iTRAQ114 and iTRAQ115 reagents, respectively.

a derivative strategy can be used, which employs the side-by-side AC from identical wild-type extracts but employs for the negative control an AC matrix which has been presaturated with a small ligand that competes with the bait protein-containing complexes for binding. In most instances, this small ligand will represent the peptide-antigen, which was originally used to raise the bait-specific antibody. The success of this approach depends on the relative affinity of bait protein and competitive ligand pair for binding to the affinity matrix. Analogous design concepts can also be employed if the AC matrix is not based on a bait-specific antibody but capitalizes on the existence of a small-molecule ligand (inhibitor, cofactor, etc.) known to strongly and selectively bind to the bait protein. We advise against the generation of negative control data based on the conjugation of unrelated antibodies, for example, generic immunoglobulins (IgGs) not known to recognize the bait protein, due to the poor discriminative power this strategy offers for the detection of crossreactivities of the bait-specific capture antibody. Regardless of the approach used, we recommend incorporation of an isotopic labeling step at this time (18) but are cognizant that further improvements to label-free quantitative mass spectrometry (MS) methods may soon eliminate the need for this extra labeling step (19). Quantitative data are indispensable for interactome studies aimed toward abundant cellular proteins where, without quantitation, it may be impossible to distinguish bona fide interactors from unspecific contaminants. Whereas for cell culture-based work metabolic isotopic labeling strategies are attractive (20), methods which conjugate isotopic tags "post-harvest" to peptides are more cost-effective and straight-forward for animal-based methods such as tcTPC (Fig. 2b). Several alternative reagents are available for this purpose and should be equally applicable (21). Most of these approaches reduce the MS analysis time (by enabling pooling of samples and controls) and reduce the risk to misinterpret performance fluctuations and run-to-run variances in the analyses of complex samples to reflect sample-to-sample differences (22).

3.2. Time-Controlled Transcardiac Perfusion Crosslinking

1. Deep anesthesia of rodents: Determine body weight of rodents and inject Deep Anesthesia Cocktail intraperitoneally to achieve a concentration of 100 mg Ketamine/10 mg Xylazine per kilogram of body weight.
2. Mounting of rodent and surgical procedures: Place deeply anesthetized animal (unresponsive to paw pinch test) on gridiron (e.g., lid of animal cage or similar grid) mounted on top of a plastic or aluminum reservoir. Stretch out limbs with hemostats and make incisions from caudal of sternum up along the sides of rib cage without injuring underlying organs using artery scissors. Secure open chest and skin with hemostats. Attach 20-gauge perfusion needle (30 mm length) with barrel tip and Luer lock hub to end of peristaltic pump tubing. Feed tubing through peristaltic pump and immerse opposite end in Purging Solution.
3. tcTPC: Grab tip of the heart with curved serrated forceps or hemostat and make 2–3 mm incision into the left ventricle (recognizable by its lighter colored appearance). Insert perfusion needle into the left ventricle (Fig. 3). Clamp the needle in place with a hemostat and secure stable position of tools. Make another incision into the right atrium to enable the

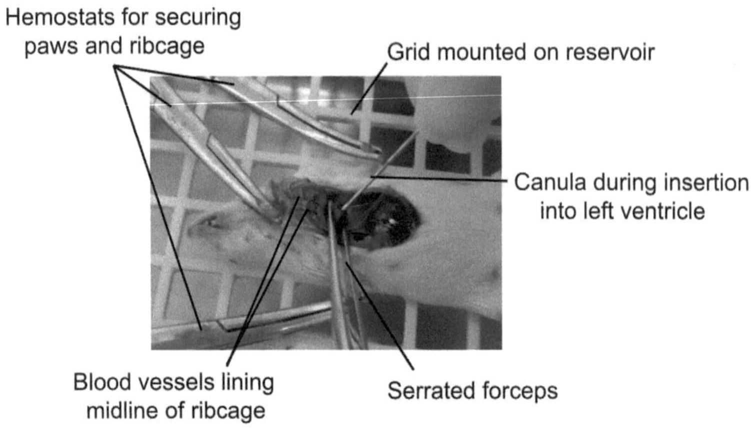

Fig. 3. Still-image of mouse captured during a critical step of tcTPC procedure. The image depicts a mouse with its chest cavity opened and mounted on grid-covered container, with perfusion canula inserted into the left ventricle of heart.

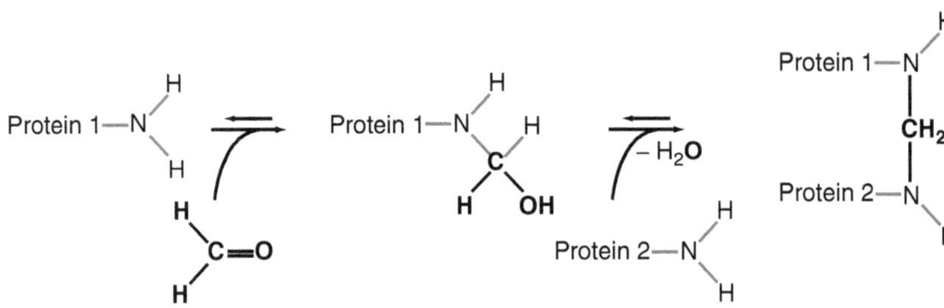

Fig. 4. Conjugation chemistry of formaldehyde crosslinking involving amino groups present within proteins. The reaction proceeds in two steps: Initially, formaldehyde mounts a nucleophilic attack toward the amino group provided by a peptide to form a Schiff-base conjugate. Subsequently, the reaction of this intermediate with a nearby primary amine is paralleled by the concomitant elimination of water and the formation of a covalent bond.

purging of blood and excess crosslinking reagent from the circulatory system during the perfusion step. Perfuse animal with PBS at 10 mL/min for 2 min to purge the circulatory system of blood. The success of this step can be observed by the decoloring of the liver and the two vessels that flank the midline of the animal's ribcage (Fig. 3). Switch perfusion solution to tcTPC crosslink solution and continue pumping at 10 mL/min for an additional 6 min. Conjugation chemistry of formaldehyde crosslinking is shown in Fig. 4 (see Notes 6 and 7).

4. Rapid dissection of brain: Remove head with sharp scalpel or large scissors. Open skin by sagittal cut from neck to nose. Remove residual muscular tissue from neck of animal. Use angled dissector scissors to cut through bone along sagittal midline of skull. Force skull open by spreading movement with back of dissector scissors. Remove brain with round-ended spatula (see Note 8). Immerse dissected tissue in tcTPC Solution (e.g., in Falcon tube) for up to 9 min (this includes the time for dissection).

3.3. Generation of Soluble Extract

All steps in this section are to be carried out at 4°C.

1. Homogenization of tissue: Pool appropriate amount of tcTPC-treated rodent tissue (see Note 9) and homogenize with rod homogenizer (e.g,. 5 × 30 s strokes) at maximum power setting using a ratio of 10:1 (w/w) of Homogenization Buffer over tissue. Prevent extracts from warming by keeping samples on ice between and following the homogenization steps.
2. Solubilization of bait protein complexes: Add Extraction Buffer (same volume as Homogenization Buffer used in previous step) to extract membrane proteins and incubate samples for 30 min with gentle agitation.
3. Removal of insoluble material: Initially, remove cellular debris from the samples by low-speed centrifugation (1,000 × *g*, 5 min). Transfer the supernatants to ultracentrifugation tubes and remove all insoluble protein material by high-speed centrifugation (100,000 × *g*, 1 h).

3.4. Generation of Affinity Matrix

The method works with a wide range of AC matrices, including chemically activated or Protein A/G-derivative matrices. Similarly, both conventional crosslinked agarose and more recent products based on improved bead-based technologies can be employed (23). The bait-specific capture antibody has to be covalently crosslinked to the matrix to avoid its inadvertent release during the protein elution step. The antibody conjugation step should follow vendor instructions for the specific matrix employed. Most commercial matrices allow the covalent conjugation of 1–5 mg of purified antibody per milliliter of matrix. A typical interactome investigation requires the matrix equivalent of 100 μg conjugated antibody per sample. The following instructions should be followed regardless of choice of AC matrix used. For steps which require the collection of AC matrix, minimize physical compression by employing gravity sedimentation, by applying minimal centrifugal force (determined by pilot sedimentation tests), or by employing magnetic bead materials.

1. Determine the amount of antibody conjugated by comparing protein concentration in purified antibody solution before and after antibody capture step, for example, with the use of BCA assay by following manufacturer instructions.
2. Prior to its use, wash the bait-specific AC matrix with at least 10 volumes each of 1× PBS, High-Salt Wash Buffer, Extraction Buffer, Pre-Elution Buffer, and Acidic-Elution Buffer (see Notes 10 and 11).
3. Finish preparative steps for AC matrix by three consecutive washes with Buffer A.

3.5. Affinity-Purification of Bait Protein Complexes

Throughout the next sample handling procedures collect and repeatedly set aside aliquots from all samples to aid with trouble-shooting if necessary. Prior to their use, rinse all tubes and reservoirs which will come in contact with samples with water and ACN to remove unwanted polymer residue and minimize dust load. We describe an implementation of this method based on HSCCs. Similar commercial products from other manufacturers can be used instead.

1. Add tissue extract obtained from high-speed centrifugation step to AC matrix and incubate for 2–24 h with gentle agitation on turning wheel (see Note 12).
2. Following the capture of bait-containing protein complexes, collect AC matrix, and save supernatant for future use.
3. Subject AC matrix to consecutive washes by briefly swirling, and subsequently sedimenting, the AC matrix in 500-fold excess (v/v) of Buffer A, High-Salt Buffer A, Buffer A, and Pre-Elution Buffer.
4. Transfer AC matrix to HSCC reservoir (with bottom plug removed) and allow the buffer to pass through the settling AC matrix leaving only enough buffer behind to just submerge the matrix (see Note 13).
5. Acidic detachment of protein complexes from AC matrix: Plug bottom of HSCC and add 300 μL of Acidic Elution Buffer. Close the screw-cap lid and gently invert HSCC to suspend beads in Elution Buffer. Incubate for 3 min (Fig. 5).

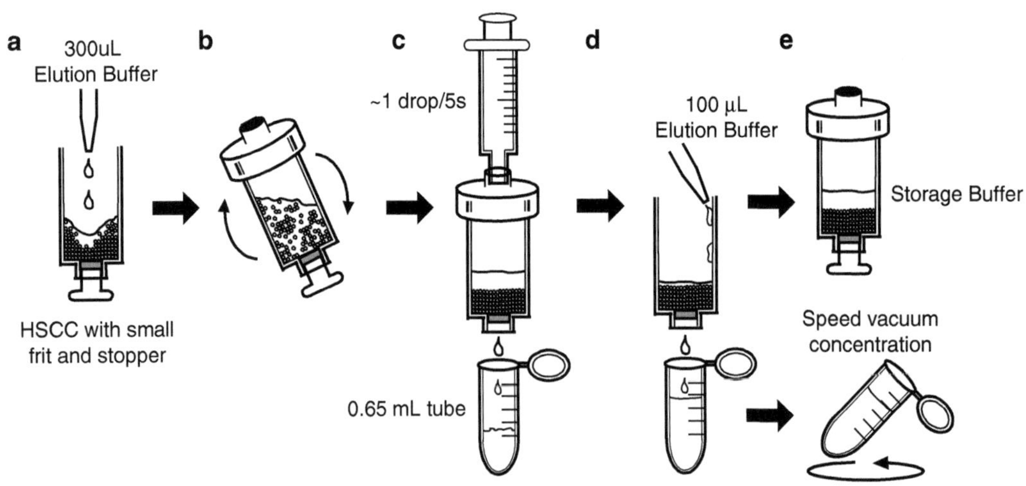

Fig. 5. Cartoon depicting steps for the elution of bait protein complexes from the AC matrix. (**a**) Add 300 μL of Elution Buffer to HSCC. (**b**) Tilt column gently to resuspend beads. (**c**) Use syringe to adjust back pressure for an elution speed of one drop/5 s and collect eluate in prerinsed 0.65 mL microcentrifuge tube. (**d**) Add 100 μL of Elution Buffer along the tube wall without disturbing the settled AC matrix. (**e**) Store HSCC for future use in Storage Buffer. Proceed with speed vacuum concentration.

6. To collect the eluate, position HSCC above 0.65 mL Safe Seal Microcentrifuge Tube, remove plug from bottom of HSCC and unscrew lid. When liquid levels in the HSCC have dropped to the front of the AC matrix, pipet an additional 100 μL of Acidic Elution Buffer into the HSCC without disrupting the settled AC matrix. Continue elution until 400 μL of eluate have accumulated in the capture tube (which should coincide with Acidic Elution Buffer reaching the AC matrix front).
7. Storage of matrix: Unplug bottom of HSCC and wash AC matrix with Storage Buffer. When pH of flow-through has returned to neutral, close HSCC at both ends and store for future use at 4°C.

3.6. Generation of iTRAQ-Labeled Tryptic Digest

Many alternative directions can be pursued for the downstream handling of eluate fractions. The following section describes one implementation of a robust method for the denaturation, reduction, alkylation, trypsinization, and iTRAQ labeling steps.

1. Reduce volume of eluate fractions in a speed vacuum concentrator to 10 μL. Add 150 μL of water and continue speed vacuum concentration to a volume of 5 μL. Repeat the preceding step once more.
2. Denaturation: Add 10 μL of Denaturation Buffer and incubate for 10 min at room temperature.
3. Test the pH by spotting a small volume (e.g., 0.1 μL) on pH paper. If pH is below 8.0, add 1 μL of pH Adjustment Buffer, gently vortex for a few seconds, and collect liquid at bottom of tube by centrifugation. Repeat preceding step until a pH of 8.0 has been reached.
4. Reduction and alkylation: Add 1 μL of Reducing Agent stock solution and incubate for 30 min at 60°C. Add 1 μL of Alkylation Agent stock solution and incubate for 1 h at room temperature in the dark. Dilute samples with water to a combined volume of 50 μL (see Note 14).
5. Trypsinization: Add 5 μL (2 μg) of Trypsin Solution to samples and incubate at 37°C for 6 h (see Note 15).
6. iTRAQ labeling: Add 70 μL of iTRAQ Reagent to digested samples (control: iTRAQ 114, sample 1: iTRAQ 115, sample 2: iTRAQ 116, etc.), vortex for 30 s and collect solution at bottom of tube by brief centrifugation. Incubate the mixtures at room temperature in the dark for 3 h with occasional brief agitation.

3.7. Tandem Mass Spectrometry

From here on the objective is to obtain an in-depth inventory of peptides present in the combined iTRAQ labeled samples. Many alternative strategies are available and the details of steps will as much depend on personal preferences as on available equipment.

Thus, rather than providing a detailed step-by-step protocol we restrict instructions to generic steps and point out possible obstacles.

Before combining iTRAQ-labeled samples and embarking on the final steps of sample preparation, it is advisable to subject a small aliquot of each sample to analytical tandem mass spectrometry (MS/MS) analysis (see Note 16).

1. Combine 5% (v/v) of individual iTRAQ-labeled samples into a fresh 0.65 mL Microcentrifuge Tube and concentrate its volume by speed vacuum concentration. Add 20 μL of water and repeat speed vacuum concentration.
2. Carry out ZipTip_{C18} (or equivalent) clean-up as described by manufacturer and subject eluate to MS/MS analysis (see Note 17).
3. Verify that iTRAQ labeling proceeded with equal efficiency by inspecting iTRAQ signature mass peak region within MS/MS spectra which can be assigned to peptides common to sample and control (e.g., trypsin auto-proteolysis peptides) (Fig. 2b).
4. Once data on the integrity, quantity, and complexity of the sample have been obtained, informed decisions can be made regarding the most suitable downstream analysis method.
5. If either the quantity or complexity of the peptide mixture is relatively low, reversed-phase separation followed by tandem analysis of the sample is recommended.
6. For samples of high complexity, it is advisable to precede the reversed-phase separation with an orthogonal fractionation either by strong cation exchange fractionation, high pH reversed-phase or isoelectric focusing.
7. Perform MS/MS analysis on an instrument suitable for the detection of low-mass iTRAQ reporter ions (e.g., a quadrupole/time-of-flight (QqTOF), a time-of-flight/time-of-flight (TOF/TOF), or an orbitrap mass spectrometer).
8. Conduct iTRAQ ratio analyses using one of several alternative software packages designed to facilitate this step (e.g., ProteinPilot (Applied Biosystems), Mascot (Matrix Science), SpectrumMill (Agilent), Warp LC (Bruker), Peaks (Bioinformatics Solutions Inc), ProteinProspector (University of California, San Francisco), and others) (see Note 18).

4. Notes

1. Clarification on the terminology of formaldehyde and paraformaldehyde: In aqueous solution, formaldehyde slowly polymerizes to paraformaldehyde, which is poorly soluble and

cannot be used as a fixative. Aqueous solutions of 4–10% formaldehyde are often referred to as formalin. Since methanol slows the spontaneous polymerization of formaldehyde, it is often added by chemical suppliers at a concentration of 10–15% (v/v). Methanol-free formaldehyde can be obtained by hydrolysis of paraformaldehyde in the presence of heat. Methanol-free formaldehyde solutions are also commercially available and are most frequently obtained in gas-sealed ampoules. Upon air exposure, these containers should only be used for up to 1 week.

2. The combination of Homogenization and Extraction buffer presented here works well for diverse bait proteins. However, its composition may need to be adjusted to accommodate biochemical idiosyncrasies of a given bait protein.
3. Deoxycholate will precipitate at a pH <8.0.
4. All subsequent procedures until the iTRAQ conjugation step avoid the use of reagents containing primary amines as their presence would interfere with the isobaric labeling chemistry. Therefore, we move to HEPES instead of Tris–HCl at this step in the procedure.
5. In aqueous solution small amounts of urea gradually decompose to ammonium and cyanate which reacts with various amino acid side chains to form their respective carbamyl derivatives. To minimize these unwanted modifications we advise preparing the 9-M urea stock immediately before its use and deplete cyanate ions by incubation with a mixed bed ion exchange resin (prior to the addition of HEPES buffer).
6. Reliable placing and securing of perfusion canula: Indicators of a successful perfusion are signs of convulsions, absence of blood in tail and limbs, and overall hardening of animal. Too deep insertions of the perfusion canula may lead to the rapid filling of lungs with perfusion liquid.
7. Purging, perfusion, dissection, and postfix crosslinking steps take a combined 17 min. To increase the throughput of tcTPC to ten animals per hour, individual perfusions can be initiated every 6 min in a workflow that requires the parallel operation of two peristaltic pumps. A team of three people is required to achieve this rate of perfusion. A convenient arrangement for subdividing tasks is to have one person carry out the intraperitoneal injections, preparative surgery, and mounting of the animals, a second person handle the perfusions, and a third person be in charge of dissection and the postfix step.
8. The tcTPC method can also be applied for the study of protein interactions occurring outside the brain. For instructions describing the rapid dissection of alternative tissues please consult the respective primary literature.

9. The amount of tissue required may need to be adjusted following a small-scale pilot investigation. Parameters that will influence the outcome are the expression-level of the bait protein, the availability of functional groups that support cross-linking and the percentage of formaldehyde used. We have collected brain interactome datasets for individual bait proteins with starting material that ranged from 1 to 100 tcTPC-treated mouse brains.
10. The washing step with Acidic Elution Buffer may need to be omitted if the antibody employed does not recover from exposure to low pH even when this step is performed rapidly.
11. To avoid physical shearing of AC matrix during transfer steps cut the end of pipet tips to generate a sufficiently wide opening.
12. The avidity and affinity of the antibody employed as well as the stability of protein complexes containing the bait protein are likely to play an important role for the capture yield. For uncrosslinked material, shorter incubation times appear to be beneficial for the capture of weak interactors (24); this is less of a factor for this method which uses covalently crosslinked biological material.
13. During this and subsequent handling steps of the HSCC the liquid flow can be manually adjusted by pressing air into the reservoir with the help of an empty syringe. To this end, the syringe can be mounted to the column with the help of the Luer lock hub present on alternative HSCC lids included with this kit.
14. The dilution with water is necessary to ensure that the concentration of urea does not exceed 1.5 M during the trypsinization step.
15. (Optional) Following trypsinization, spike a known amount of a synthetic peptide (e.g., 1 pmol (Glu1)-Fibrinopeptide B (GluFib) (Sigma-Aldrich, St. Louis, MO)) to samples. Equal iTRAQ labeling can then be confirmed by verifying equal intensities of 114:115:116:117 iTRAQ signature mass peaks upon fragmentation of the GluFib parent ion.
16. This is to verify that AC, digestion, and iTRAQ labeling steps were successful, to assess the complexity and quantity of the peptide mixture and to confirm that the sample is not inadvertently dominated by common contaminants such as polymers or keratins.
17. The sample needs to be acidified and depleted of organic solvent prior to ZipTip clean-up or reversed-phase separation steps. If the complexity of the sample is to be reduced prior to reversed-phase capture, fractionation by strong cation exchange is recommended as it does not require the prior removal of ethanol.

18. The relative quantitation works well for the filtering of candidate interactors in a qualitative sense (as long as relative iTRAQ ratios exceed a practical discriminatory threshold of ~1.5-fold) but tends to underestimate abundance ratios in instances when one of the samples contains much higher levels of a given protein. Biological and/or technical repetitions of the experiment are frequently accompanied by changes in the relative ratios of peptides but are not expected to alter the qualitative assignment of iTRAQ ratios. Inconsistent iTRAQ ratios for different peptides derived from the same protein may indicate that this protein exists in multiple isoforms which differ in their propensity to be retained during the AC step.

Acknowledgments

Work on this project was funded through an operating grant MOP-74734 of the Canadian Institute for Health Research (CIHR). AJ was supported by an award from The Scottish Rite Charitable Foundation of Canada and GS received support from the W. Garfield Weston Foundation.

References

1. Sobott, F., and Robinson, C. V. (2002) Protein complexes gain momentum, *Curr Opin Struct Biol 12*, 729–734.
2. Aloy, P., and Russel, R. B. (2002) The third dimension for protein interactions and complexes, *Trends Biochem Sci 12*, 633–638.
3. Neubauer, G., and Wilm, M. (1999) From genome to function: analysis of multi-protein complexes and protein phosphorylation, *Curr Opin Mol Ther 1*, 695–701.
4. Blackstock, W., and Rowley, A. (1999) Probing cellular complexity with proteomics, *Curr Opin Mol Ther 1*, 702–711.
5. Gavin, A. C., Aloy, P., Grandi, P., Krause, R., Boesche, M., Marzioch, M., Rau, C., Jensen, L. J., Bastuck, S., Dumpelfeld, B., Edelmann, A., Heurtier, M. A., Hoffman, V., Hoefert, C., Klein, K., Hudak, M., Michon, A. M., Schelder, M., Schirle, M., Remor, M., Rudi, T., Hooper, S., Bauer, A., Bouwmeester, T., Casari, G., Drewes, G., Neubauer, G., Rick, J. M., Kuster, B., Bork, P., Russell, R. B., and Superti-Furga, G. (2006) Proteome survey reveals modularity of the yeast cell machinery, *Nature 440*, 631–636.
6. Charbonnier, S., Gallego, O., and Gavin, A. C. (2008) The social network of a cell: recent advances in interactome mapping, *Biotechnol Annu Rev 14*, 1–28.
7. Berggard, T., Linse, S., and James, P. (2007) Methods for the detection and analysis of protein-protein interactions, *Proteomics 7*, 2833–2842.
8. Drewes, G., and Bouwmeester, T. (2003) Global approaches to protein-protein interactions, *Curr Opin Cell Biol 15*, 199–205.
9. Gingras, A. C., Aebersold, R., and Raught, B. (2005) Advances in protein complex analysis using mass spectrometry, *J Physiol 563*, 11–21.
10. Burgess, R. R., and Thompson, N. E. (2002) Advances in gentle immunoaffinity chromatography, *Curr Opin Biotechnol 13*, 304–308.
11. Reisinger, V., and Eichacker, L. A. (2008) Solubilization of membrane protein complexes for blue native PAGE, *J Proteomics 71*, 277–283.
12. Hooker, B. S., Bigelow, D. J., and Lin, C. T. (2007) Methods for mapping of interaction networks involving membrane proteins, *Biochem Biophys Res Commun 363*, 457–461.
13. Schmitt-Ulms, G., Hansen, K., Liu, J., Cowdrey, C., Yang, J., DeArmond, S., Cohen, F. E., Prusiner, S. B., and Baldwin, M. A.

(2004) Time-controlled transcardiac perfusion cross-linking for the study of protein interactions in complex tissues, *Nat Biotechnol 22*, 724–731.
14. Wells, J., and Farnham, P. J. (2002) Characterizing transcription factor binding sites using formaldehyde crosslinking and immunoprecipitation, *Methods 26*, 48–56.
15. Jackson, V. (1999) Formaldehyde cross-linking for studying nucleosomal dynamics, *Methods 17*, 125–139.
16. Orlando, V., Strutt, H., and Paro, R. (1997) Analysis of chromatin structure by in vivo formaldehyde cross-linking, *Methods 11*, 205–214.
17. Fragoso, G., and Hager, G. L. (1997) Analysis of in vivo nucleosome positions by determination of nucleosome-linker boundaries in cross-linked chromatin, *Methods 11*, 246–252.
18. Ong, S. E., and Mann, M. (2005) Mass spectrometry-based proteomics turns quantitative, *Nat Chem Biol 1*, 252–262.
19. Zhu, W., Smith, J. W., and Huang, C. M. Mass spectrometry-based label-free quantitative proteomics, *J Biomed Biotechnol 2010*, 840518.
20. Zhang, G., and Neubert, T. A. (2009) Use of stable isotope labeling by amino acids in cell culture (SILAC) for phosphotyrosine protein identification and quantitation, *Methods Mol Biol 527*, 79–92, xi.
21. Chen, X., Sun, L., Yu, Y., Xue, Y., and Yang, P. (2007) Amino acid-coded tagging approaches in quantitative proteomics, *Expert Rev Proteomics 4*, 25–37.
22. Rudnick, P. A., Clauser, K. R., Kilpatrick, L. E., Tchekhovskoi, D. V., Neta, P., Blonder, N., Billheimer, D. D., Blackman, R. K., Bunk, D. M., Cardasis, H. L., Ham, A. J., Jaffe, J. D., Kinsinger, C. R., Mesri, M., Neubert, T. A., Schilling, B., Tabb, D. L., Tegeler, T. J., Vega-Montoto, L., Variyath, A. M., Wang, M., Wang, P., Whiteaker, J. R., Zimmerman, L. J., Carr, S. A., Fisher, S. J., Gibson, B. W., Paulovich, A. G., Regnier, F. E., Rodriguez, H., Spiegelman, C., Tempst, P., Liebler, D. C., and Stein, S. E. Performance metrics for liquid chromatography-tandem mass spectrometry systems in proteomics analyses, *Mol Cell Proteomics 9*, 225–241.
23. Yingyongnarongkul, B. E., How, S. E., Diaz-Mochon, J. J., Muzerelle, M., and Bradley, M. (2003) Parallel and multiplexed bead-based assays and encoding strategies, *Comb Chem High Throughput Screen 6*, 577–587.
24. Cristea, I. M., Williams, R., Chait, B. T., and Rout, M. P. (2005) Fluorescent proteins as proteomic probes, *Mol Cell Proteomics 4*, 1933–1941.

Part IV

Ligand Discovery

Chapter 17

Ligand Discovery Using Small-Molecule Microarrays

Dominick E. Casalena, Dina Wassaf, and Angela N. Koehler

Abstract

Genome-wide association studies and genetic linkage studies have created a growing list of proteins related to disease. Small molecules can serve as useful probes of function for these proteins in a cellular setting or may serve as leads for therapeutic development. High-throughput and general binding assays may provide a path for discovering small molecules that target proteins for which little is known about structure or function or for which conventional functional assays have failed. One such binding assay involves small-molecule microarrays (SMMs) containing compounds that have been arrayed and immobilized onto a solid support. The SMMs can be incubated with a protein target of interest and protein–small molecule interactions may be detected using a variety of fluorescent readouts. Several suitable methods for manufacturing SMMs exist and different immobilization methods may be more or less preferable for any given application. Here, we describe protocols for covalent capture of small molecules using an isocyanate-coated glass surface and detection of binding using purified protein.

Key words: Small-molecule microarrays, Ligand discovery, Isocyanate, Rapamycin, FK506, FKBP12

1. Introduction

Genome-wide association studies have produced an increasing list of proteins that may play a key role in disease. Small molecules may be useful tools in understanding the function of these proteins in a cellular context and in some cases may serve as therapeutic leads. Designing small molecule ligands for proteins in the absence of knowledge about structure or function is a significant challenge. Developing high-throughput functional assays (e.g., enzymatic, DNA-binding, etc.) is also a challenge for protein targets that lack assigned molecular function. High-throughput binding assays may provide a general route toward identifying small molecule probes

Gerard Drewes and Marcus Bantscheff (eds.), *Chemical Proteomics: Methods and Protocols*, Methods in Molecular Biology, vol. 803, DOI 10.1007/978-1-61779-364-6_17,

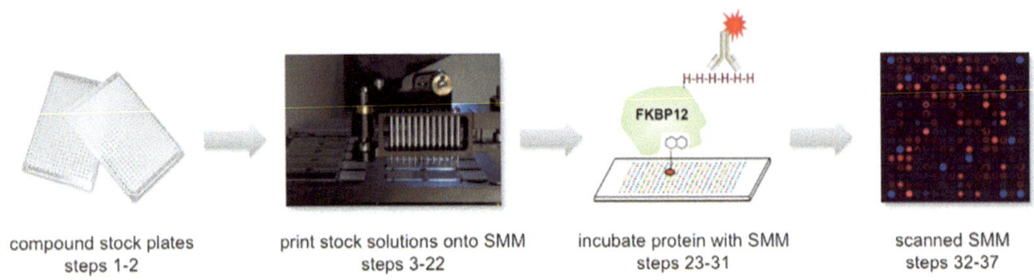

Fig. 1. General scheme for printing and screening small-molecule microarrays (SMMs).

for such targets (1, 2). The small-molecule microarray (SMM) platform has proven to be a robust technology for discovering protein–small molecule interactions (1). The microarray format is particularly attractive for binding assays due to the miniaturized and parallel nature of the format. Typically, more than 10,000 small molecules are evaluated for their ability to bind a protein of interest on a single microarray. Very little compound is required (<200 pL per feature) to make the arrays and tens of thousands of compounds may be screened with as little as 20 μg of protein. SMMs have been used to identify functional probes of many proteins to date including various "druggable" enzymes (3–6), "undruggable" transcription factors (7), and extracellular growth factors (8), as well as nonprotein targets such as pathogenic intron RNA (9).

Several covalent attachment approaches have been developed to immobilize small molecules onto solid substrates such as glass or silicon and have been reviewed elsewhere (1, 10, 11). Most of these capture strategies take advantage of either latent functionalities or specific appendages on the small molecules that react with the modified surface. These approaches can result in either homogenous (5–7, 12) or heterogenous (13–15) display of small molecules. In this chapter, we present an updated step-by-step protocol for manufacture of SMMs using isocyanate-mediated capture and detection of protein–small molecule interactions (Fig. 1). The isocyanate capture approach for small molecules presented here was originally developed by our laboratory in an effort to expand the compatibility of the SMM approach with a wider set of molecules (13, 14). The surface is reactive to compounds containing a variety of nucleophilic functional groups and has been used to immobilize complex natural products, products of diversity-oriented synthesis, drug-like molecules, and known bioactives from commercial collections, and FDA-approved drugs (8, 13). Many compounds contain multiple isocyanate-reactive functionalities, giving rise to the potential for multiple modes of display on the surface. This heterogeneous surface may allow proteins to sample multiple binding modes for any given immobilized small molecule in the array. The surface is also compatible with screens involving pure proteins or proteins residing in complexes or clarified cell lysates. The SMM fabrication and screening protocols described here serve as a blueprint for building a ligand discovery platform that is general for most types of compounds and proteins.

2. Materials

2.1. Compound Printing and Array Manufacture

1. Amine-functionalized glass slides, GAPSII-Coated Slides (Corning Life Sciences, Inc.).
2. Aushon 2470 Micro-Arrayer (Aushon Biosystems) outfitted with a 48-pin print head and 85 μm diameter solid pins. Other arrayers using split pins ("quill pins") may also be used.
3. Contact Angle Goniometer (VCA Optima XE, AST Products, Inc.) or equivalent.
4. Vacuum Desiccator (VWR International) or equivalent.
5. Glass staining dishes with cover (Wheaton Scientific), or equivalent.
6. Stainless steel slide racks (Wheaton Scientific), or equivalent. Glass slide racks may also be used.
7. Gas filter gun with 0.01 μm PVDF filter (Wafergard GN, Entegris Inc.).
8. Nitrogen gas, ultrahigh purity grade (Airgas, Inc.).
9. Dimethyl sulfoxide, DMSO.
10. Solutions of small molecules (~2.5–10 mM in DMSO) in 384-well V-bottom polypropylene plates (Abgene or Greiner). Typical purity requirements for compounds are ≥ 90%.
11. Dimethylformamide, DMF.
12. FMOC 8-amino-3,6-dioxaoctanoic acid (Chem-Impex International, Inc.). Polyethylene glycol spacers of varying lengths ($n = 2$–10 ethylene glycol units) have been successfully used with this protocol.
13. (Benzotriazole-1-yloxy)tripyrrolidinophosphonium hexafluorophosphate, PyBOP (EMD Chemicals, Inc.).
14. *N*,*N*-Diisopropylethylamine, DIPEA.
15. Piperidine, redistilled.
16. 1,6-Diisocyanatohexane (Sigma-Aldrich).
17. Tetrahydrofuran, THF.
18. Pyridine.
19. (PEO)3-monoamine (Molecular Biosciences, Inc.).
20. Deionized water.
21. HiLyte Fluor 488 amine, TFA salt, fluorescent dye for sentinels (Anaspec).

22. Rapamycin, control ligand to FKBP12 (LC Laboratories).
23. FK506, control ligand to FKBP12 (LC Laboratories).

2.2. Screens with Pure Proteins

1. FKBP12 protein fused to a His tag, for control screen (Abcam).
2. Anti-His antibody, Penta-His Alexa Fluor647 Conjugate (Qiagen).
3. 1× TBST: 50 mM Tris–HCl, 150 mM sodium chloride, 0.1% Tween-20, pH 7.5.
4. 4-Well dish untreated ST lid (Nalge Nunc International).
5. 1-Well dish untreated ST lid (Nalge Nunc International).
6. Rocking platform shakers (VWR).
7. Orbital shaker, works orbital shaker (IKA).
8. Gas filter gun with 0.01 μm PVDF filter (Wafergard GN, Entegris Inc.).
9. Nitrogen gas, ultrahigh purity grade (Airgas Inc.).
10. Distilled water.
11. Microarray scanner, Axon Genepix 4300A Scanner (GENEPIX 4300, Molecular Devices), or suitable equivalent.

3. Methods

Here, we describe an updated protocol for identifying protein–small molecule interactions using SMMs. The protocol was first reported in two manuscripts from our laboratory in 2006 (13, 14). Since that time, we have further optimized selected steps in the SMM manufacture protocol. We also provide updated sources for selected instruments and materials used for both manufacture and screening of SMMs. In this specific example, compounds known to bind to the FKBP12, including the commercially available natural products FK506 and rapamycin, are printed onto isocyanate-coated glass slides. We provide a protocol for detecting interactions of these printed ligands with purified and epitope-tagged FKBP12 as a tutorial for screening (Fig. 1). These interactions are typically used for quality control studies when we manufacture arrays containing screening collections. The ligands described here are typically placed throughout the forty-eight 16 × 16 subarrays on SMMs containing 12,288 printed features. While the protocol described here focuses on the use of a pure protein in the screen, proteins residing in cell lysates may be screened as well. For guidance on screening proteins from cell lysates the reader is referred to the protocol in (14).

3.1. Compound Printing/Array Manufacture

3.1.1. Glass Surface Activation (Fig. 2)

1. Prepare stock solutions of compounds for printing, including dyes for sentinels, FK506, and rapamycin, by dissolving them in DMSO. Typical stocks are prepared in a concentration range of 2.5–10 mM depending upon solubility. Stock solutions are stored at –20°C.
2. Transfer 5 μL of each compound stock solution to individual wells in a V-bottom 384-well polypropylene plate. Plates may be centrifuged before use to eliminate air pockets inside the wells, and insure compound solutions are in the well bottoms.
3. Clean amine-functionalized glass slides prior to printing by using a stream of filtered ultrahigh purity grade nitrogen gas to remove any particulates from the surface. Static contact angle of deionized water on this surface should be 40–45°. Rinse the water from the contact angle test slide with DMF and dry under a stream of ultrapure nitrogen gas.
4. Load amine-functionalized slides carefully into a slide rack. Submerge each rack in a glass staining dish containing fresh PEG linker solution: FMOC-8 amino-3,6-dioxaoctanoic acid (1 mM), PyBOP (2 mM), DIPEA (20 mM) in DMF. The solution should

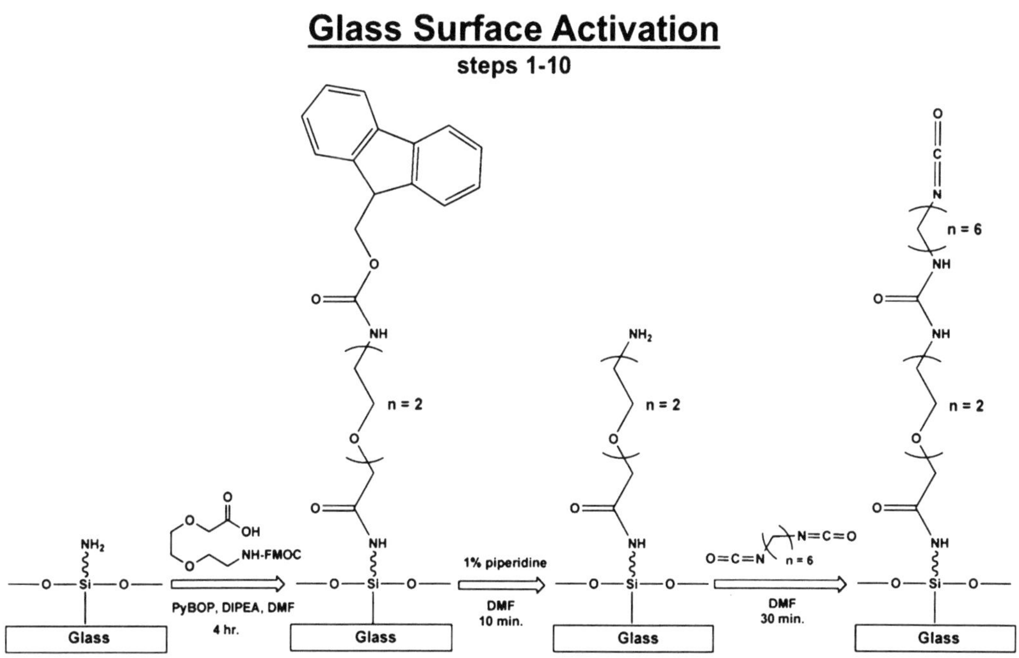

Fig. 2. Scheme for installing a reactive isocyanate group on the surface of a glass slide. Gamma aminopropyl silane (GAPS) slides are coated with an Fmoc-protected polyethylene glycol linker. The protecting group is removed using piperidine, and 1,6-diisocyanatohexane is coupled to the surface to provide the isocyanate glass substrate used in SMM printing. *PyBOP* (benzotriazol-1-yloxy)tripyrrolidinophosphonium hexafluorophosphate; *DIPEA* *N,N*-diisopropylethylamine; *DMF* *N,N*-dimethylformamide.

completely cover the slides. Incubate the slides in PEG linker solution for at least 4 h (up to 24 h) at room temperature with stirring. Avoid forming a vortex in the stirring solution. Cover the staining dish.

5. Remove one slide from the PEG linker solution, rinse thoroughly with DMF from a squirt bottle and dry under ultrapure nitrogen gas. Static contact angle of deionized water on this surface should be 55–60°. Rinse the water from the contact angle test slide with DMF from a squirt bottle and return to the slide rack.
6. Remove the rack from the PEG linker solution. Briefly rinse slides with a gentle stream of DMF from a squirt bottle, then submerge the rack in a clean staining dish that contains FMOC deprotection solution: 1% (v/v) piperidine in DMF. The solution should completely cover the slides. Incubate the slides in the FMOC deprotection solution for at least 10 min (up to 24 h) at room temperature with stirring. Avoid forming a vortex in the stirring solution. Cover the staining dish.
7. Remove the rack from the FMOC deprotection solution. Briefly rinse slides with a gentle stream of DMF from a squirt bottle, then submerge the rack in a clean staining dish containing DMF, and wash for 1 min with stirring. Avoid forming a vortex in the stirring solution.
8. Remove the rack from the wash solvent, then submerge the rack in a clean staining dish containing diisocyanate solution: 1% (v/v) 1,6-diisocyanatohexane in DMF. The solution should completely cover the slides. Incubate the slides in diisocyanate solution for at least 30 min (up to 2 h) at room temperature with stirring. Avoid forming a vortex in the stirring solution. Cover the staining dish.
9. Remove the rack from the diisocyanate solution. Briefly rinse slides with a gentle stream of DMF from a squirt bottle, then submerge the rack in a clean staining dish containing DMF, and wash for 3 min with stirring. Avoid forming a vortex in the stirring solution. Repeat with fresh DMF for 3 min. Remove the rack from DMF and submerge in THF with stirring and wash for 2 min.
10. Remove the rack from the wash solvent and dry the slides using a stream of filtered ultrahigh purity grade nitrogen gas. Static contact angle of deionized water on this surface should be 40–45°. Rinse the contact angle test slide with THF and dry thoroughly.

3.1.2. Small Molecule Printing (Fig. 3)

1. Isocyanate functionalized slides should either be used for array manufacture immediately or stored under nitrogen or argon gas and desiccated to avoid ambient moisture or other chemical vapors.

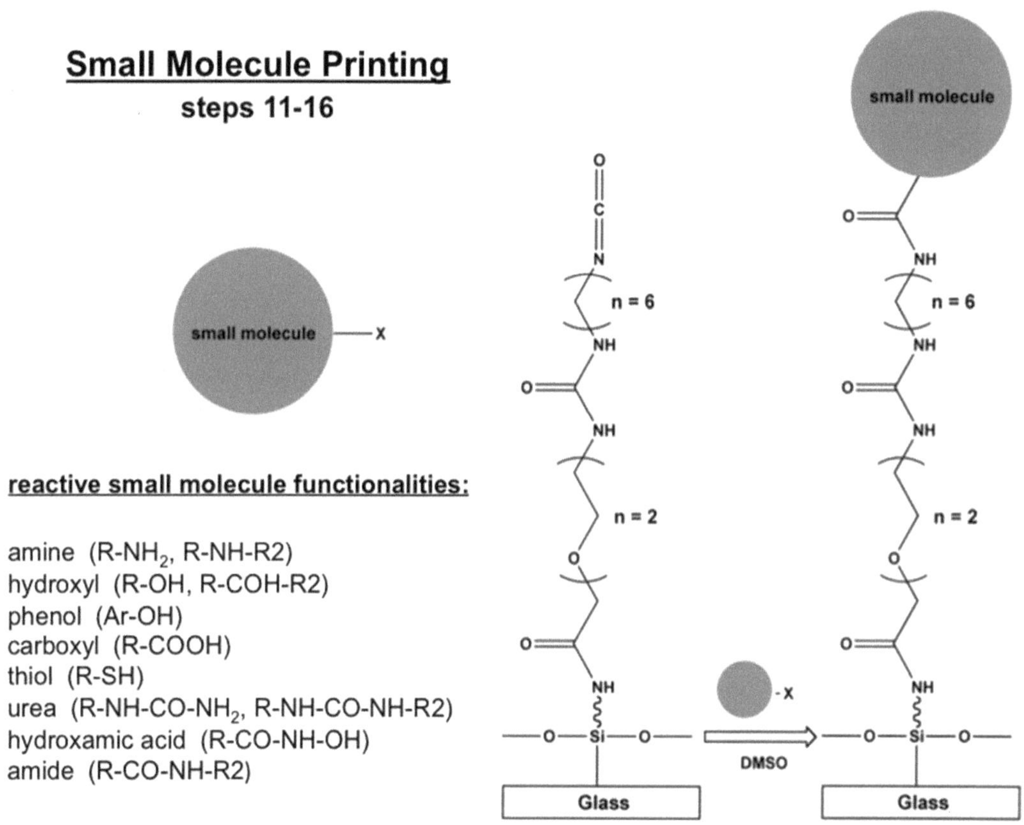

Fig. 3. Activated isocyanate slides are loaded onto the microarrayer platen for small molecule printing. Selected functional groups will react immediately with the glass surface (e.g., amines, thiols, primary alcohols) while less reactive groups require pyridine as catalyst.

2. 384-well V-bottom stock plates containing small molecule solutions should be at room temperature.
3. Arrayer printing pins should be cleaned by sonication in solvents appropriate for removing any trace remains from the last molecule library printed. Typically pins are sonicated in a 50/50 solution of DMSO/water for 20 min followed by 20 min sonication in DMF then 20 min sonication in acetone. Thoroughly dry the pins with a stream of nitrogen prior to printing.
4. Carefully place isocyanate functionalized slides and small molecule stock plates onto the microarrayer platform.
5. Print microarrays using microarrayer and corresponding software. DMSO solutions typically yield 160 μm diameter spots on the isocyanate functionalized glass surface. Typical array feature parameters are set at 270–300 μm feature-to-feature spacing to avoid spots touching or blending together. Pins are washed between each pick up of unique compounds by rinsing

in DMSO five times for 5 s, then touched to bibulous paper to remove the DMSO drop from each pin.

6. When the arrayer run is complete, allow the slides to remain in the arrayer for at least 10 min to allow the DMSO to evaporate from the slide surface.
7. Carefully return the slides to the slide rack. Extra care must be used to not touch or scratch the printed slide surface (see Notes 1–3).

3.1.3. Catalysis and Quench (Fig. 4)

1. Place an open scintillation vial containing 3–5 mL of pyridine inside a vacuum desiccator. Place the slide rack containing arrayed slides in the vacuum desiccator and close. Attach a vacuum

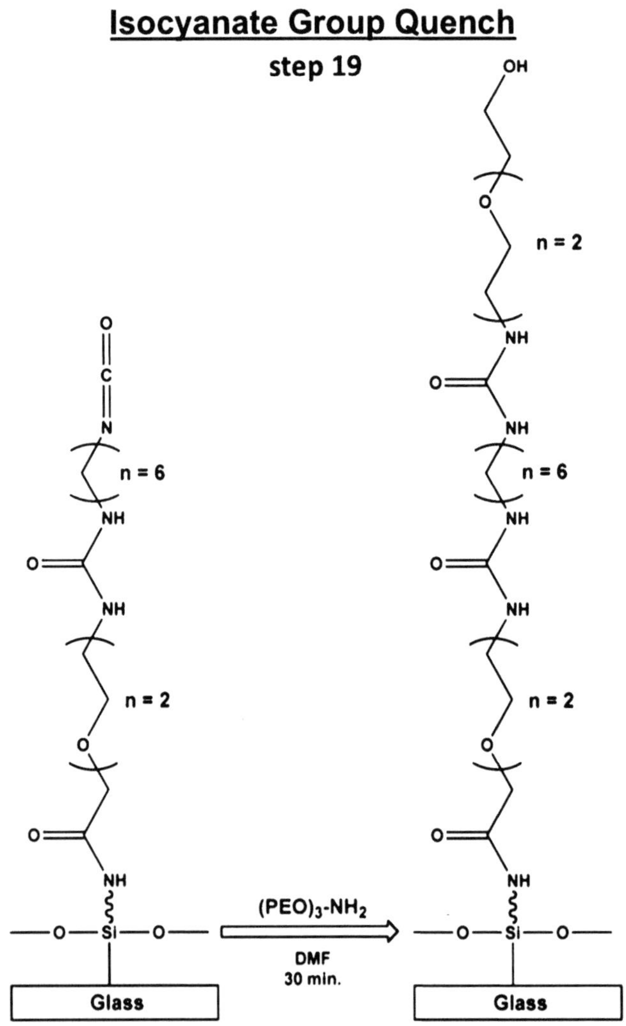

Fig. 4. Unreacted isocyanate groups on the slide surface are quenched with (PEO)3-monoamine to avoid covalent attachment of proteins during a screen. Failure to create an inert background surface may result in very high-background fluorescence on the slide.

hose and pump and evacuate the desiccator. This will create a pyridine vapor inside the desiccator chamber. Allow the slides to remain in the vacuum chamber exposed to the pyridine vapor for 24 h. Pyridine catalyzes the covalent attachment of functional groups that have low reactivity to isocyanate. The vapor pressure of pyridine at room temperature is 18 mmHg.

2. Remove the slide rack from the desiccator and submerge in a clean staining dish containing EG3 quench solution: (PEO)3-monoamine (10 mM) in DMF. The solution should completely cover the slides. Incubate the slides in EG3 quench solution for at least 30 min at room temperature with stirring. Avoid forming a vortex in the stirring solution. Cover the staining dish.
3. Remove the rack from the EG3 quench solution. Briefly rinse slides with a gentle stream of DMF from a squirt bottle, then submerge the rack in a clean staining dish containing DMF, and wash for 1 h with stirring. Avoid forming a vortex in the stirring solution. Remove the rack from DMF and submerge in THF with stirring and wash for 3 min. Repeat with fresh THF for 3 min.
4. Remove the rack from the wash solvent and dry the slides using a stream of filtered ultrahigh purity grade nitrogen gas. Static contact angle of deionized water on this surface should be 40–45°. Rinse the contact angle test slide with THF and dry thoroughly (see Notes 1 and 2).
5. Dry arrayed slides may be stored under inert gas in sealed containers at −20°C for up to 6 months if they will not be used in binding assays immediately.

3.2. Screens with Pure Proteins: Detecting Known Interactions with Fkbp12

1. Prepare 6 mL of FKBP12 protein solution (1 μg/mL) for each microarray slide in TBST. Most proteins are typically screened at concentrations in the 0.1–2 μg/mL range. TBST is a good general binding buffer for many proteins but we typically choose an appropriate buffer for each protein on a case-by-case basis. For example, buffers should contain specific cofactors to maintain appropriate protein conformation or activity when necessary. Care should be taken to avoid autofluorescent additives to any buffers used in the protocol. Insert the microarray slide, printed face up, in one well of the four-well dish.
2. Add 6 mL of the protein solution from the bottom right corner of the well while manually rocking the dish back and forth. Take caution to avoid air bubbles.
3. Place dish with slide on a rocking platform and allow protein binding to occur for 30 min at room temperature. Note that thermally sensitive proteins may be incubated with the microarrays at 4°C for 60 min. Be sure to use chilled buffers for all applications involving thermally sensitive proteins.

4. After the incubation, remove slide from the four-well dish and place it in a single well dish containing TBST buffer. Place on orbital shaker for 1 min and repeat the rinse two times.
5. Prepare a 6-mL solution of a 1:1,000 dilution of anti-His antibody in TBST, and dispense solution in one of the wells of a four-well dish.
6. Place washed slide in antibody solution, printed face up, and incubate while on a rocking platform. Allow antibody binding to occur for 30 min at room temperature. As mentioned previously in step 4, alternative conditions may be used for thermally sensitive proteins.
7. After incubation, remove slide from the four-well dish and single well dish containing TBST buffer. Place on an orbital shaker for 2 min. Repeat the wash step twice using TBST. Perform one final wash in TBS buffer.
8. Perform a final brief 10-s rinse in distilled water with gentle manual agitation by gently rocking the dish back and forth.
9. Dry slides by centrifugation or under a gentle stream of nitrogen gas. Ideally slides are scanned for fluorescence immediately after drying. When using most fluor-conjugated proteins or antibodies, dried slides may be stored at room temperature and in the dark for up to 2 days prior to scanning without significant deterioration of fluorescent signal.
10. Place dried microarray in scanner. Scan the array using the 635 and 488 nm lasers. Adjust the PMT voltages appropriately to avoid features with saturated signal. See Fig. 5 for a scanned array image (see Notes 4 and 5).
11. Align the corresponding GAL file for the microarray print run onto the scanned image using the GenePix Software. Printed fluorescent sentinels (488 nm) are used to generally align the grid and the diameters of the grid features are fit to the fluorescent features using the automated feature in the software. Manually inspect the fitted GAL grid to be sure that the software has sized the features appropriately (see Note 6).
12. Using the Genepix software, generate a GPR file that contains information about fluorescent intensity, signal-noise-ratios, spot diameter, and other types of information for the scanned microarray.
13. Examine GPR data to determine whether known ligands to FKBP12 have been detected. Check to see whether compounds have carried over to the next sample pickup resulting in contamination of the next printed feature.

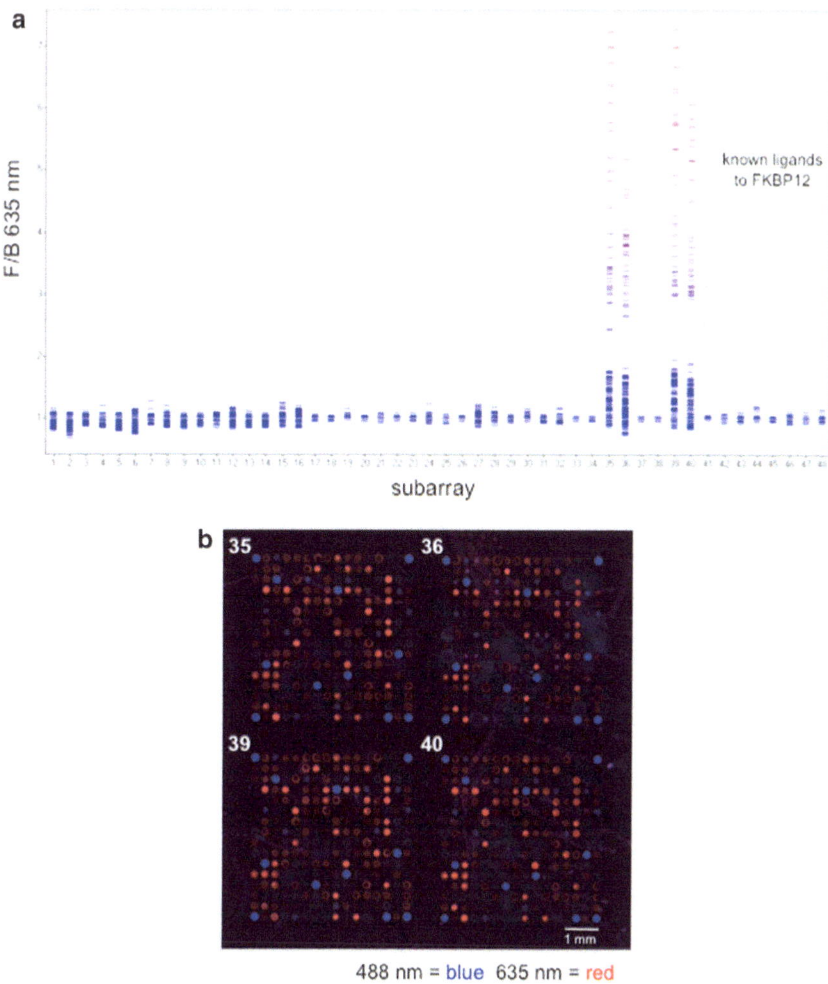

Fig. 5. Small-molecule microarray (SMM) probed with pure FKBP12-His6 followed by an Alexa647-labeled anti-His antibody. (**a**) Plot of foreground signal over local background signals (F/B) for compounds printed in each of the 48 subarrays. Each feature has a local background set to three times larger than the feature diameter. The typical SMM contains 48 subarrays, each with 256 printed features in a 16 × 16 array configuration. Subarrays 35, 36, 39, and 40 contain positive control ligands to FKBP12. All of the ligands show higher F/B values than the other printed features. (**b**) *Inset* image of subarrays 35, 36, 39, and 40. Signal at 488 nm corresponding to the control fluorescein dye features used for aligning the GAL file grid is false-colored *blue*. Signal at 635 nm corresponding to features where the Alexa647-labeled antibody is bound to the SMM is false-colored *red*.

14. Using data contained within the GPR file, compute foreground signal over local background (F/B) values where each feature has a local background that is set to three times the feature diameter (Fig. 5).
15. Compare data from replicate screens. Most screens are run on three replicate slides.

4. Notes

1. In case of a problem with array features blending together:
   (a) Source plate contaminated with solvent. Solvent contamination (e.g., methanol, ethanol, etc.) in the DMSO solution source plates can lead to changes in spot morphology and typically results in spreading of the spot diameter.
   (b) Arrayer pins are not clean. Contamination of pins with dust or precipitated compound may result in pin tips with artificially greater surface area, resulting in pick up and deposition of larger volumes of compound stock solution. Thoroughly clean pins by sonication as described, or remove contaminated pins and gently clean with a cotton swab and solvent. *It is critical not to use cotton swabs that contain adhesive or plastic.*
   (c) Isocyanate surface activation reaction errors. If static water contact angles are not within the range described in the protocol, it is likely the surface chemistry activation reactions have failed. Use fresh reagents, and insure no water or other contamination occurs to the reagents.
   (d) Arrayed slide is not allowed to thoroughly dry. Printed arrays must be allowed to thoroughly dry before removal from the arrayer. The volume of each printed spot should be small enough that 10 min is enough time to allow the plate to dry after printing. No artificial humidification should be supplied to the printing atmosphere.
   (e) Printed features are too close together. Aim to leave at least 100 μm spacing between spots, larger volume spots may require more distance. Also program the arrayer to place successive spots at least two rows and columns away from the most recent printed spot if possible.
   (f) Arrayed slides kept in pyridine vapor too long or vacuum pressure too low. Some compound spots may increase in diameter if kept under vacuum too long or under vacuum that nears the vapor pressure of the compound. The pyridine vapor catalysis step should be done under vacuum not below 17 mmHg.
2. In case of a problem with array spots missing:
   (a) Stock solution not delivered to the glass slide. Check that the source plate contains solution at the bottom of the wells in sufficient volume and that the source plate solutions have not evaporated. Ensure that the arrayer print head is adjusted in the *z*-axis so that pins are contacting the source plate solutions as well as the glass surface.

Inspect and clean any pins that show contamination; "quill" type pins may become obstructed with compounds that have dried on the pins. Also inspect pins to insure they are not damaged or bent.

(b) Compound is not reactive. Selected compounds may not contain a functional group that covalently reacts with an isocyanate. Nonreactive compounds typically rinse away during the EG3 quench reaction, and those spots will react with the (PEO)3-monoamine quenching reagent. Selected compounds that lack an isocyanate-reactive group will nonspecifically stick to the surface. In our experience, some flat and hydrophobic compounds that resemble dyes will stick nonspecifically.

(c) Pyridine catalysis step fails. Ensure the vacuum desiccator is completely sealed and vacuum is maintained to create a pyridine vapor atmosphere.

3. In case of a problem with compound solution carryover from one sample pick up to the next:

(a) Pins are not clean. Increase the duration and number of pin wash cycles between each unique sample pick up.

(b) Pin blotter is contaminated. If a blotter is used to remove solvent from the pins after cleaning, change or clean the blotter between print runs.

(c) Wash solution is contaminated. Replace wash solution with fresh solvent.

4. In case of high-background fluorescence across array:

(a) Avoid autofluorescent additives in buffers. The reagents may form a fluorescent film across the slide that obscures signal from binding interactions. When in doubt, a plain glass slide may be incubated in the buffer of interest, dried and scanned for fluorescence.

(b) Protein concentration is too high. High concentrations of protein (>5 μg/mL) may lead to film formation on the surface. Reduce the amount of protein or evaluate a series of concentrations to optimize F/B values.

(c) Protein is denatured. In our experience, denatured protein samples often display higher background binding to the array surface relative for native protein samples.

(d) Increase duration of washes after final antibody incubation. Extra washes may reduce binding to array background as well as printed small molecules.

(e) Preblock arrays prior to incubation with protein. Although not required for most proteins, blocking agents such as bovine serum albumin (BSA) or SynBlock (AbD Serotec) may be used to reduce background binding. Slides may be preblocked in SynBlock solution 0.1% (v/v) BSA in the appropriate incubation buffer for 60 min prior to incubating with protein. BSA may also be used in subsequent incubation steps.

5. In case that scanned features appear white: The saturation limit of the scanner has been reached. Lower the PMT voltage on the scanner and rescan.
6. Genepix Pro does not properly autofit each printed feature using the GAL file. Selected printed features of irregular shape or size may require manual fitting of the GAL file to the printed feature image.

Acknowledgments

The authors would like to thank Michelle Palmer, Yan-Ling Zhang, Gil Walzer, Hong Chen, Jacob Asiedu, Lisa Marcaurelle, Michael Foley, James Bradner, Olivia McPherson, and Stuart Schreiber for materials, technical support, or advice that was relevant to developing this updated protocol. Work described herein was funded with Federal funds from the National Cancer Institute's Initiative for Chemical Genetics, National Institutes of Health, under contract no. N01-CO-12400.

References

1. Vegas, A. J., Fuller, J. H., and Koehler, A. N. (2008) Small-molecule microarrays as tools in ligand discovery, *Chem Soc Rev 37*, 1385–1394.
2. Astle, J. M., Simpson, L.S., Huang, Y., Reddy, M.M., Wilson, R., Connell, S., Wilson, J., Kodadek, T. (2010) Seamless bead to microarray screening: rapid identification of the highest affinity protein ligands from large combinatorial libraries, *Chem Biol 7*, 38–45.
3. Kwon, S. J., Lee, M. Y., Ku, B., Sherman, D. H., and Dordick, J. S. (2007) High-throughput, microarray-based synthesis of natural product analogues via in vitro metabolic pathway construction, *ACS Chem Biol 2*, 419–425.
4. Uttamchandani, M., Lee, W. L., Wang, J., and Yao, S. Q. (2007) Quantitative inhibitor fingerprinting of metalloproteases using small molecule microarrays, *J Am Chem Soc 129*, 13110–13117.
5. Vegas, A. J., Bradner, J. E., Tang, W., McPherson, O. M., Greenberg, E. F., Koehler, A. N., and Schreiber, S. L. (2007) Fluorous-based small-molecule microarrays for the discovery of histone deacetylase inhibitors, *Angew Chem Int Ed Engl 46*, 7960–7964.
6. Urbina, H. D., Debaene, F., Jost, B., Bole-Feysot, C., Mason, D. E., Kuzmic, P., Harris, J. L., and Winssinger, N. (2006) Self-assembled small-molecule microarrays for protease screening and profiling, *Chembiochem 7*, 1790–1797.
7. Koehler, A. N., Shamji, A. F., and Schreiber, S. L. (2003) Discovery of an inhibitor of a transcription factor using small molecule microarrays and diversity-oriented synthesis, *J Am Chem Soc 125*, 8420–8421.
8. Stanton, B. Z., Peng, L. F., Maloof, N., Nakai, K., Wang, X., Duffner, J. L., Taveras, K. M., Hyman, J. M., Lee, S. W., Koehler, A. N., Chen, J. K., Fox, J. L., Mandinova, A., and Schreiber, S. L. (2009) A small molecule that binds Hedgehog and blocks its signaling in human cells, *Nat Chem Biol 5*, 154–156.
9. Labuda, L. P., Pushechnikov, A., and Disney, M. D. (2009) Small molecule microarrays of

RNA-focused peptoids help identify inhibitors of a pathogenic group I intron, *ACS Chem Biol 4*, 299–307.

10. Duffner, J. L., Clemons, P. A., and Koehler, A. N. (2007) A pipeline for ligand discovery using small-molecule microarrays, *Curr Opin Chem Biol 11*, 74–82.

11. Uttamchandani, M., Wang, J., and Yao, S. Q. (2006) Protein and small molecule microarrays: powerful tools for high-throughput proteomics, *Mol Biosyst 2*, 58–68.

12. Kohn, M., Wacker, R., Peters, C., Schroder, H., Soulere, L., Breinbauer, R., Niemeyer, C. M., and Waldmann, H. (2003) Staudinger ligation: a new immobilization strategy for the preparation of small-molecule arrays, *Angew Chem Int Ed Engl 42*, 5830–5834.

13. Bradner, J. E., McPherson, O. M., Mazitschek, R., Barnes-Seeman, D., Shen, J. P., Dhaliwal, J., Stevenson, K. E., Duffner, J. L., Park, S. B., Neuberg, D. S., Nghiem, P., Schreiber, S. L., and Koehler, A. N. (2006) A robust small-molecule microarray platform for screening cell lysates, *Chem Biol 13*, 493–504.

14. Bradner, J. E., McPherson, O. M., and Koehler, A. N. (2006) A method for the covalent capture and screening of diverse small molecules in a microarray format, *Nature Protocols 1*, 2344–2352.

15. Miyazaki, I., Okumura, H., Simizu, S., Takahashi, Y., Kanoh, N., Muraoka, Y., Nonomura, Y., and Osada, H. (2009) Structure-affinity relationship study of bleomycins and Shble protein by use of a chemical array, *Chembiochem 10*, 845–852. 2

Chapter 18

Working with Small Molecules: Preparing and Storing Stock Solutions and Determination of Kinetic Solubility

Andrea Wolf, Satoko Shimamura, and Friedrich B.M. Reinhard

Abstract

The handling of organic compounds in the laboratory requires the use of organic (co-) solvents to mediate solubility. Advantages and disadvantages of the widely used solvent dimethylsulfoxide (DMSO) are discussed, and guidelines for dissolution and storage of compounds are given. Finally, nephelometry is introduced as a fast method to determine the kinetic solubility of a compound.

Key words: Dissolution, Compound handling, Compound storage, DMSO, Kinetic solubility

1. Introduction

All methods and procedures using small organic molecules described in this book require the chemical substances to be present in solution. As organic molecules all show a certain degree of lipophilicity, an organic solvent with suitable properties has to be identified. Organic compound solutions are usually tolerated if not preferable for experiments like immobilisation of the compound on a non-biological surface or any other organic-chemical reaction. However, if the compound is to be used in an aqueous environment either as a ligand (e.g. an enzyme inhibitor) or as a reactant, the number of possible options is somewhat limited. As most organic compounds used in chemical proteomics cannot be dissolved in water easily it is advisable to use an organic solvent as an intermediate step. An optimal solvent for this application has to have a high dielectric constant to allow the disruption of dipole–dipole interactions typically found in compounds designed to interact with proteins. It should not contain any hydrogen donor–acceptor network which would have to be

Gerard Drewes and Marcus Bantscheff (eds.), *Chemical Proteomics: Methods and Protocols*, Methods in Molecular Biology, vol. 803, DOI 10.1007/978-1-61779-364-6_18,

disrupted by the compound during the dissolution process. Additionally, it should be readily mixable with water and non-toxic when applied to cells. Finally, it should be non-denaturing when used with protein solutions in typical concentration ranges present after the final step of diluting the compound with water of buffer. An excellent example for this type of solvent is dimethylsulfoxide (DMSO) which is the reason for its widespread usage for this purpose: the dielectric constant is with 33 quite high, no H-donor groups are present, it can be mixed with water to all percentages and cultured cells will typically tolerate 0.5% DMSO without effects on viability. Protein solutions often tolerate between 2 and 10% DMSO, sometimes even more, without substantial protein denaturation.

However, DMSO also has several disadvantages, and consequently several precautions should be taken. DMSO is a solvent with a very high hygroscopy. The reason for this are strong interactions of DMSO and water using H-bonding which leads to a liquid phase with higher structure than DMSO alone, peaking in a well-defined complex of one DMSO and two water molecules (1, 2). At this ratio, the physicochemical properties of DMSO are largely altered: the viscosity increases and the solubility of lipophilic compounds in DMSO decreases. The consequence for compounds stored as DMSO solutions is that dissolved compounds start to precipitate, leading to lower true compound concentrations in solution. This precipitation is further accelerated by freeze/thaw cycles. Additionally, the melting point of the DMSO–water mixture decreases from 18°C (pure DMSO) to −78°C (3), resulting in unfrozen samples even at −20°C. Several studies conducted on the effect of different storage conditions on compound stability (4–7) identified water as having the most deleterious effect and resulted in the recommendation of −20°C and dry environment as the best way of storing compound solutions in DMSO while keeping the number freeze/thaw cycles low. However, for long-term storage (more than 1 year) samples should be stored as powder or film.

As most reactions or drug discovery experiments assume the compound being completely solubilized, the knowledge of the maximum solubility of a substance under the used conditions is essential. Therefore, several methods have been implemented in the past years to measure compound solubility in a convenient way (8). A kinetic solubility assay is the method of choice for determination of the maximum solubility of a compound under most reaction or assay conditions. In contrast to a thermodynamic/equilibrium solubility assay where the compound is added as solid to the aqueous solution and therefore the dissolution of compound in the presence of excess solid is measured, in the kinetic solubility assay a small volume of a DMSO stock solution of the compound is added to the buffer system. Hence, this assay is mimicking the typical use of a compound better then the equilibrium assay. The most common method is the nephelometric kinetic assay (9–11). A serial dilution

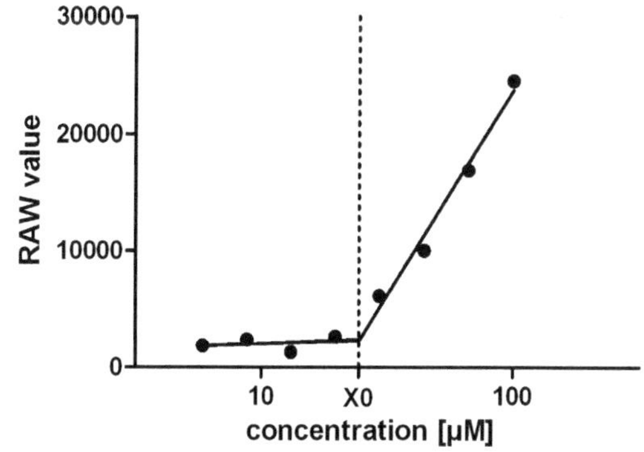

Fig. 1. Segmental regression analysis of light scattering data of positive control pyrene. The *intersection* of the *two lines* represents the determined concentration of the kinetic point of solubility of the compound under these conditions, in this case 25 μM.

of a stock solution of compound is added to a well of a 96- or 384-well plate filled with buffer. A laser beam at typically 640 nm illuminates the well and any light scattering caused by compound precipitates is measured and quantified. The light scattering signal is plotted versus the concentration and two lines are fitted to the data (Fig. 1). The intersection point is used as the point of solubility (10). As microplate readers supporting this type of measurement are available, this method allows the determination of solubility of several compounds in a variety of conditions at a high throughput.

2. Materials

2.1. Dissolution of and Storage of Compounds

1. DMSO (dimethylsulfoxid), e.g. "purissimum" grade DMSO stored over molecular sieve.
2. Glass vials for compounds storage (Zinsser Analytics, Qualyvials).
3. Polypropylene tubes (e.g. Thermo Fisher Scientific, Pittsburgh, PA, Screenmates in 96-tube latch rack).
4. Seals for polypropylene tubes (e.g. Thermo Fisher Scientific, Pittsburgh, PA, SepraSeal mats #4463).
5. Tube or microplate shaker (e.g. Vortex Genie, Scientific Industries, Inc., Bohemia, NY, or Titramax 100, Heidolph Instruments GmbH & Co. KG, Schwabach, Germany).

2.2. Kinetic Solubility Measurement

1. 96-Well dilution plates: V-bottom polypropylene plates (e.g. Greiner Bio-One, Frickenhausen, Germany).
2. 384-Well dilution plates: V-bottom polypropylene plates (e.g. Greiner Bio-One, Frickenhausen, Germany).

3. 96-Well assay plates: Black, clear-bottom plates (e.g. µClear®, Greiner Bio-One, Frickenhausen, Germany).
4. 384-Well assay plates: Black, clear-bottom plates (e.g. µClear®, Greiner Bio-One, Frickenhausen, Germany).
5. Phosphate-buffered saline (PBS): Dissolve 8 g NaCl, 0.2 g KCl, 1.44 g Na_2HPO_4, 0.24 g KH_2PO_4 in 800 ml of distilled H_2O. Adjust the pH to 7.4 with 1 M hydrochloric acid. Add H_2O to 1 l. Filter with 0.2-µm filter to sterilise and remove any floating particles that might cause light scattering signals.
6. Control compounds: it is advisable to use a compound with a known and well-defined solubility limit in the concentration range of interest as positive control, e.g. pyrene (Sigma-Aldrich, Germany). In this set-up, the solubility of pyrene was determined to be in the range of 18–30 µM. Additionally, a negative control, defined as a compound with no solubility limitations in the concentration range of interest, should be used as well, e.g. corticosterone (Fluka, Germany) or DMSO alone. Dissolve pyrene and corticosterone to a concentration of 10 mM in DMSO and perform eight steps of 1:1.5 serial dilutions in DMSO to get to a 100× concentrated stock solution assuming a desired final DMSO concentration of 1% (adjust concentrations if different percentages of DMSO are required). Prepare several aliquots of these dilutions in plates, seal the plates, and store the dilutions at −20°C. Aliquots should be thawed and used once.
7. Sample compound preparation: sample compounds with unknown solubility are dissolved to a sensible concentration (e.g. 10 or 30 mM). Serial dilution and storage are performed as for the control compounds.
8. Centrifuge containing microplate holders (e.g. Multifuge 3S-R, Heraeus).
9. Automated liquid handling workstation: serial dilutions in dilution plates and compound addition to assay plates should be done with robotic liquid handling for optimal reproducibility (e.g. Biomek FXp with Biomek software, Beckman Coulter, Inc., Brea, CA, or Janus™ Automated Workstation with WinPREP Software (Version 4.3.0.0016), PerkinElmer, Inc., Waltham, MA).
10. NepheloStar Galaxy 504 reader with Galaxy software (BMG Labtech GmbH, Offenburg, Germany).
11. Reagent bulk dispenser (e.g. Multidrop combi or Wellmate, Thermo Fisher Scientific, Pittsburgh, PA).
12. Data analysis software, e.g. GraphPad Prism V5.0, GraphPad Software, Inc., La Jolla, CA.

3. Methods

3.1. Dissolution and Storage of Compounds

1. Ideally, compound handling is performed in an environment with low humidity and inert gas atmosphere (Argon).
2. Weigh out the required amount of your solid compound sample in either a glass vial or a polypropylene tube with the appropriate volume to dissolve to the desired concentration (see Note 1).
3. Add the calculated volume of fresh and dry DMSO to the container and close it with the screw cap (glass vials) or the SepraSeal tube sealing (polypropylene tubes).
4. Agitate the sealed container at room temperature using the tube/microplate stirrer until the compound is completely dissolved. However, avoid extending the time the compound is kept at room temperature to more than 24 h (see Note 2).
5. Transfer aliquots of the compound solution into new containers in quantities that are typically required per experiment.
6. Seal the containers and freeze them at −20°C unless they are used immediately (see Note 3).

3.2. Kinetic Solubility Measurement

1. Fill 384-well assay plates with 99 μl of filtered PBS solution using a bulk dispenser. Transfer 1 μl of the compound solution in DMSO to the assay plates followed by three mixing steps with 20-μl volume. Accordingly, 96-well assay plates should be filled with 198 μl of buffer and 2 μl of compound solution. Centrifuge the plates at 350 × *g* for 1 min to remove air bubbles. Seal the top of the assay plate with an adhesive seal to avoid evaporation and to prevent entering of dust (see Note 4).
2. Incubate replicate of plates for 2 h at 37°C to obtain stable results (see Note 5). Place plates in centrifuge and spin again at 350 × *g* for 1 min. Remove adhesive seal and proceed to measurement.
3. Instrument setup: 30 min prior to reading, the Nephelostar reader is set to the incubation temperature of 37°C. Optimal parameters for the instrument settings are plate and buffer dependent and should therefore be optimized. However, a laser beam focus of 2 mm for 96-well assay plates and 1 mm for the 384-well assay plates used in this assay along with a gain setting of 80% and a laser intensity setting of 90% proved to yield reproducible results.
4. The whole microplate is scanned and the raw data are exported as text file (see Note 6).
5. Data analysis is performed in GraphPad Prism: create a new XY-data table and paste the log transformed concentrations as

X values and the Nephelostar raw data as *Y* values. Analyse the data using segmental regression as a fit model. This will fit two lines to the data with an interception at $X = X_0$, the solubility point (see Fig. 1).

4. Notes

1. When calculating the volume of DMSO to dissolve a compound check if the solid sample contains any salt (e.g. as counter ion) and consider the additional molecular weight of the salt.
2. Never store samples dissolved in DMSO in the refrigerator as water vapour will diffuse slowly through the walls of polypropylene tubes.
3. Keep freeze/thaw cycles of compound solutions to the absolute minimum. Avoid thawing and handling in a humid environment to keep DMSO as dry as possible.
4. Always wear gloves while handling the clear bottom plates as fingerprints will interfere with the nephelometric measurement. Likewise, carefully avoid any scratches as nephelometer readings are affected by scratches and dust, often yielding an incorrectly low solubility value. Wipe off the dust on the bottom surface with ethanol soaked Kim towel or paper towel, and dry the bottom before measurement.
5. Other temperatures might be chosen as well. A prolongation of the incubation time did not result in any significant changes.
6. If temperatures below room temperature are used for the incubation of the kinetic solubility be sure to bring plate to room temperature to avoid fogging of the plate bottom.

References

1. Catalán J, Díaz C and García-Blanco F (2001) Characterization of Binary Solvent Mixtures of DMSO with Water and Other Cosolvents. *J. Org. Chem.* *66*, 5846–5852.
2. Lipinsky CA (2006) Samples in DMSO: What an end user needs to know. *LRIG presentations.* http://lab-robotics.org/Presentations/Lipinski%200602/Lipinski_LRIG%20Feb%202,%202006.pdf. Accessed 15 December 2010
3. Rasmussen DH and MacKenzie AP (1968) Phase Diagram for the System Water-Dimethylsulphoxide. *Nature.* *220*, 1315–1317.
4. Kozokowski BA, Burt TM, Tirey DA, Williams LE, Kuzmak BR, Stanton DT, Morand KL and Nelson SL (2003) The Effect of Room-Temperature Storage on the Stability of Compounds in DMSO. *J. Biomol. Screen.* *8*, 205–209.
5. Kozokowski BA, Burt TM, Tirey DA, Williams LE, Kuzmak BR, Stanton DT, Morand KL and Nelson SL (2003) The Effect of Freeze/Thaw Cycles on the Stability of Compounds in DMSO. *J. Biomol. Screen.* *8*, 210–215.
6. Cheng X, Hochlowski J, Tang H, Hepp D, Beckner C, Kantor S and Schmitt R (2003) Studies on Repository Compound Stability in DMSO under Various Conditions. *J. Biomol. Screen.* *8*, 292–304.

7. Ilouga PE, Winkler D, Kirchhoff C, Schierholz B and Wöcke J (2007) Investigation of 3 Industry-Wide Applied Storage Conditions for Compound Libraries. *J. Biomol. Screen.* *12*, 21–32.

8. Kerns EH, Di L and Carter GT (2008) In Vitro Solubility Assays in Drug Discovery. *Curr. Drug. Metab.* *9*, 879–885.

9. Dehring KA, Workman HL, Miller KD, Mandagere A and Poole SK (2004) Automated robotic liquid handling/laser-based nephelometry system for high throughput measurement of kinetic aqueous solubility. *J. Pharm. Biomed. Anal.* *36*, 447–456.

10. Fligge TA and Schuler A (2006) Integration of a rapid automated solubility classification into early validation of hits obtained by high throughput screening. *J. Pharm. Biomed. Anal.* *42*, 449–454.

11. Thomas S, Brightman F., Gill, H, Lee, S and Pufong B (2008) Simulation Modelling of Human Intestinal Absorption Using Caco-2 Permeability and Kinetic Solubility Data for Early Drug Discovery. *J. Pharm. Sci.* 97, 4557–4574.

Chapter 19

A Database for Chemical Proteomics: ChEBI

Paula de Matos, Nico Adams, Janna Hastings, Pablo Moreno, and Christoph Steinbeck

Abstract

Chemical proteomics is concerned with the identification of protein targets interacting with small molecules. Hence, the availability of a high quality and free resource storing small molecules is essential for the future development of the field. The Chemical Entities of Biological Interest (ChEBI) database is one such database. The scope of ChEBI includes any constitutionally or isotopically distinct atom, molecule, ion, ion pair, radical, radical ion, complex, conformer, etc., identifiable as a separately distinguishable entity. These entities in question are either products of nature or synthetic products used to intervene in the processes of living organisms. In addition, ChEBI contains a chemical ontology which relates the small molecules with each other thereby making it easier for users to discover data. The ontology also describes the biological roles that the small molecules are active in. The ChEBI database also provides a central reference point in which to access a variety of bioinformatics data points such as pathways and their biochemical reactions; expression data; protein sequence and structures.

Key words: Chemical compound, Chemical nomenclature, InChI, InChIKey, IUPAC, Molecular entity, Ontology, Substructure search, Similarity search, Web Services

1. Introduction

Small molecules play an important role in chemical proteomics which is concerned with the identification of protein targets interacting with small molecules. Hence, the availability of a high quality and free resource storing small molecules is essential for the future development of the field. The Chemical Entities of Biological Interest (ChEBI) (1) database is one such database. The scope of ChEBI includes any constitutionally or isotopically distinct atom, molecule, ion, ion pair, radical, radical ion, complex, conformer, etc., identifiable as a separately distinguishable entity. These entities in question are either products of nature or synthetic products used

Gerard Drewes and Marcus Bantscheff (eds.), *Chemical Proteomics: Methods and Protocols*, Methods in Molecular Biology, vol. 803, DOI 10.1007/978-1-61779-364-6_19, © Springer Science+Business Media, LLC 2012

to intervene in the processes of living organisms. In addition, ChEBI contains a chemical ontology which relates the small molecules with each other, thereby making it easier for users to discover data. The ontology also describes the biological roles that the small molecules are active in. The ChEBI database also provides a central reference point in which to access a variety of bioinformatics data points such as pathways and their biochemical reactions; expression data; protein sequence and structures. In the following sections, we give a brief outline of the database, its features, and how it can be used.

2. Materials

The following materials are required to use the ChEBI service.

2.1. ChEBI Web Access

A computer with an internet browser (e.g. Internet Explorer or Mozilla Firefox) and Java version 6 or higher is required to access ChEBI.

2.2. ChEBI Programmatic Access

A computer with over 20 GB of disk space. In addition, depending on what programmatic access is required the following software should be installed.

- Internet browser.
- An RDBMS such as Oracle 11g, MySQL 5, or PostgreSQL.
- A spreadsheet application.
- A compression/decompression utility that can handle gzip-compressed files.
- Java version 1.6.

3. Methods

In this section, we give a detailed description of the ChEBI database and the data contained within it. The following sections focus on how to access its data.

3.1. ChEBI Database

The ChEBI database contains a number of diverse chemical entities. Each entity is manually annotated by an expert chemist and contains a number of data fields. Entities in ChEBI include chemical classes, groups, and chemical compounds. The database consists of two types of entities namely, three and two star entities, which mainly distinguish the level of annotation assigned to each.

- Three star entities have been manually annotated by ChEBI annotators with annotators assigning a recommended name, additional nomenclature, references to the literature and classification within the chemical ontology.
- Two star entities have been loaded from the ChEMBL database (2), which is also manually annotated, or have been submitted by external users.

3.2. Data Fields

A typical ChEBI entity can contain a number of data fields. In this section, a description of each data field is given.

3.2.1. ChEBI ID

A unique and stable identifier for the entity, for example, CHEBI:16236. It has no chemical significance and should be used solely for identification purposes. If a redundant entity is discovered in ChEBI then the two entities will be merged and one of the identifiers will be deemed a secondary identifier. Both the identifiers will continue to be stable and maintained.

3.2.2. ChEBI Names

The ChEBI Name is a name for an entity recommended for use by the biological community. The ChEBI Name is unique within the ChEBI database and ChEBI names are selected to be unambiguous. The ChEBI Name is also provided in ASCII format if the original includes special characters, which require a Unicode presentation.

3.2.3. Definition

A short textual definition is included in some entries.

3.2.4. Stars

All entries are rated using a star system as follows:

- 3 stars: The entity has been manually annotated by the ChEBI team.
- 2 stars: The entity has been manually annotated by the ChEMBL project or by a ChEBI submitter.
- 1 star: The entity is a Preliminary Entry loaded automatically from a data source but which has not been manually annotated.

An absence of stars indicates that the entity is either deleted or obsolete.

3.2.5. Chemical Structure Representations

- *MDL Molfiles.* ChEBI stores the two-dimensional or three-dimensional structural diagrams as connection tables in MDL Molfile (3) format. All other structural information is generated from this format. An entity can have one or more connection tables though there is always a primary or default structure assigned. The additional structures are shown to provide alternative views of the same entity.
- *IUPAC International Chemical Identifier (InChI) and InChIKey.* The InChI (4) is a non-proprietary identifier for

chemical substances that can be used in printed and electronic data sources thus enabling easier linking of diverse data compilations. It expresses chemical structures in terms of atomic connectivity, tautomeric state, isotopes, stereochemistry, and electronic charge in order to produce a sequence of machine-readable characters unique to the respective molecule.

The InChIKey is a 25-character hashed version of the full InChI, designed to allow for easy web searches of chemical compounds. It is also available from ChEBI and is generated from the ChEBI Molfile. The InChI and InChIKey are generated from the default MDL Molfile stored in the ChEBI database.

- *SMILES* (simplified molecular input line entry system) (5) is a simple but comprehensive chemical line notation, created in 1986 by David Weininger and further extended by Daylight Chemical Information Systems, Inc. SMILES specifically represents a valence model of a molecule and is widely used as a data exchange format. The SMILES are generated from the default MDL Molfile using the MarvinBeans ChemAxon software package.

3.2.6. Formula

Where possible, formulae are assigned for entities and groups. For compounds consisting of discrete molecules, this is generally the molecular formula, a formula according with the relative molecular mass (or the structure).

3.2.7. Charge and Mass

For ions the magnitude of the charge is given in Arabic numerals preceded by the sign of the charge. For neutral molecules, the charge is indicated as a numerical zero. Relative molecular, atomic, and ionic masses are shown for molecular, atomic, and ionic entities, respectively. The relative masses are calculated from tables of relative atomic masses (atomic weights) published by IUPAC.

3.2.8. Nomenclature

The ChEBI database was initially created as chemical nomenclature database and hence allows users to store different types of nomenclature.

- *IUPAC names.* An IUPAC name is a name provided for an entity based on current recommendations of the International Union of Pure and Applied Chemistry (IUPAC). It need not be fully systematic as it makes use of "retained names." In most cases, a single IUPAC Name is provided for a molecular entity or a group.
- *INN.* In cases where an entity is a pharmaceutical substance, an International Nonproprietary Name (INN) may be shown. The INN is the official non-proprietary or generic name given to a pharmaceutical substance, as designated by the World

Health Organisation (WHO). INNs may appear in ChEBI in English, Latin, Spanish, and French language versions.

- *Synonyms*. Alternative names for an entity which either have been used in EBI or external sources or have been devised by the curators based on recommendations of IUPAC, Nomenclature Committee of the International Union of Biochemistry and Molecular Biology (NC-IUBMB) or their associated bodies, are included in ChEBI as synonyms. The source of each synonym is clearly identified. Systematic names may also be included in this section. In addition to English-language synonyms, versions may be shown in French, German, Spanish, and Latin, the language being indicated by a flag.

3.2.9. Database Links

Database links are direct links to the entries for an entity in the databases cited. These links may be manually annotated by a ChEBI annotator or they may be automatically generated by external databases who use ChEBI identifiers. ChEBI has an extensive collection of links including links to pathway and reaction databases, bioactivity databases, interaction databases, and enzyme databases. The ever expanding list of databases that ChEBI links to can be found on the ChEBI web site in the user manual.

3.2.10. Registry Numbers

Registry numbers are essentially identifiers which are assigned when a molecule enters a certain type of database. ChEBI currently stores three types of registry numbers, namely Chemical Abstracts Service (CAS), Beilstein, and Gmelin.

3.2.11. Citations

Publications which cite the entity are listed as citations, along with hyperlinks to the PubMed entry via CiteXplore (6), a web application of the EBI for the exploration of literature related to biological research and bioinformatics.

3.3. The ChEBI Ontology

Although ChEBI is designed as a relational database, it makes extensive use of ontology annotations, which allows the faceted search and browsing of ChEBI content. Ontologies are computable "formal explicit specifications of a shared conceptualisation" of a knowledge domain (7). In other words, ontologies define both entities and the relationships between entities and express this knowledge in a formal and computable way. Ontologies are in widespread use in bio- as well as medical informatics, but also in software engineering, knowledge representation, artificial intelligence research, and applications and are an integral component of the semantic web.

Usually, ontologies consist of several distinct building blocks, namely instances or individuals, sets of instances (often referred to as classes or, under special circumstances as "universals") and relations. Individuals are instances of classes (the single benzene molecule, which is being observed in an IR spectroscopy experiment in matrix isolation is an instance of the class "benzene molecule," which is the set of all benzene molecules in the world). Ontologies aim to make statements about a domain at a class, rather than an instance level. Classes are organised in the form of a taxonomy, which specifies the subsumption or "is a" relationship structure ("primary amine is a amine"). Often, ontologies contain further relationships, which specify other relationships between classes such as, for example, disjoints or mereological relationships.

The following section discusses both the origin and structure of the present ChEBI ontology as well as future developments. ChEBI terms are cited using the ChEBI identifier.

3.3.1. Origins of the ChEBI Ontology

The development of the ChEBI Ontology is a consequence of the development of the Gene Ontology (GO) (8) and the increasing need to annotate chemical entities involved in biological processes both with chemical information as well as biological roles. This is complementary to the goals of the Gene Ontology, namely to develop a dynamic controlled vocabulary, that can be used to annotate the roles and functions of genes and therefore also of gene products, which may be proteins or (small) molecules. Very early incarnations of the Gene Ontology, therefore, contained descriptions of these small molecules and their functions, which, in subsequent revisions, were factored out into a "Chemical Ontology," which together with the KEGG Compound (9) and IntEnz (10) databases, form the foundations of the modern version of ChEBI.

3.3.2. ChEBI Ontology Structure

The ChEBI ontology consists of three subontologies, describing molecular structure (CHEBI:24431), roles (CHEBI:50906), and subatomic particles (CHEBI:36342). It has the structure of a directed acyclic graph (DAG), meaning that ChEBI allows multiple parentage. Each class in the ChEBI ontology has, in accordance with OBO guidelines, a numerical term ID, a short ChEBI name, and a full IUPAC name together with a number of synonyms from different sources. We also provide Aristotelian definitions for classes and chemical structures (11). The molecular structure ontology describes chemical entities in terms of their structural characteristics as well as composition (Fig. 1).

The molecular structure ontology is subdivided into smaller ontologies of functional groups, atoms, and various types of molecular entities. The description of molecular entities in terms of structure or functional groups has been established in the chemical community for over 100 years now and is the primary "lens" through which a chemist will typically view and describe a chemical entity. Biomedical researchers, by contrast, often find a description of chemical entities in terms of roles or functions more useful and in accordance with their model of the natural world. ChEBI provides for this through the provision of a role ontology. The latter describes approximately 500 roles, which molecular entities may

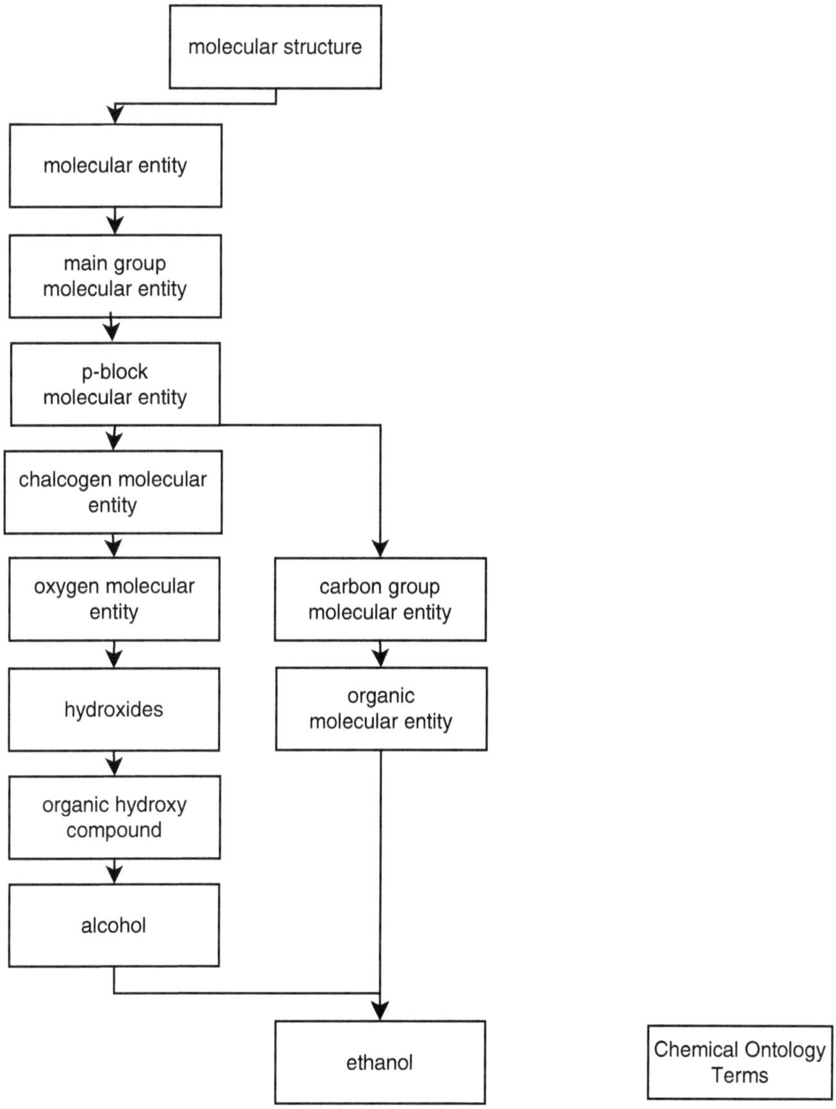

Fig. 1. Snapshot of the DAG for ethanol (ChEBI:16236) showing the position in the molecular structure ontology.

assume under a given set of circumstances. The roles are subdivided into biological roles (CHEBI:24432), chemical roles (CHEBI:51086), and application roles (CHEBI:33232). Typical biological roles, for example are "immunomodulator" (CHEBI: 50846), "neurotransmitter agent" (CHEBI:35942), or "hormone antagonist" (CHEBI:49020). Chemical roles include terms such as "acid" (CHEBI:51086), "chelator" (CHEBI:38161), or "chromophore" (CHEBI:23240). Application roles, finally, encompass terms such as "fragrance" (CHEBI:48318), "chemical tracer" (CHEBI:35205), or "fertilizer" (CHEBI:33287).

The subatomic particle ontology describes particles smaller than atoms and is useful for the description of many chemical concepts such as atoms (atoms have a parthood relationship with electrons).

3.3.3. Relationships in ChEBI

Terms in the ChEBI ontology are linked to each other via relationships. We have already discussed the subsumption ("is a") relationship, which is supplemented by other relations as detailed in Table 1.

Using these added relations, both the chemical as well as biological roles of a molecule as well as its position in multiple taxonomies can now be comprehensively described (Fig. 2).

3.3.4. Future Work

ChEBI is part of the OBO Foundry community, which is currently in the process of a significant restructuring of its ontological assets in order to ensure maximum interoperability and expressivity. This also requires a major restructuring of the ChEBI ontology, which, though useful as a standalone ontology, has room for improvement in interoperating with other Foundry ontologies. Specifically, this means that ChEBI will, in the medium term, be aligned to other Foundry ontologies through the adoption of the Basic Formal Ontology and the Relationship Ontology, followed by extensive annotation of ChEBI content after adoption of these foundational ontologies.

3.4. ChEBI Website

The ChEBI Website provides public access to the ChEBI data and includes searching, browsing, and visualisation facilities.

3.4.1. Searching

ChEBI can be searched via a simple or advanced text search, or via a chemical structure search.

Simple Search

The simple search is the main entry point into the ChEBI data for the user. It is available from the main ChEBI home page via a search box centrally located in the page, and additionally a search may be performed from any ChEBI entry page via the search box in the top right of the screen. The search query may be any data associated with an entity, such as names, synonyms, formulae, CAS or Beilstein Registry numbers, or InChIs.

When there are multiple results returned from a search, the search takes you to a search results page, which displays the results in a grid format together with the name, identifier, and structure (where applicable) of the search hit. Search results may be downloaded for import into other applications, in a flat file, tab delimited, or chemical structure SDF file (for those search results which contain structures only). When there is only one result found the search takes you directly to the entity result page, bypassing the search results table.

Wildcards are available for both the simple and the advanced search. The wildcard character is "*." A wildcard character allows you to find compounds by typing in a partial name. The search engine will then try to find names matching the pattern you have specified.

Table 1
The relationships of the ChEBI ontology

ChEBI relation	Usage/definition
has part	The "has part" relationship is used to denote parthood relationships between a part and a whole [e.g. potassium tetracyanonickelate$^{(2-)}$ (CHEBI:30071) has part tetracyanonickelate$^{(2-)}$ (CHEBI:30025)]
is conjugate base of	The "is conjugate base of" relationship connects an acid with its conjugate base and, together with the "is conjugate acid of" relationship leads to a set of cyclic relationships. The relationship is often used for disambiguation purposes: biologists, for example, often confuse "acetate" with "acetic acid". This relationship assists disambiguation by making the distinction explicit
is conjugate acid of	See above
is tautomer of	Tautomers are chemical entities, which can rapidly interconvert through a chemical reaction known as tautomerism. The cyclic "is tautomer of" relationship is used to show the relationship between two chemical entities with significantly different structural representation to warrant their separate inclusion in ChEBI
is enantiomer of	Enantiomers are compositionally identical molecules, which are structurally non-superimposable and are, in fact mirror images of one-another. The relationship is cyclic
has functional parent	The "has functional parent" relationship is used to connect two chemical entities, which can be derived from each other and where one of the chemical entities contains a characteristic arrangement of atoms in space, which defines a structural class of molecules, e.g. 16α-hydroxy-progesterone (CHEBI:15826) has functional parent progesterone (CHEBI:17026) (the former can be derived from the latter via 16α hydroxylation)
has parent hydride	The "has parent hydride" relationship relates a chemical entity to its parent hydride (defined by IUPAC as "an unbranched acyclic or cyclic structure or an acyclic/cyclic structure having a semisystematic or trivial name to which only hydrogen atoms are attached"). For example, naphthalene (CHEBI:16482) has parent hydride 1,4-naphthoquinone (CHEBI:27418)
is substituent group from	The "is substituent from" relationship indicates the relationship between a parent molecular entity and a functional group (or an atom), which has been derived from the parent entity via loss of one or several atoms. For example, l-valino group (CHEBI:32854) is substituent group from l-valine (CHEBI:16414)
has role	The "has role" relationship indicates a particular role a chemical entity may assume under a given set of circumstances. For example morphine (CHEBI:17303) has role opioid analgesic (CHEBI:35482)

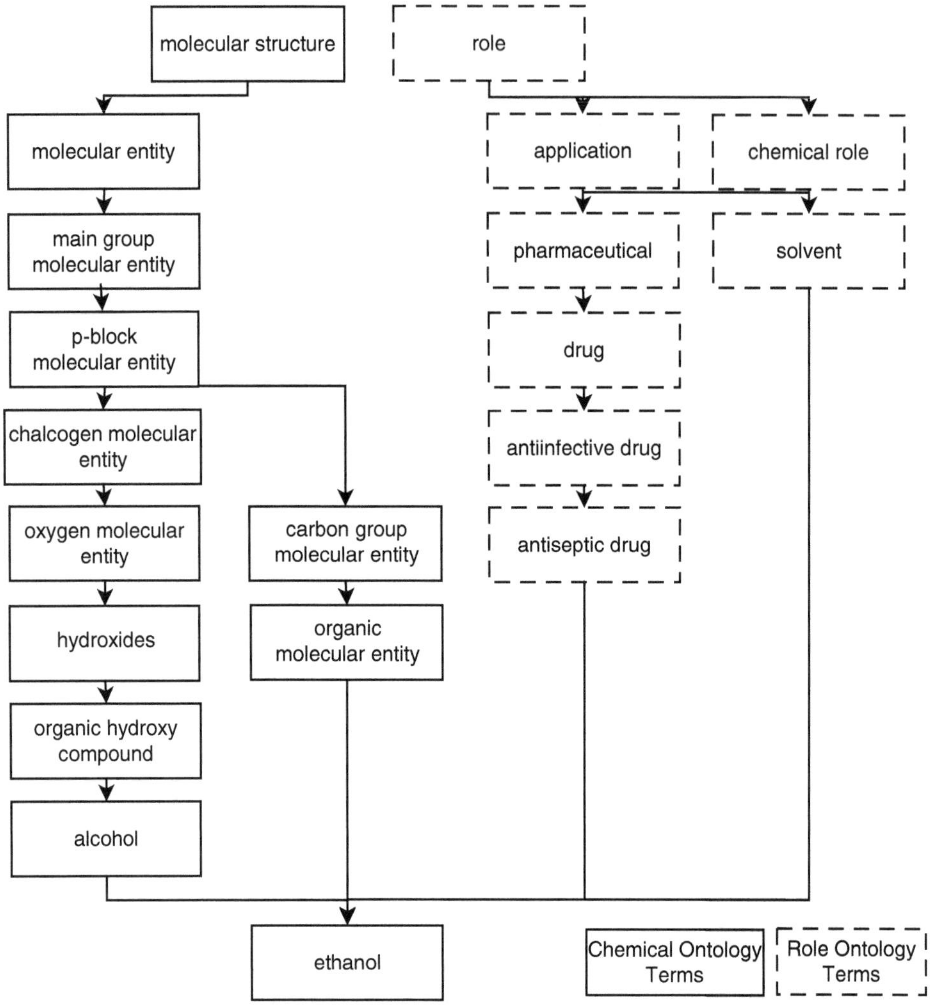

Fig. 2. Snapshot of the DAG for ethanol (ChEBI:16236) including role definitions.

- To match words *starting with* your search term add the wildcard character to the end of your search term. For example, searching for *aceto** will find compounds such as *acetochlor*, *acetophenazine*, and *acetophenazine maleate*.
- To match words *ending with* your search term add the wildcard character to the start of your search term. For example, searching for **azine* will find compounds such as *2-(pentaprenyloxy) dihydrophenazine*, *acetophenazine*, and *4-(ethylamino)-2-hydroxy-6-(isopropylamino)-1,3,5-triazine*.
- To match words *containing* a search term, add the wildcard character to the start and the end of your search term. For example, searching for **propyl** will find compounds such as *(R)-2-hydroxypropyl-CoM*, *2-isopropylmaleic acid*, and *2-methyl-1-hydroxypropyl-TPP*.

Any number of wildcard characters may be used *within* a search term. Data containing Unicode characters may be searched for by using Unicode UTF-8 characters or by using their ASCII representation. For example, when searching for subscripts such as H_2O, the ASCII representation is H_2O.

Advanced Text Search

The advanced text search provides for searches with additional granularity and specificity by allowing specific categories to be searched in, as well as by providing combinations of searches using Boolean operators. The advanced search page also contains the structure-based search which may be combined with the text-based search. To access the advanced search screen, select the "Advanced Search" link from the left-hand menu. Alternatively, if a simple search fails to return any results, you will be taken there automatically.

The text search facility of the Advanced Search allows users to search all the data or to filter a search by any of the available categories, which are:

- All – allows you to search all the categories (i.e. the same behaviour as the simple text search).
- ChEBI Identifier – allows searching for specific ChEBI identifiers.
- ChEBI Name – will search only for ChEBI Names matching your search term.
- Definition – allows you to search within the Definition field.
- IUPAC Name – will search for IUPAC Names.
- All Names – will search for ChEBI Names, IUPAC Names and Synonyms.
- InChI/InChIKey – will search for InChIs and InChIKeys matching the search criteria.
- SMILES – will search for SMILES strings matching the search criteria.
- Cross references – allows searching for accession numbers from other linked sources.
- Registry numbers – will search for CAS, Beilstein, and Gmelin Registry Numbers matching the search criteria.
- Citations – will search within the Citations for matches with the search criteria.
- Formula – will search within the Formula field.
- Mass – will search within the Mass field.

Chemical formulae can be searched for, and this search parses the formula (which must be specified with the correct case) so as to determine the difference between Co (Cobalt) and CO (Carbon-Oxygen). Mass and charge can be searched within ranges: for example, one can search for all entities with a mass of between 150 and 300 atomic mass units. Furthermore, searches can be filtered by database: for example, one can search for entities used in the NMRShiftDB (12) or PubChem (13) databases.

Users also have the ability to filter on the ChEBI ontology. This functionality allows one to retrieve all the children of a specific entity based on the relationship given. For example, all cofactors (CHEBI:23357) can be retrieved by entering the term cofactors using the "has role" relationship and this will retrieve not only its direct children such as "pantothenic acids" (CHEBI:25848) but also further entities in the graph related via an "is a" relationship, such as NADPH (CHEBI:16474). It also allows retrieval of only those entities with chemical structures by ticking the "Filter by chemical structure" checkbox.

As in the Quick search, the asterisk (*) is provided as the wildcard character. Furthermore, all of these searches can be combined by using the logical operators AND, OR, and BUT NOT.

Structure Search

Also available on the ChEBI advanced search page is the structure search facility. The chemical structure search algorithm which is implemented is provided by the OrChem Oracle chemical cartridge (14). The user is able to perform the structure search by drawing or pasting a structure into the JChemPaint (15) applet, an editor and viewer for 2D chemical structures. The specified structure is then converted into fingerprints, each fingerprint representing the occurrence of a particular structural feature. It is important to remember that fingerprints have limitations: they are good at indicating that a particular structure feature is not present but they can only indicate a structure feature's presence with some probability.

The different types of structure searches that may be performed are the substructure, similarity, and identity search, each of which is described below. Structure searches may also be combined (using the standard Boolean operators) with any of the advanced text search queries available lower down on the screen.

- *Substructure search.* Fingerprints are used to reduce the size of the search space in substructure searching, by eliminating candidates for which it can be determined on the basis of the fingerprints that they cannot be a superstructure of the search structure. If molecule A is a substructure of molecule B, then all bits set ("on") in the fingerprint of molecule A should also be set in molecule B. Note, however, that it does not follow that if all bits are set in molecule A that are set in molecule B, then molecule A is necessarily a substructure of molecule B (although it may well be). So, once this initial screening is performed, the potential substructure candidates are subjected to a more rigorous and computationally demanding subgraph isomorphism matching algorithm to determine whether molecule A is genuinely a substructure of molecule B. To perform a substructure search in ChEBI draw your chemical structure using the JChemPaint applet. Then select the "Chemical Structure Search" option "Substructure" and click "Search." If your substructure is found within the database the results will be displayed with relevant links to the entities found.

- *Similarity search.* Similarity searching is performed by calculating the Tanimoto coefficient (16) for each structure within the database against the query structure. The Tanimoto coefficient calculates how many structural features two chemical structures have in common based on the fingerprint described above. A Tanimoto score of 1.0 indicates that all fingerprint bits are shared between the two structures, which implies that the two structures are highly similar. However, as the fingerprints are calculated on a chemical structure bond path depth of eight, it means that some structures will have similar fingerprints and high similarity scores even though they might not be very structurally similar.
- *Identity search.* Identity searching is performed using the InChI chemical identifier. The structure which is specified as a search structure in the applet is converted into an InChI representation, and the database is then scanned for molecules with that identical InChI.

3.4.2. Browsing

The ChEBI ontology public website may be browsed via the interrelationships between entities in the ontology, and via the atomic constitution of the entities in terms of the periodic table. Additionally, ontology browsing functionality is provided via external websites, including the Ontology Lookup Service (17), which is described below.

- *Interrelationships between entities.* As discussed above in the section on the ChEBI ontology, each ChEBI entity is interrelated by the relationships between entities specified in the ChEBI ontology. These ontology relationships are displayed on the entry page of the public website together with a clickable link to the target of the relationship. Additionally, the ChEBI ontology tree view allows the user to show all paths from the entity they are currently looking at, to the root of the ontology, and each entity along the path to the root is clickable for navigation to the entry page for that entity.
- *Periodic table.* The ChEBI Periodic Table allows browsing of the dataset by the familiar periodic table. The periodic table may be browsed for molecular entities containing those elements, or for the elements themselves. Clicking one of the links on the molecular entities periodic table browser takes you to the entry page for the particular molecular entity class within ChEBI. For example, clicking the link on sodium (Na) takes you to the "sodium molecular entity" entry page, which is a class within the ChEBI ontology. You can navigate further within the data by clicking one of the links in the ChEBI ontology section of

the entry page. Similarly, clicking one of the links on the elements periodic table takes you to the ChEBI entity for the atom of that type, such as the sodium atom.

- *Ontology lookup service (OLS)*. The Ontology Lookup Service is a facility which provides a centralised query interface for ontology and controlled vocabulary lookup. The link to activate the Ontology Lookup Service for ChEBI, a link is available under the "Browse" menu on the left-hand menu bar of the ChEBI public website, which opens the OLS website in a new window or tab, with the ChEBI ontology pre-loaded. The OLS provides both a tree-based visualisation and navigation facility for the ontology, details of data such as synonyms and cross-references associated with each entity, and a graph visualisation of the ontology around the selected node.

3.5. Programmatic Access

3.5.1. Web Services

The ChEBI Web Service provides users with programmatic access to the ChEBI public data and ontology. This access allows users to integrate ChEBI with their own applications. Currently, the Web Service allows users to search for small molecules through a number of options, retrieve data, and explore the ChEBI ontology. The service is SOAP based, and through the published WSDL definition file, clients for most programming languages can be generated by the user with appropriate software. Alternatively, packaged clients for Java and Perl are available for download through the ChEBI Web Services home page (18).

Web Service Methods

The ChEBI Web Service implements seven methods that allow users to search, retrieve data, and explore the ChEBI database. Each method specifies certain parameters which need to passed to the programme and there is a predefined return type for each method.

getLiteEntity Method

This is the main search entry point as it allows free text search as well as through other categories. In its default case (search category and stars set to "ALL"), it is equivalent to the simple search defined in Subheading 3.4.1.

The parameters that need to be provided to this method are listed below.

- Search: The text to be searched against the database. This field allows the use of an asterisk as a wildcard and any Unicode character. It operates in the same way as the simple search in the web page.
- Search category: This defines which data field of the database will be searched using the search text. Only one category can be selected. The default case "All" behaves as the simple search. The list of available categories to select from are:
  - All
  - ChEBI Id

 - ChEBI Name
 - Definition
 - All Names
 - IUPAC Name
 - Database Link/Registry Number/Citation
 - Formula
 - Mass
 - InChI/InChI Key
 - SMILES

- Maximum results: Defines the maximum number of results that the search can return. The maximum results that the system can return by default is 5,000, however, this can be reduced by setting this parameter.
- Stars: Minimum number of stars (0–3) that the ChEBI entities must have in order to be eligible for the search.
- This method retrieves a list of "lite" entity objects. Each "lite" entity object represents an entity in the ChEBI database. These objects contain the following data items.
- ChEBI ASCII name: The ChEBI recommended name in ASCII format.
- ChEBI Id: The ChEBI database identifier.
- Search score: The text search score for this hit.
- Stars: The number of stars.

getCompleteEntity Method

This method returns the complete entity available in the ChEBI database for a ChEBI identifier. Once a particular ChEBI entity has been identified through the getLiteEntity search or other way, this method receives that ChEBI identifier and returns a complete entity, which contains most of the data fields available in the ChEBI database. The only parameter required for this method is:

- ChEBI Id: The complete ChEBI Identifier of the entity to retrieve, e.g. CHEBI:15377.
- The result, a complete entity object, contains the following data fields for the desired entity (for more detailed descriptions of the data fields, see Subheading 3.2):
- ChEBI Identifier: The ChEBI identifier for the entity.
- ChEBI Name: The ChEBI recommended name in ASCII format for that entity.
- Definition: A short free text description of the entity.
- Status: It has four possible states; *checked* – the entity was checked by a curator; *ok* – the entity was loaded but has not

been checked; *deleted* – the entity was deleted by a curator; *obsolete* – the entity has been obsoleted by the system.

- SMILES: The SMILES string computed for the molecule.
- InChI: The InChI calculated for the molecule.
- InChI Key: 25 Character key computed from the InChI.
- Charge: The electrostatic charge of the molecule as deposited.
- Mass: The expected atomic mass of the molecule.
- Stars: The number of ChEBI stars of the entity.
- Secondary ChEBI Identifiers: Sometimes a ChEBI entity is referenced by more than one ChEBI ID. In such a case, a collection of those ChEBI IDs is stored in this field.
- Synonyms: This is a collection of synonyms for the chemical entity.
- IUPAC Names: Standard chemical name according to the naming conventions of the International Union of Applied Chemistry.
- Formulae: The empirical chemical formulae of the entity in ASCII format. It includes the source of this data.
- Registry Numbers: CAS, Beilstein, and Gmelin registry numbers when available.
- Citations: A collection of references to the literature that cite that entity.
- Chemical Structures: A collection of available chemical structures, in MDL mol format, for the entity.
- Database Links: A number of different database links which are relevant to this specific entity.
- Ontology Parents: The outgoing relationships for the entity.
- Ontology Children: The incoming relationships for the entity.

getComplete EntityByList Method

This method produces a list of "complete" entity objects for a list of ChEBI Identifiers given. The complete entity object is the same as the one returned by the getCompleteEntity method. The maximum size of the list to be retrieved is 50 "complete" entities. As such the only parameter given is:

- ChEBI Identifiers List: A list of ChEBI identifiers to retrieve for example, CHEBI:15377, CHEBI:25805.
- The result retrieved contains a list of "complete" entity objects.
- Complete Entity List: A list of "complete" entity objects, one complete entity per provided ChEBI Identifier in the initial list.

  In simple terms, this method is equivalent to calling the getCompleteEntity method for several different ChEBI identifiers one after the other.

getOntologyParents and getOntologyChildren Methods

Given a ChEBI identifier, these methods retrieve either all the direct parents or direct children for that identifier within the ChEBI Ontology, regardless of the relation type. For every parent or child, the type of relation such as "is a," "has part," and "has role," that connects it to the queried entity, is included in the result. These methods are not recursive, thus only the first level of parents or children are provided. The only parameter given to the method is:

- ChEBI Identifier: The ChEBI identifier (example CHEBI:15377) of the entity whose parents or children one is looking for.

  The method returns a list of pairs composed of entities and relationships.
- ChEBI Identifier: The ChEBI identifier of a child or parent of the queried entity.
- Relationship Type: The relationship (as described in Subheading 3.3.4) that connects the queried ChEBI entity with the parent or child ChEBI entity of this pair.

getAllOntologyChildrenInPath Method

This method is similar to the getOntologyChildren method, but returns all the children obtained by recursively traversing along the ontology relationship connections to the leaf ontology nodes. The relationship type that is being retrieved has to be specified, and the ontology will be searched for children in "is a" paths which are connected to the given search entity via the given search relationship. The search can additionally be constrained to consider only ChEBI entities that have a chemical structure. The output is a list of child ChEBI identifiers. Parameters for this method are as follows.

- ChEBI Identifier: The ChEBI identifier of the entity whose children one is looking for.
- Relationship Type: The relationship (as described in Subheading 3.3.4) that is to be considered for the search.
- Structure Only: True or false value stating whether the search should be limited only to ChEBI entities that have a chemical structure.

  The method returns a list of ChEBI identifiers
- Children List: A list of ChEBI identifiers that are direct or recursive children of the provided ChEBI identifier satisfying the relationship type specified and structure constraints.

getStructureSearch Method

This method provides chemical structure search capability. The allowed chemical structure search types are substructure, similarity and identity, which are explained in Subheading 3.4.1. The method receives as input a structure, which needs to be provided by the tuser in MDL Molfile format, Chemical Markup Language (CML) (19) format, or SMILES format. Additional to the structure, options need to be selected, such as the structure search type, the maximum amount of results to be retrieved, and the Tanimoto cutoff score, which only applies for similarity search. Parameters for this method are as follows.

- Structure: Text representing the chemical structure to use as query. This can be written in MDL Mol format, Chemical Markup Language (CML), or SMILES.
- Type: The format type that was used for describing the query molecule in the previous field. This is either MDL Molfile, CML, or SMILES.
- Structure Search Category: The type of structural search to perform with the query which can be Similarity, Substructure, or Identity.
- Total Results: An integer limiting the maximum number of results that will be retrieved.
- Tanimoto Cutoff: A floating point number between 0 and 1 that will be taken as the minimum Tanimoto score to be returned. Values close to 1 represent a high level of similarity.

The result delivered by the method is a list of objects, having one object for each ChEBI entity. Each object consists of the following data items.

- ChEBI Identifier: ChEBI Identifier for the entity.
- ChEBI ASCII Name: ChEBI recommended name for the entity.
- Entity Star: The star rating that the resulting ChEBI entity has (0–3).

Example Workflow for Retrieving the Functional Children of an Entity

This workflow is based on the results using ChEBI Release 65 (see Note 1). It takes as input a KEGG Compound database identifier and retrieves a list of ChEBI entities and their KEGG Compound identifiers corresponding to compounds that have a functional parent in the initial query. The workflow can be followed in Fig. 3 and the steps are described below:

1. A search using method getLiteEntity is conducted using as input “C00022”, which is the KEGG Compound identifier for pyruvate (pyruvic acid), “ALL” as the search category and “THREE ONLY” for the stars field. Failing to set up the stars numbers would change the results radically. This search yields two results, CHEBI:15361 for pyruvate and CHEBI:32816 for pyruvic acid. ChEBI distinguishes between the two conjugates.
2. The two resulting ChEBI identifiers are fed into the method getOntologyChildren, one at the time. For pyruvate, CHEBI: 15361, 12 children result of invoking getOntologyChildren, of which 10 have the relationship “has functional parent”.

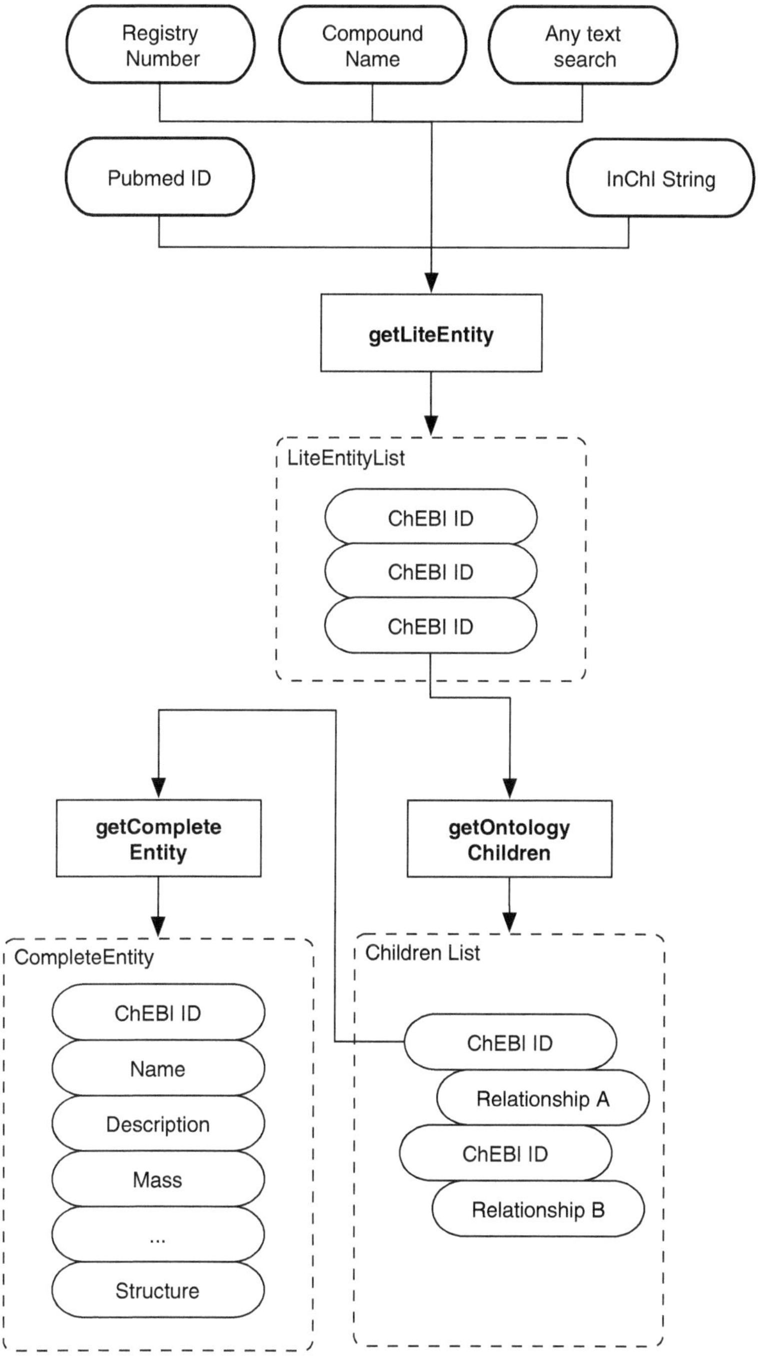

Fig. 3. An example workflow using the web service.

This means that there are ten compounds in ChEBI that can be derived from pyruvate. Among those ten functional children, you should see 3-hydroxypyruvate CHEBI:17180, 3,4-dihydroxyphenylpyruvate CHEBI:29055 and keto-phenylpyruvate CHEBI:18005 to name a few.

3. Invoking the method getCompleteEntity with the identifier of keto-phenylpyruvate, CHEBI:18005, yields its complete entity, which contains in the "DatabaseLink" field C00166 for the KEGG Compound database, which corresponds to keto-phenylpyruvate in KEGG Compound.
4. Repeating step 3 iteratively with the results from step 2 will produce results such as those listed in Table 2.

This workflow could be somewhat simplified by replacing method getOntologyChildren by getOntologyChildrenInPath, and adding to this method's invocation the necessary relationship type "has functional parents."

3.5.2. Downloads

For access to the complete data set deposited in ChEBI, a number of download alternatives are provided through the ChEBI downloads web page (20).

SDF File

The SDF file should be the format of choice if the main interest is the chemical structure data. The MDL SDF file format is one of the most widely used standards for storing chemical structures for a set of molecules. SDF is essentially a concatenation of MDL Mol files (each Molfile houses one chemical structure). When it comes to ChEBI, four collections are available as SDF files for download:

- ChEBI Lite: Contains the chemical structure, ChEBI identifier, ChEBI ASCII name, and ChEBI Stars for every ChEBI entity. This data is written in the [Data Header] and [Data] fields of the standard SDF file, and as such, they should be accessible to any software that reads compliant MDL SDF format.
- ChEBI Lite 3 Stars: Same as ChEBI Lite, but only containing ChEBI entities with 3 stars, in other words those entities which have been manually annotated by ChEBI curators.
- ChEBI Complete: Contains all chemical structures and associated data for every ChEBI entity. The data fields included in the format are listed in Subheading 3.2.
- ChEBI Complete 3 Stars: Same as ChEBI Complete, but limited only to the ChEBI entities with 3 stars.

MDL SDF file format can be easily read by most chemical aware software packages such as BioEclipse (21) and chemoinformatics programming libraries such as the Chemistry Development Kit (22). Note that the SDF file does not contain any ontological information.

OBO File

If the main interest is the extraction of the ChEBI ontology then the Open Biomedical Ontologies (OBO) file format should be used. The OBO file format is one of the standards for biomedical ontology storage and exchange. ChEBI provides the ChEBI ontology

Table 2
The results from implementing and running the example workflow using the KEGG Compound identifier C00022 as an input in ChEBI Release 65

Functional child entity	Child ChEBI name	KEGG compound link
CHEBI:15361 pyruvate		
CHEBI:17180	3-Hydroxypyruvate	C00168
CHEBI:18110	3-Phosphonatooxypyruvate(3-)	C03232
CHEBI:19602	2-Hydroxy-3-carboxybenzylidenepyruvate	No link found
CHEBI:19603	2-Hydroxy-3-methylbenzylidenepyruvate	No link found
CHEBI:27040	Trans-2-carboxybenzylidenepyruvate	No link found
CHEBI:36242	3-(4-Hydroxyphenyl)pyruvate	C01179
CHEBI:17271	3-Phosphonatopyruvate(3-)	C02798
CHEBI:17468	(4-Bromophenylsulfanyl)pyruvate	C04264
CHEBI:29055	3,4-Dihydroxyphenylpyruvate	No link found
CHEBI:18005	Keto-phenylpyruvate	C00166
CHEBI:32816 pyruvic acid		
CHEBI:16894	3-Sulfopyruvic acid	C05528
CHEBI:18007	3-[Hydroxy(oxido)phosphoranyl] pyruvic acid	C06368
CHEBI:30841	3-Hydroxypyruvic acid	C00168
CHEBI:30933	3-Phosphonooxypyruvic acid	C03232
CHEBI:8934	(4-Bromophenylsulfanyl)pyruvic acid	C04264
CHEBI:27597	3-(5-Hydroxyindol-3-yl)pyruvic acid	No link found
CHEBI:29750	3-(Indol-3-yl)pyruvic acid	C00331
CHEBI:17406	3-(Imidazol-5-yl)pyruvic acid	C03277
CHEBI:27501	3-(Imidazol-1-yl)pyruvic acid	No link found
CHEBI:19610	2-Hydroxy-4-hydroxymethylbenzylidenepyruvic acid	No link found
CHEBI:1665	3-Sulfinylpyruvic acid	C05527
CHEBI:30935	3-Phosphonopyruvic acid	C02798
CHEBI:16208	3-Mercaptopyruvic acid	C00957
CHEBI:19891	3,4-Dihydroxyphenylpyruvic acid	No link found
CHEBI:15999	(4-Hydroxyphenyl)pyruvic acid	C01179
CHEBI:28128	3,5-Dibromo-4-hydroxyphenylpyruvic acid	C04285
CHEBI:28025	3,5-Dichloro-4-hydroxyphenylpyruvic acid	No link found
CHEBI:17131	(3,5-Diiodo-4-hydroxyphenyl)pyruvic acid	C01244
CHEBI:28039	4-Hydroxy-3-iodophenylpyruvic acid	C03987
CHEBI:27981	3,5-Dinitro-4-hydroxyphenylpyruvic acid	No link found
CHEBI:18184	3,5,3′-Triiodothyropyruvic acid	C03832
CHEBI:18184	3,5,3′-Triiodothyropyruvic acid	C11138
CHEBI:30851	Keto-phenylpyruvic acid	C00166
CHEBI:51850	Methyl pyruvate	No link found
CHEBI:1447	3-Acylpyruvic acid	No link found
CHEBI:52989	4-Hydroxy-3-iodophenylpyruvate	C03987

in OBO format version 1.2. The OBO file can be explored with the freely available OBO-Edit tool (23), which provides a graphical user interface to view and edit ontologies in this format.

Flat File Tab-Delimited

ChEBI is stored in a relational database and all of its tables can be downloaded as an independent flat-file tab delimited format. For each table, two versions are available.

1. 3 Stars Version: Contains data only for the 3 Stars ChEBI entities, and files are names < table_name > _3stars.tsv.
2. All Stars Version: Contains data for all the entities in ChEBI, and files are named < table_name > .tsv. This of course includes the 3 Stars Version.

All these files can be opened with spreadsheet applications, manipulated, or read by unix standard text tools or used to produce a local replicate of the database. If the interest is of general data manipulation, this should be the format of choice. To recreate the database within a local relational database management system (RDBMS), probably a better choice would be to download the Database table dumps.

Database Table Dumps

ChEBI can be downloaded in amenable format for recreating the relational database in the user's local RDBMS. For this aim, two additional download options are provided.

- *Oracle binary table dumps.* ChEBI provides an Oracle binary table dump that can be imported into an Oracle relational database. The binary table data dumps are available in two versions.

1. 3 Stars version: Light version that contains only those entities in ChEBI with 3 ChEBI Stars (file named CHEBI_ASCII_3STAR.dmp.gz).
2. Complete version: Contains all the entities in ChEBI (file named CHEBI_ASCII_ALL_STAR.dmp.gz). This of course includes the 3 Stars version.
3. In both cases, the dump file which provide the instructions for generating the table's structures and the parameters files (ending with ".par") need to be downloaded.

   Once downloaded, the chosen dump file (either All Stars or 3 Stars) can be imported into Oracle using the "imp" command. The parameter file, either import.par (for the 3 Stars version) or import_all_star.par (for the All Stars version), should reside in the same directory as the dump when the import is done. The correct command to execute is:

   imp<database_name>/<database_password>@<Instance_name>PARFILE=import.par

   imp<database_name>/<database_password>@<Instance_name>PARFILE = import_all_star.par

- *Generic SQL table dumps.* ChEBI provides a generic SQL dump which consists of generic SQL insert statements (meaning that the SQL dialect used is supposed to be RDBMS independent). The dump is provided in two versions:

1. 3 Stars Version: Contains only those entities in ChEBI with 3 Stars (file named generic_dump_3stars.zip).
2. Complete version: Contains all the entities in ChEBI (file named generic_dump_allstar.zip).

   The archive files called generic_dump_3star.zip and generic_dump_allstar.zip consists of 12 files each which contain SQL table insert statements of the entire database. The file called compounds.sql should always be inserted first in order to avoid any constraint errors. Included in the folder are a MySQL and PostgreSQL create table scripts as an example for other users of the database. These insert statements should be usable in any database which accepts SQL as its query language but has only been tested on MySQL.

4. Note

1. All the example data used in this text was taken from ChEBI Release 65 dated 3 February 2010.

References

1. de Matos, P., Alcantara, R., Dekker, A., Ennis, M., Hastings, J., Haug, K., Spiteri, I., Turner, S., and Steinbeck, C. (2010) Chemical Entities of Biological Interest: an update, *Nucl. Acids Res. 38*, D249–254.
2. ChEMBL DB. http://www.ebi.ac.uk/chembldb. Accessed 1 March 2010.
3. Symyx CTFile formats. http://www.symyx.com/solutions/white_papers/ctfile_formats.jsp. Accessed 1 March 2010.
4. Stein, S. E., Heller, S. R., and Tchekhovskoi, D. (2003) An Open Standard for Chemical Structure Representation - The IUPAC Chemical Identifier, *International Chemical Information Conference*, 131–143.
5. Weininger, D., Weininger, A., and Weininger, J. L. (1989) SMILES. 2. Algorithm for generation of unique SMILES notation, *Journal of Chemical Information and Computer Sciences 29*, 97–101.
6. CiteXplore database. http://www.ebi.ac.uk/citexplore. Accessed 1 March 2010.
7. Studer, R., Benjamins, V. R., and Fensel, D. (1998) Knowledge engineering: Principles and methods, *Data & Knowledge Engineering 25*, 161–197.
8. Ashburner, M., Ball, C. A., Blake, J. A., Botstein, D., Butler, H., Cherry, J. M., Davis, A. P., Dolinski, K., Dwight, S. S., Eppig, J. T., Harris, M. A., Hill, D. P., Issel-Tarver, L., Kasarskis, A., Lewis, S., Matese, J. C., Richardson, J. E., Ringwald, M., Rubin, G. M., and Sherlock, G. (2000) Gene ontology: tool for the unification of biology. The Gene Ontology Consortium, Nature Genetics 25, 25–29.
9. Kanehisa, M., and Goto, S. (2000) KEGG: Kyoto Encyclopedia of Genes and Genomes, *Nucl. Acids Res. 28*, 27–30.
10. Fleischmann, A., Darsow, M., Degtyarenko, K., Fleischmann, W., Boyce, S., Axelsen, K., Bairoch, A., Schomburg, D., Tipton, K., and Apweiler, R. (2004) IntEnz, the integrated relational enzyme database, Nucl. Acids Res. 32, D434–437.

11. Schober, D., Smith, B., Lewis, S., Kusnierczyk, W., Lomax, J., Mungall, C., Taylor, C., Serra, P., and Sansone, S. (2009) Survey-based naming conventions for use in OBO Foundry ontology development, *BMC Bioinformatics 10*, 125.
12. Steinbeck, C., and Kuhn, S. (2004) NMRShiftDB - compound identification and structure elucidation support through a free community-built web database, *Phytochemistry 65*, 2711–2717.
13. Wang, Y., Xiao, J., Suzek, T., Zhang, J., Wang, J., and Bryant, S. (2009) PubChem: a public information system for analyzing bioactivities of small molecules, *Nucl. Acids Res. 37*, W623–633.
14. Rijnbeek, M., and Steinbeck, C. (2009) OrChem - An open source chemistry search engine for Oracle(R), *Journal of Cheminformatics 1*, 17.
15. Krause, S., Willighagen, E., and Steinbeck, C. JChemPaint - Using the Collaborative Forces of the Internet to Develop a Free Editor for 2D Chemical Structures, *Molecules*, 93–98.
16. Tanimoto TT (1957) IBM Internal Report.
17. Cote, R., Jones, P., Apweiler, R., and Hermjakob, H. (2006) The Ontology Lookup Service, a lightweight cross-platform tool for controlled vocabulary queries, *BMC Bioinformatics 7*, 97.
18. ChEBI Web Services. http://www.ebi.ac.uk/chebi/webServices.do
19. Murray-Rust, P., Rzepa, H., and Wright, M. (2001) Development of chemical markup language (CML) as a system for handling complex chemical content, New J. Chem. 25, 618–634.
20. ChEBI Downloads. http://www.ebi.ac.uk/chebi/downloadsForward.do. Accessed 1 March 2010.
21. Spjuth, O., Helmus, T., Willighagen, E., Kuhn, S., Eklund, M., Wagener, J., Rust, P., Steinbeck, C., and Wikberg, J. (2007) Bioclipse: an open source workbench for chemo- and bioinformatics, *BMC Bioinformatics 8*, 59.
22. Steinbeck, C., Hoppe, C., Kuhn, S., Floris, M., Guha, R., and Willighagen, E. (2006) Recent Developments of the Chemistry Development Kit (CDK) - An Open-Source Java Library for Chemo- and Bioinformatics, *Current Pharmaceutical Design 12*, 2111–2120(2110).
23. Day-Richter, J., Harris, M., Haendel, M., The Gene Ontology, O. B. O., and Lewis, S. (2007) OBO-Edit an ontology editor for biologists, *Bioinformatics 23*, 2198–2200.

Chapter 20

Working with Small Molecules: Rules-of-Thumb of "Drug Likeness"

Ming-Qiang Zhang

Abstract

Based on analyses of existing small organic drug molecules, a set of "rules-of-thumb" have been devised to assess the likeness of a small molecule under study to those existing drugs in terms of physicochemical and topological properties. These rules can be used to estimate the likelihood of a small molecule to possess the desired efficacy, pharmacokinetic/pharmacodynamic properties, and toxicity profiles to eventually become a drug, and therefore, whether it justifies further experimental work and development. These rules are particularly useful when selecting a chemical starting point for a given project or choosing a chemical series to focus when multiple series are available. Caution should be paid, however, not to overly rely on these rules for decision-making, since these rules are restricted by knowledge of existing drugs. Novel chemotypes and/or targets may be exceptions.

Key words: Drug-likeness, Drug properties, Chemotype

1. Introduction

Drug-likeness is a qualitative estimate of chemical tractability of a given small molecule hit/lead or a preclinical development candidate to become a successfully launched drug. It is a collection of empirical guidelines that are based on algorithm translation of experienced medicinal chemists' intuition. Since each individual medicinal chemist has slightly different intuitive understanding of the characteristics that make a compound look like a drug (1), some of the guidelines may have imprints of different individual and organizational experience and interpretation (2). This chapter attempts to list the most commonly used "rules-of-thumb" for evaluating drug-likeness, since these are the easiest to implement,

Gerard Drewes and Marcus Bantscheff (eds.), *Chemical Proteomics: Methods and Protocols*, Methods in Molecular Biology, vol. 803, DOI 10.1007/978-1-61779-364-6_20,

particularly for nonspecialists. For more in-depth coverage and computational treatment of the topic, the reader is referred to special monographs cited in ref. (3).

All "rules-of-thumb" for assessing drug-likeness are derived from the analysis of existing drugs and/or drug candidates such as those listed in the databases of MDL Drug Data Report, Derwent World Drug Index, Comprehensive Medicinal Chemistry (CMC), etc. As such, practice according to drug-likeness will inevitably follow a power law, i.e., the structures or structural frameworks appeared in existing drugs will be most likely used in new drug discovery programs, resulting in very narrow structural diversity. A recent analysis of structural scaffold content of all organic compounds in the Chemical Abstract Service (CAS) Registry found that 17.2% of the total 24 million organic compounds in the registry are represented by only 30 structural frameworks (4). Therefore, to broaden the druggable chemical space and increase the structural diversity of small molecules, it may be beneficial to work occasionally beyond the rules of drug-likeness.

2. Efficacy

2.1. Structural Frameworks

The use of structural frameworks to analyze drug-likeness relies on topological similarity of a given molecule to those of existing drugs. For this purpose, structural frameworks are defined as ring systems that are either alone or connected by chemical bonds, ignoring the branching side-chains or substituents (Fig. 1a, ref. (5)). Hydrogen atoms are excluded for simplicity. Of the 5,120 drugs recorded in the CMC database (version 94.1), there are 1,170 structural frameworks found, but when atoms and bonds are considered equivalent, 50% of these drugs can be represented by 32 frameworks (Fig. 1b). Interestingly, the top 17 frameworks found in all organic compounds (representing ~7 million out of 24 million compounds) in CAS registry are within these 32 drug frameworks (4), suggesting considerable overlap between the most common shapes of drugs and the most common shapes of organic compounds in general.

Not surprisingly many drugs sharing the same structural framework do not necessarily act on the same biological target or have the same biological activity or therapeutic utility. Such a single molecular framework able to provide ligands for diverse biological targets is also referred to as a "privileged structure" (Fig. 2, ref. (6)).

Privileged structures are particularly useful when designing screening libraries, since potential hit rate may be higher than those of random libraries. Several commercial providers of screening libraries, for example, ChemBridge (http://www.chembridge.com) and BioFocus DPI (http://www.biofocus.com), have taken advantages of

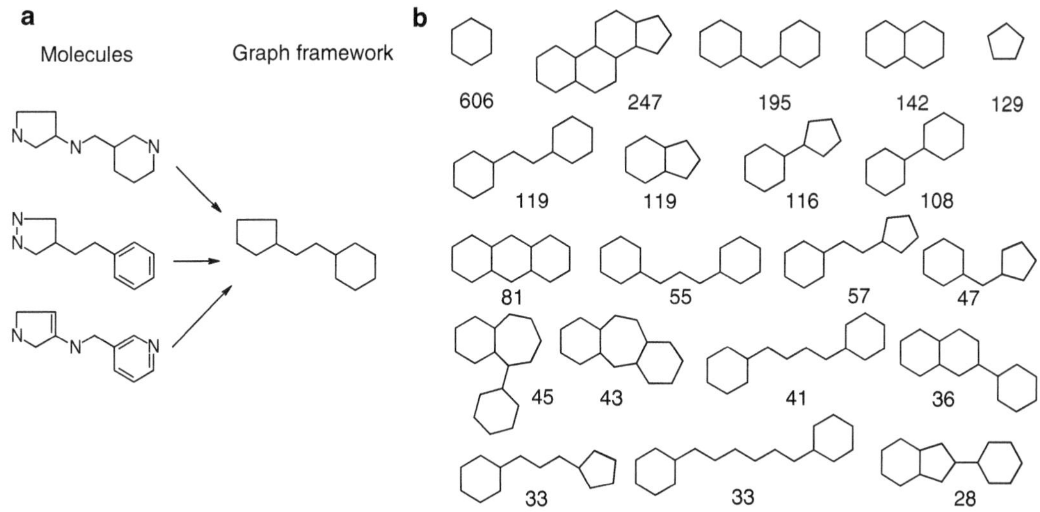

Fig. 1. (**a**) Definition of graph framework; (**b**) the top 20 graph frameworks most frequently found in 5,120 drug molecules (5). Numbers indicate frequency of occurrence.

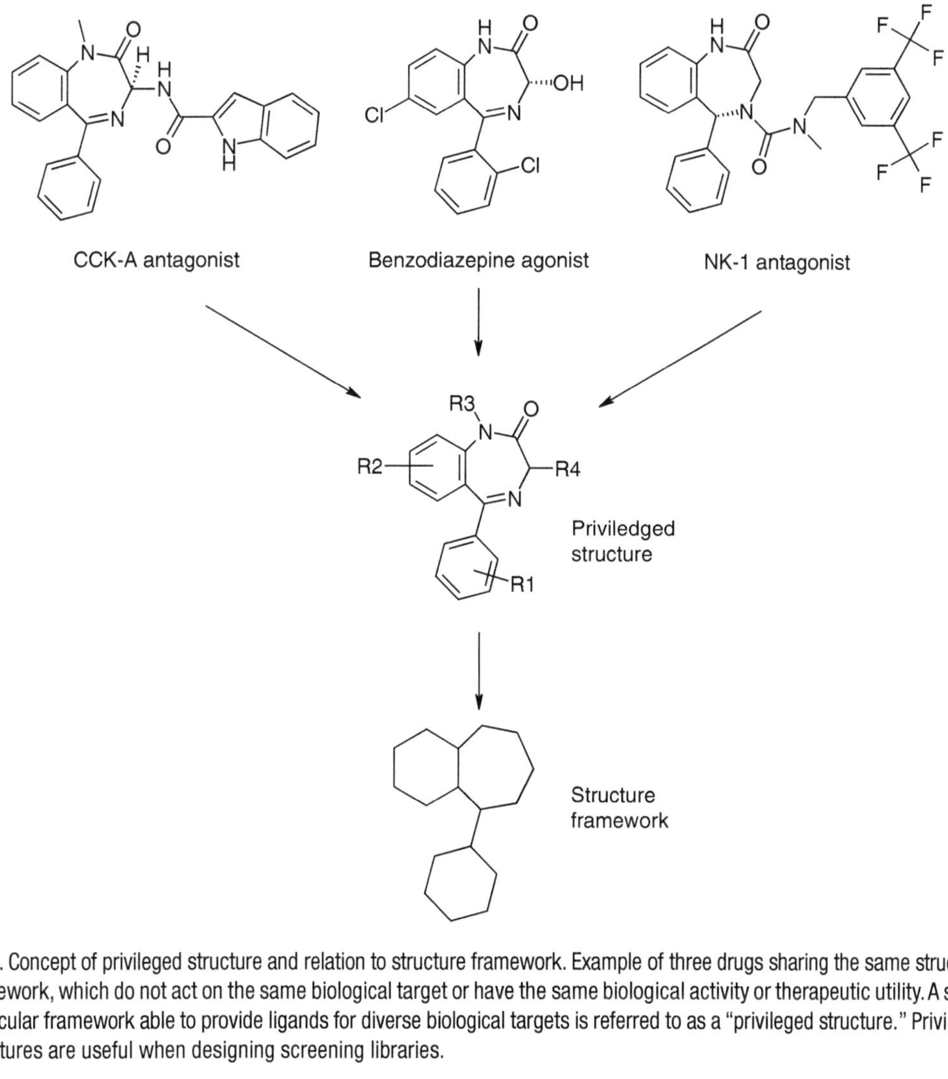

Fig. 2. Concept of privileged structure and relation to structure framework. Example of three drugs sharing the same structural framework, which do not act on the same biological target or have the same biological activity or therapeutic utility. A single molecular framework able to provide ligands for diverse biological targets is referred to as a "privileged structure." Privileged structures are useful when designing screening libraries.

privileged structures and some target-specific SAR in producing target-focused libraries, e.g., GPCR libraries, kinase libraries, etc. These should be valuable when searching for new hits for a given target.

2.2. Ligand Efficiency

Ligand efficiency (LE) measures the contribution per heavy (non-hydrogen) atom to the overall potency of a molecule. It is defined as:

$$\mathrm{LE} = \frac{\mathrm{Potency}}{N} \tag{1}$$

where N denotes the number of heavy atoms in the molecule. Potency can be defined by pIC_{50}, pKi, or ΔG.

Since molecule size tends to increase during lead optimization, LE is a useful parameter to assess the impact of additional groups and structural fragments on potency. Molecules that achieve a given potency with fewer heavy atoms are by definition more efficient. Taking into account of the impact of molecule size on other drug-like properties (see below for further discussion), a molecule with an LE of ≥ 0.3 kcal/heavy atom (using ΔG as the measure of potency (see Note 1)) is considered more tractable and therefore drug-like.

Several modified/improved LE measurements are proposed or in use. These include the use of molecular weight instead of number of heavy atoms as the measure of molecule size (7), since MW is superior in dealing with the contribution of heteroatoms from different rows in the periodic table (e.g., fluorine vs. iodine). Size-independent measure of LE has also been developed to take into account of the maximum potency a molecule can achieve and hence the size dependency of LE (8, 9). Equation (1) can therefore be modified as follows (9):

$$\mathrm{LE} = \frac{\mathrm{Potency}}{N^{0.3}}. \tag{1a}$$

2.3. Enthalpy and Entropy

Evaluation of a compound potency or affinity alone may not be sufficient in assessing its tractability, since compounds with similar potency may have very different thermodynamics in their interaction with the protein target. In thermodynamics, the Gibbs free energy gain (ΔG) from binding of a ligand to its target is determined by both enthalpic (ΔH) and entropic (ΔS) gains from the interaction, as expressed in the following equation (see Note 1):

$$\Delta G = \Delta H - T\Delta S = -RT\mathrm{Ln}K_{\mathrm{B}} \tag{2}$$

where R is the gas constant, T is the absolute temperature, and K_{B} is the binding constant. Specific noncovalent bonds such as hydrogen-bonding, electrostatic attraction, and van der Waals interactions favor the enthalpic gain (ΔH), whereas nonspecific hydrophobic interactions (solvent entropy) and gains in freedom of molecule and bond motions favor entropic gain (ΔS).

Synthetic small molecules often have proportionately greater favorable entropic contribution (ΔS) to the gain in free energy (ΔG) than natural, biological ligands (10). This is due to the tendency of medicinal chemists to add hydrophobic groups/fragments in order to increase potency or affinity during lead optimization. To design specific noncovalent bonds between a small molecule and its protein target (so to enhance enthalpic gain ΔH) is very difficult, presumably because of the limit to the number of noncovalent bonds that can be introduced in a given binding pocket (10). On the other hand, studies of marketed statins and HIV protease inhibitors have shown that enthalpically optimized compounds are often the best-in-class (11). Therefore, compounds with better bonding complementarity with the protein target and hence more favorable ΔH values signify better and more tractable compounds.

2.4. Target Resident Time

In addition to energetic/thermodynamic analysis, kinetic characterization of ligand-target interaction is also important in evaluating the tractability of a given compound. Typically, most in vitro biological/biochemical assays are performed under conditions in which the target is exposed to an invariant concentration of the compound throughout the period of the measurement. This poorly mimics the temporal concentrations and duration of exposure that the compound has against its target in a dynamic in vivo system. Therefore, it is often observed that apparent in vitro potency/ affinity does not correlate (at least not quantitatively) with in vivo efficacy, especially when the duration of pharmacological effects is an important component of overall in vivo efficacy (12).

Ligand-target residence time is a measure of the duration in which a ligand is bound to its target. It is often expressed as the time required for 50% dissociation of the remaining ligand–target complex ($t_{1/2}$) (see Note 2).

Whether a compound under study should have long ($t_{1/2}$=h or days) or short ($t_{1/2}$<1 s to min) target residence time is dependent on its mechanism of action and intended indication. For example, compounds designed to target nonhuman targets for antiinfective treatment or mutated oncogenic targets for anticancer treatment may have long target residence time, whereas compounds designed to modulate ion channels may have short target residence time. Interestingly, most anticancer kinase inhibitors that are approved for clinical use or under active development tend to have longer plasma half time (an imperfect surrogate for cellular target residence time) than those of which development has been discontinued (13).

A balanced target residence time lies in-between the desired efficacy and mechanism-based toxicity. Target selectivity is also influenced by relative target residence times between the primary and secondary targets. A ligand may have higher binding affinity to its primary target than its secondary target but its selectivity would be impaired if its residence time with the secondary target is longer than that with its primary target (12).

3. ADME and Toxicity Properties

3.1. "Rule-of-Five"

An analysis of 2,287 compounds that were believed to have entered Phase II clinical development, judging from their registration at USAN (United States Adopted Name) or INN (International Non-proprietary Name), indicated that poor oral absorption are more likely when a compound has:

- MW > 500 Da
- Clog P > 5 (see Note 3)
- Number of H-bond donors (HBD) > 5
- Number of H-bond acceptors (HBA) > 10

and that molecules should be flagged for poor absorption or permeability if any two of the above criteria are exceeded (14).

Several variations and further work on the above simple "rule-of-thumb" for drug oral absorption have been published (15). Two additional physicochemical parameters worth of noting are: number of rotatable bonds (RB) and polar surface area (PSA). It is often cited that for good oral absorption a compound should also have:

- Number of RB ≤ 10
- PSA ≤ 120 Å^2 (see Note 3)

Similar rules-of-thumb have also been proposed for good CNS penetration (16). CNS penetration is likely if a compound has:

- MW ≤ 400
- CLog P ≤ 5
- Number of HBD ≤ 3
- Number of HBA ≤ 7

suggesting for CNS penetration the tolerated physicochemical properties have a smaller range than oral absorption.

It is important to note that inter-relationships exit among some of these physicochemical parameters. For example, molecular weight is positively correlated with CLog P, whereas PSA is negatively correlated with CLog P. To achieve a 50% probability of attaining high permeability (e.g. P_{app} > 100 nm/s), a log D of >1.7 is required for compounds with molecular weight of 350–400, >3.1 for molecular weight of 400–450, >3.4 for molecular weight of 450–500 and >4.5 for molecular weight of >500 (17).

3.2. Number of Aromatic Rings

Since aromatic rings are one of the most common structural features that are encountered when working with small molecules, it is important to note that too many aromatic rings in a molecule can result in low developability. Based on analysis of the

Table 1
Minimum acceptable solubility for human dosing in relation to permeability

Human dose (mg) (MAD)	7	7	70	70	700	700
Human dose (mg/kg)	0.1	0.1	1	1	10	10
Permeability (K_a, min^{-1})	0.003 (low)	0.03 (high)	0.003 (low)	0.03 (high)	0.003 (low)	0.03 (high)
Minimum acceptable solubility (mg/ml)	0.035	0.0035	0.35	0.035	3.5	0.35

GlaxoSmithKline corporate collection of compounds and pipeline drug candidates, it was concluded that the fewer the number of aromatic rings contained in an oral drug candidate, the more developable that candidate is likely to be; specially, more than three aromatic rings in a molecule correlates with poorer compound developability and, therefore, an increased risk of compound attrition (18).

3.3. Aqueous Solubility (see Note 4)

After oral administration, a drug must dissolve in the small intestine (water volume ~250 ml) and diffuse to the surface of the intestinal epithelium (small intestinal transit time ~270 min) to be absorbed into systemic circulation. Insoluble compounds tend to have incomplete absorption and therefore low oral bioavailability.

The minimum aqueous solubility required for a drug molecule is dependent on its permeability and potency (i.e., dose needed to exert therapeutic effects). The more potent and more permeable the compound, the lower the solubility required to achieve complete absorption. Table 1 summarizes the relationship among dose, permeability, and minimum solubility for a given drug molecule (19).

To assist quick assessment of drugability of a molecule, the following classification ranges are suggested for medicinal chemists (20):

- <10 μg/ml low solubility
- 10–60 μg/ml moderate solubility
- >60 μg/ml high solubility

Compounds with low solubility are flagged out for probability of low oral absorption.

3.4. Toxicity

3.4.1. Reactive Functionalities

Reactive functional groups such as electrophiles α,β-unsaturated carbonyl compounds, quinones/quinone imines, epoxides, alkyl sulfates, halides, acryl carbonium, nitrenium ions, benzylic carbonium ions, and alkyl carbonium ions can form covalent bond with nucleophilic groups of biological macromolecules such as proteins, nucleic acids and small peptides such as glutathione, causing nonspecific cytotoxicity.

Reactive functionalities should therefore be avoided, unless cytotoxicity is desired, for instance, as anticancer agents and targeted delivery can be achieved, such as antibody-directed enzyme prodrug therapy (ADEPT).

3.4.2. Metabolic Activation (21)

Some structural motives, although themselves nonreactive, are prone to be activated through metabolism. These include catechols, hydroquinones, anilines, thiophenes, thioureas, etc. For example, the thionocarbonyl function of thioureas can be metabolized by flavin-containing monooxygenases and cytochrome P450 enzymes to reactive sulphenic, sulphinic, and sulphonic acids which can alkylate proteins. Another example is the aniline group which can be oxidized in the liver to hydroxylamines and further products such as nitroso-compounds. The sulphonamides such as sulfamethoxazole are a group of drugs that contain an aniline moiety, and they can cause a variety of unpredictable drug reactions (fever, skin rashes, hepatitis, nephritis, and blood dyscrasias) due to their bioactivation to hydroxylamines and nitroso-derivatives (22). It is believed that metabolism-induced activation is at least partially responsible for the so-called idiosyncratic toxicity, for example, through immune responses caused by protein–metabolite adducts.

3.4.3. Log P

In general, more lipophilic compounds are more promiscuous in binding to biological targets and the more promiscuous compounds are likely to be more toxic. Various studies have indicated that promiscuity increases dramatically for compounds with log $P > 3$ (23). Specifically, it was found that lipophilicity contributes strongly to binding to hERG, a cardiac potassium channel known for its promiscuous binding to various chemotypes and their cardiac side-effects. For example, to achieve a >70% chance of hERG $IC_{50} > 10\ \mu M$ requires a log D of <3.3 for a neutral compound and <1.4 for a basic compound.

Drug-induced phospholipidosis is also related to lipophilicity. It was suggested that the risk of a compound causing phospholipidosis increases if the sum of log P^2 and pKa^2 is >90 (24). For example, for a basic molecule with a pKa of 9, a log P of <3 is desirable to minimize potential phospholipidosis, whereas for a weaker base the log P limit could be higher.

It is suggested that there is an optimum range of log P 1–3 (24), too low impairing the permeability and too high impairing solubility and potentially causing nonspecific toxicity. This optimum range may be different within different chemical series, especially when other parameters are improved simultaneously, e.g., reduction of molecular weight, removal of aromatic rings, increased saturation or added chirality, etc.

In summary, the following rules-of-thumb can be used to assess the drug-likeness of a small molecule compound, although no strict need to comply with every of them:

1. Structural framework looks like those commonly found in existing drugs.
2. Ligand efficiency >0.3 kcal/atom.
3. Binding affinity reasonably driven by ΔH.
4. A balanced target resident time compatible to the target.
5. "Rule-of-5" compliant.
6. Log P lies within the optimum range of 1–3.
7. Number of aromatic rings <3.
8. Water solubility >60 μg/ml.
9. No reactive functional groups.
10. No groups that are prone to bio-activation, e.g., thiourea or aniline.

4. Notes

1. The change in Gibbs free energy (ΔG) is a thermodynamic measure of energy gained from binding of a small molecule to its target. It is typically obtained by an isothermal titration calorimetry (ITC) experiment, which measures the heat generated (exothermic) or absorbed (endothermic) by the interaction of a small molecular ligand with its binding target. Measurement of this heat allows accurate determination of binding constants (K_B), reaction stoichiometry (n), enthalpy (ΔH), and entropy (ΔS), thereby providing a complete thermodynamic profile of the molecular interaction in a single experiment. Because ITC goes beyond binding affinities and can elucidate the mechanism of the molecular interaction, it has become the method of choice for characterizing biomolecular interactions. Typical experimental procedures involve successive injection of a solution of one binding partner (e.g., small molecular ligand) in the syringe into the cell that contains the other binding partner (e.g., a protein target) (see Microcal web site for details http://www.microcal.com/technology/itc.asp). It is extremely important that both binding partners are dissolved in strictly identical buffer (salts, pH, etc.) or vehicle. ΔS and ΔG are deduced from observed K_B and ΔH: $\Delta G = -RT \log K_B = \Delta H - T\Delta S$ (25).
2. There are a number of experimental methods for the measurement of dissociation half-life ($t_{1/2}$) (12). One of the common methods is based on changes of surface plasmon resonance

(SPR), such as Biacore (http://www.biacore.com/). Here, a binding partner (either ligand or protein) is immobilized to a surface which is flowed over with a solution containing the other binding partner. The formation of the binary complex causes changes in SPR or refractive index (RI) until equilibrium is established. The rate of association and the final equilibrium concentration of the binary complex can thereby be determined. Once equilibrium is reached, the flow solution can be replaced by the blank vehicle to perturb the equilibrium. The rate of return of the RI value to the prebinding level is followed to define the dissociation rate.

3. Log *P* is the logarithm of a compound's partition coefficient between octanol and water. The most frequently used log *P* values are obtained by calculation with a plethora of different computational methods, although errors (ranging from 0.5 to 1 log unit) do exist in these calculated values compared to the experimentally measured values. The Interactive Laboratory (http://www.acdlabs.com/ilab) is a commercial product from Advanced Chemistry Development (ACD), which provides online calculation of log *P* and log *D* (partition coefficient between octanol and buffer at a given pH for an ionisable compound). A free interactive web service is available from Molinspiration Chemoinformatics (http://www.molinspiration.com/cgi-bin/properties) for the calculation of log *P* and PSA and other molecular properties.

4. Kinetic solubility can be utilized to assess compound solubility during biological screening and this experiment can be performed in a high-throughput fashion. The measurement is routinely performed by adding small amount of test compound stock solution into physiological relevant aqueous solution in the presence of small percentage of organic solvent, i.e., 1% DMSO. After constantly agitating for several hours (e.g., 4 h), the mixture can be analyzed using direct measurement such as nephelometry. Alternatively, the mixture can be separated using filtration or centrifugation method. The dissolved compound concentration is often quantified by analytical methods such as UV absorbance, HPLC/UV, or LC/MS. Equilibrium solubility represents the intrinsic solubility of a drug substance under specific condition such as temperature, ionic strength, and pH. The equilibrium solubility is often measured in a specific solution such as water or other physiological aqueous solution in absence of co-solvent or other types of organic modifiers. The sample is prepared with the addition of excessive amount of test compound in solution and mixing time normally lasts for at least 24 h. After equilibrium is reached, the mixture is segregated by filtration or centrifugation. The solubility is determined using back calculation from a standard curve.

References

1. Takaoka, Y., Endo, Y., Yamanobe, S., Kakinuma, H., Okubo, T., Shimazaki, Y., Ota ,T., Sumiya, S., and Yoshikawa, K. (2003) Development of a method for evaluating drug-likeness and ease of synthesis using a data set in which compounds are assigned scores based on chemists' intuition. *J. Chem. Inf. Comput. Sci.* 43, 1269–1275.
2. Leeson, P.D., and Springthorpe, B. (2007) The influence of drug-like concepts on decision-making in medicinal chemistry. *Nat. Rev. Drg Disc.* 6, 881–890.
3. Young, D.C. Computational Drug Design: A Guide for Computational and Medicinal Chemists, Wiley-Interscience (February 12, 2009) ISBN-10: 047012685X; Tudor I. Oprea (Editor). Chemoinformatics in Drug Discovery. Wiley-VCH (May 6, 2005) ISBN-10: 3527307532 and other volumes in the book series of Methods and Principles in Medicinal Chemistry, Mannhold R., Kubinyi H., and Folkers G (Eds).
4. Lipkus, A.H., Yuan, Q., Lucas, K.A., Funk, S.A., Bartelt III, W. F., Schenck, R. J., and Trippe, A.J. (2008) Structural diversity of organic chemistry. A scaffold analysis of the CAS registry. *J. Org. Chem.* 73, 4443–4451.
5. Bemis, G.W., and Murcko, M.A. (1996) The properties of known drugs. 1. Molecular frameworks. *J. Med. Chem* 39, 2887–2893.
6. Duarte, C.D., Barreiro, E.J., and Fraga, C.A. (2007) Privileged structures: a useful concept for the rational design of new lead drug candidates. *Mini Rev. Med. Chem.* 7, 1108–1119.
7. Abad-Zapatero, C., and Metz, J.T. (2005) Ligand efficiency indices as guideposts for drug discovery. *Drug Disc Today* 10, 464–469.
8. Bembenek, S.D., Tounge, B.A., and Reynolds, C.H. (2009) Ligand efficiency and fragment-based drug discovery. *Drug Disc. Today* 14, 278–283.
9. Nissink, J.W.M. (2009) Simple size-independent measure of ligand efficiency. *J. Chem. Inf. Model* 49, 1617–1622.
10. Olsson, T.S.G., Williams, M.A., Pitt, W.R., and Ladbury, J.E. (2008) The thermodynamics of protein-ligand interaction and salvation: insights for ligand design. *J. Mol. Biol.* 384, 1002–1017.
11. Freire, E. (2008) Do enthalpy and entropy distinguish first-in-class from best-in-class? *Drug Disc. Today* 13, 869–874.
12. Copeland, R.A., Pompliano, D.L, and Meek, T.D. (2006) Drug-target residence time and its implications for lead optimization. *Nat. Rev Drug Disc.* 5, 730–739.
13. LoRusso, P.M., and Eder, J.P. (2008) Therapeutic potential of novel selective-spectrum kinase inhibitors in oncology. *Expert Opin. Invest. Drugs* 17, 1013–1028.
14. Lipinski, C.A., Lombardo, F., Dominy, B.W., and Feeney, P.J. (1997) Experimental and computational approaches to estimate solubility and permeability in drug discovery and development settings. *Adv. Drug Deliv. Rev.* 23, 3–25.
15. van de Waterbeemd, H., Smith D.A., Beaumont K., and Walker D.K. (2001) Property-based design: Optimization of drug absorption and pharmacokinetics. *J. Med. Chem.* 44, 1313–1333.
16. Pajouhesh, H., and Lenz, G.R. (2005) Medicinal chemical properties of successful central nervous system drugs. *NeuroRx* 2, 541–553.
17. Waring, M.J. (2009) Defining optimum lipophilicity and molecular weight ranges for drug candidates – Molecular weight dependent lower logD limits based on permeability. *Bioorg. Med. Chem. Lett.* 19, 2844–2851.
18. Ritchie, T.J., and Macdonald, S.J.F. (2009) The impact of aromatic ring count on compound developability – Are too many aromatic rings a liability in drug design? *Drug Discovery Today* 14, 1011–1020.
19. Kerns, E.H., and Di, L. (2008) Drug-like properties: Concepts, structure, design and methods, from ADME to toxicity optimization. ISBN 978-0-1236-9520-8, Academic Press, Burlington, MA 01803, USA, pp64.
20. Lipinski, C.A. (2000) Drug-like properties and the cause of poor solubility and poor permeability. *J. Pharmacol. Toxicol. Methods* 44, 235–249.
21. Smith, D.A., van de Waterbeemd, H., and Walker, D.K. (2001) Pharmacokinetics and metabolism in drug design. ISBN 3-527-30197-6, Wiley-VCH Verlag GmbH, Weinheim, Germany, pp 99–122.
22. Williams, D.P., and Park B.K. (2003) Idiosyncratic toxicity: The role of toxicophores and bioactivation. *Drug Discovery Today* 8, 1044–1050.
23. Waring, M.J. (2010) Lipophilicity in drug discovery. *Expert Opin. Drug Discov.* 5, 235–248.
24. Ploemen, J.P., Kelder, J., Hafmans, T., van de Sandt, H., van Burgsteden J.A., Saleminki P.J., and van Esch E. (2004) Use of physicochemical calculation of pKa and CLogP to predict phospholipidosis-inducing potential: a case study with structurally related piperazines. *Exp Toxicol Pathol.* 55, 347–55.
25. Wiseman, T., Williston, S., Brandts, J.F., and Lin, L.N. (1989) Rapid measurement of binding constants and heats of binding using a new titration calorimeter. *Anal Biochem.* 179, 131–137.

Index

Gerard Drewes and Marcus Bantscheff (eds.), *Chemical Proteomics: Methods and Protocols*, Methods in Molecular Biology, vol. 803, DOI 10.1007/978-1-61779-364-6, © Springer Science+Business Media, LLC 2012

N

O

P

Q

R

S

T

U

X

MIX
Papier aus verantwortungsvollen Quellen
Paper from responsible sources
FSC® C105338

If you have any concerns about our products,
you can contact us on
ProductSafety@springernature.com

In case Publisher is established outside the EU,
the EU authorized representative is:
Springer Nature Customer Service Center GmbH
Europaplatz 3, 69115 Heidelberg, Germany

Printed by Libri Plureos GmbH
in Hamburg, Germany